Fuzzy-Model-Based Sliding Mode Control of Complex Nonlinear Systems

This book presents recent theoretical findings on Sliding Mode Control (SMC) issues for various classes of complex nonlinear systems based on Takagi-Sugeno (T-S) fuzzy models. Research results and controller design methods on SMC of complex nonlinear systems through T-S fuzzy models are presented and show how the methods can be used in various practical systems, such as control of nonlinear gas absorption, water stream heating, and air-drying processes in some chemical production systems.

Fuzzy-Model-Based Sliding Mode Control of Complex Nonlinear Systems discusses new fuzzy SMC methods for some typical classes of complex nonlinear systems and offers new stability analysis results on sliding motion based on more advanced T-S fuzzy affine models. New convexification techniques are presented to handle the couplings in the sliding surface gain with the Lyapunov matrices and/or system matrices and extra freedom of the existence conditions for the sliding surface gains. The book provides new SMC laws for both one-dimensional and two-dimensional systems that tend to be more complicated and widely used in many industrial processes. New types of sliding surface and dynamic sliding mode controller design methods are also covered.

This book finds its audience as a reference to researchers and engineers that are engaging in the areas related to fuzzy modelling and control and SMC with its applications.

Engineering Mathematics and Operations Research

Tools, Techniques, Theory, and Applications Used in Manufacturing and Management Science

Series Editor: Aliakbar Montazer Haghighi

The main aim of this new book series is to publish books by qualified authors in the areas related to mathematics and its applications in engineering and operations research areas. Thus, the series will publish rigorous and quality books related to a variety of areas of mathematics and science related to engineering. All books will include some applications, tools, and theories as well as the latest research and development of mathematics and its applications in engineering, manufacturing, and management sciences areas. Such topics that will be covered in this series are: vector analysis, linear algebra, fuzzy mathematics, sequences, series, probability, statistics, reliability, numerical analysis, approximation, teaching of mathematics for engineers, complex analysis, ordinary differential equations, differential equations, partial differential equations, stochastic processes, random walk, point process, Brownian motions, birth and death processes, and queueing theory.

Books in this series will be of interest to undergraduate and graduate students as course textbooks as well as reference books.

Optimization Techniques and Associated Applications
Edited by Sandeep Singh and Sandeep Dalal

Foundation Engineering Mathematics
By Faridon Amdjadi and Dharminder Singh

Fuzzy-Model-Based Sliding Mode Control of Complex Nonlinear Systems

By Wenqiang Ji, Jianbin Qiu and Heting Zhang

For more information about this series, please visit: www.routledge.com/Engineering-Mathematics-and-Operations-Research/book-series/CRCEMORTTTA

Fuzzy-Model-Based Sliding Mode Control of Complex Nonlinear Systems

Wenqiang Ji, Jianbin Qiu and Heting Zhang

CRC Press
Taylor & Francis Group
Boca Raton London New York

CRC Press is an imprint of the
Taylor & Francis Group, an **informa** business

Cover design: Shutterstock

First edition published 2026
by CRC Press
2385 Executive Center Drive, Suite 320, Boca Raton, FL 33431, U.S.A.

and by CRC Press
4 Park Square, Milton Park, Abingdon, Oxon, OX14 4RN

CRC Press is an imprint of Taylor & Francis Group, LLC

ISBN: 978-1-032-69466-5 (hbk)
ISBN: 978-1-032-69570-9 (pbk)
ISBN: 978-1-032-69571-6 (ebk)

DOI: 10.1201/9781032695716

Typeset in Nimbus Roman font
by KnowledgeWorks Global Ltd.

Publisher's note:This book has been prepared from camera-ready copy provided by the authors.

Contents

Preface

Many real-world systems and industrial processes inevitably contain instinctive nonlinearities. All these inherent nonlinearities exist in various forms and always introduce much difficulty in stability analysis and controller synthesis. From this perspective, research on nonlinear systems is generally a problematic issue. Based on fuzzy logic control theory, the Takagi-Sugeno (T-S) fuzzy model has been viewed as an effective method in modelling and control of highly complex systems subject to various types of nonlinearities. The overall dynamical features of the T-S fuzzy model can be described by a set of local linear/affine models, which are in smooth connection by a family of fuzzy IF-THEN rules. In fact, the T-S fuzzy system is a multi-model method wherein sub-models (or known as local linear or affine models) are "blended" via fuzzy membership functions to characterize the global dynamical properties of a nonlinear system. It has been shown that in any convex compact set, the T-S fuzzy model can serve as a technically efficient tool to approximate smooth nonlinear systems to arbitrary degrees of accuracy.

On another research front, sliding mode control (SMC) has been known as an effective robust control method, as it has appealing features such as strong robustness against perturbations or uncertainties, good transient performance and fast response. In view of these merits, the SMC technique has drawn increasing research interests from both academic study and practical engineering, and has been widely utilized in industrial applications such as spacecrafts, power systems, industrial manipulators, electrical motors and so on. Due to the universal function approximation competence of the T-S fuzzy models, some valuable works on SMC have been reported for nonlinear systems based on the T-S fuzzy models. It is worth mentioning that when investigating SMC problem for nonlinear systems through T-S fuzzy models, conventional stability and performance analysis methods and general sliding mode controller synthesis approaches do not usually meet some practical requirements. It should be also noted early fuzzy SMC results were mainly obtained based on full state feedback and a common quadratic Lyapunov function, while the system states are usually not fully measurable for various practical situations, and the analysis and design conservatism also needs to be further reduced. This monograph attempts to propose innovative stability and robust performance analysis methods to sliding motion and develop a series of new state feedback and output feedback (dynamic) sliding mode controller synthesis approaches for complex nonlinear systems based on the T-S fuzzy (affine) models with local linear/affine sub-models under a unified convex optimization framework. Researchers investigating the fields of SMC of nonlinear systems based on T-S fuzzy models can also find valuable reference materials in this monograph.

In this monograph, the stability and robust performance analysis of sliding motion, state feedback and output feedback sliding mode controller design, fuzzy modelling problems are thoroughly investigated for various classes of complex nonlinear systems including one-dimensional (1-D) systems, two-dimensional (2-D) systems, and singular perturbed systems based on T-S fuzzy models with local linear/affine sub-models. Fresh new techniques are used to these systems which include state-input augmentation method, singular piecewise affine system method, auxiliary matrix multiplier method, and so on. The content of this monograph is organized as eight chapters. Chapter 1 introduces the background, fuzzy modelling and sliding mode control problem statements, which can be viewed as the fundamental description of fuzzy sliding mode control. Chapter 2-Chapter 5 sufficiently investigate the state feedback and output feedback sliding mode control for nonlinear 1-D systems subject to both matched and mismatched parameter uncertainties and external disturbance through T-S fuzzy affine models. The main contents include 1) model approximation for continuous-time and discrete-time nonlinear 1-D systems by T-S fuzzy affine models 2) stability and robust performance analysis of sliding motion; 3) state feedback and output feedback SMC law design; 4) state feedback and output feedback dynamic sliding mode controller design; 5) output feedback adaptive sliding mode controller design. Considering the multiple time-scale dynamical phenomena, Chapter 6 focuses on the fuzzy integral sliding mode control for nonlinear singularly perturbed systems. Topics include 1) a new type of integral sliding surface design; 2) stability analysis of nominal closed-loop system and sliding motion; 3) dynamic sliding mode controller design. In view of the complex structure and evolution properties of 2-D systems, the theories and techniques developed in the previous part are extended to tackle the continuous-time and discrete-time nonlinear 2-D systems in Chapter 7 and Chapter 8 based on T-S fuzzy affine models. The main contents include 1) stability and robust performance analysis of sliding motion based on 2-D piecewise quadratic Lyapunov functions; 2) fuzzy state feedback and output feedback SMC design.

In general, this monograph presents the most recent theoretical findings on sliding mode control issues for various classes of complex nonlinear systems based on T-S fuzzy (affine) models. This monograph can serve as a reference for researchers and engineers engaging in the areas related to fuzzy modelling and control, and sliding mode control with its applications. Some prerequisites for reading this monograph include linear system theory, fuzzy control theory, matrix theory, and so on.

Harbin, China *Wenqiang Ji*
Harbin, China *Jianbin Qiu*
Harbin, China *Heting Zhang*

Jul. 2025

Acknowledgements

The authors would like to dedicate their gratitude to numerous individuals who have made direct and indirect contribution to this monograph. Special thanks go to Professor Hak-Keung Lam from King's College London, Professor Mohammed Chadli from University of Paris-Saclay, and Professor Ligang Wu from Harbin Institute of Technology, for their valuable suggestions, constructive comments and supports. The authors are especially grateful to their families for their encouragement and great support.

The writing of this book was supported in part by the National Natural Science Foundation of China under Grant U21B6001, 62273121, 12301636, 62473128, and U25A204591, and in part by the Fundamental and Interdisciplinary Disciplines Breakthrough Plan of the Ministry of Education of China under Grant JYB2025XDXM109, and in part by the China Postdoctoral Science Foundation under Grant 2022M711639, 2023T160320, and 2024M752396.

List of Figures

List of Tables

Notations and Acronyms

$\blacksquare$	end of proof
$:=$	is defined as
$\in$	belongs to
$\subset$	is included in
$\forall$	for all
$\sum$	sum
$\Re$	field of real numbers
$\Re^n$	space of n-dimensional real vectors
$\Re^{n\times m}$	space of $n \times m$ real matrices
$\mathbf{I}$	identity matrix
$\mathbf{I}_n$	$n \times n$ identity matrix
$\mathbf{0}$	zero matrix
$\mathbf{0}_{n\times m}$	zero matrix of dimension $n \times m$
X^{T}	transpose of matrix X
X^{-1}	inverse of matrix X
Sym–X″	$\mathtt{Sym}\{X\} = X + X^{\mathrm{T}}$
$X > (<) 0$	X is real symmetric positive (negative) definite
$L_2\{[0,\infty),[0,\infty)\}$	space of square integrable functions on $\{[0,\infty),[0,\infty)\}$ (continuous case)
$l_2\{[0,\infty),[0,\infty)\}$	space of square summable sequences on $\{[0,\infty),[0,\infty)\}$ (discrete case)
$L_2[0,\infty)$,	space of square integrable functions on $[0,\infty)$ (continuous case)
$l_2[0,\infty)$	space of square summable sequences on $[0,\infty)$ (discrete case)
$\lVert\cdot\rVert$	Euclidean matrix norm (spectral norm)
$\lVert\cdot\rVert_2$	L_2-norm: $\sqrt{\int_0^\infty \lvert\cdot\rvert^2 dt}$ (continuous case) l_2-norm: $\sqrt{\sum_0^\infty \lvert\cdot\rvert^2}$ (discrete case)

CQLF	common quadratic Lyapunov function
FLC	fuzzy logic control
LMI	linear matrix inequality
MF	membership function
PQLF	piecewise quadratic Lyapunov function
SMC	sliding mode control
T-S	Takagi-Sugeno

1 Introduction

Immense physical systems and industrial processes frequently encounter complex nonlinear dynamics, which bring much difficulty in system analysis and controller synthesis. Though local linearization is well known as a classical analysis and synthesis technique for nonlinear systems, it is noted that local linearization can only ensure the local stability for the original nonlinear system. From this perspective, the resulting local linearization model is available only for a certain range of operating conditions. Fuzzy logic control (FLC), which is known as a simple and powerful scheme to control complex nonlinear systems or even nonanalytic systems, has been extensively applied in various industrial processes [1, 2, 3, 4, 5, 6, 7, 8, 9, 10]. The basic structure for a fuzzy system is composed of four parts, that is, knowledge base, inference engine, fuzzification interface, and defuzzification interface [16]. With research on FLC proceeding rapidly, FLC has been suggested to be an alternative technique to conventional control methods. Nevertheless, it is also noted that as mathematical models are lacking for conventional fuzzy controller synthesis, the conventional fuzzy logic control approaches also have some drawbacks, for instance, the difficulty of inconstant performance and systematic design.

In the past few decades, the fuzzy-model-based method has become a hot research area due to its unique advantages in complex nonlinear system modelling, analysis, and control. Among various fuzzy control approaches, the Takagi-Sugeno (T-S) fuzzy-model-based method has been verified to be an effective model-based nonlinear control method and has been widely utilized in the control of complex nonlinear dynamic systems during past years [11, 12, 13, 14, 15, 16, 17, 18, 19, 20, 21, 22, 23, 24]. The key procedure of this model is to utilize a group of fuzzy IF-THEN rules to characterize a nonlinear plant in the form of a family of local affine or linear models, which are connected via fuzzy membership functions (MFs) smoothly. It has been well-known that T-S fuzzy models can approximate smooth nonlinear functions to arbitrary degrees of accuracy in any convex compact region [25]. Due to the universal approximation capability of T-S fuzzy models, numerous remarkable stabilization and filtering design results on T-S fuzzy systems have been reported [26, 27, 28, 29, 30, 31, 32, 33, 34, 35, 36, 37, 38, 39, 40, 41, 42, 43, 44, 45, 46, 47].

1.1 T-S FUZZY AFFINE MODELS

A general affine nonlinear system is given in a compact form as follows,

$$\begin{cases} \delta x(t) = f_1(x(t)) + g_1(x(t))u(t) + h(x(t))w(t) \\ y(t) = f_2(x(t)) \\ z(t) = f_3(x(t)) + g_2(x(t))u(t) \end{cases} \tag{1.1}$$

where $\delta x(t) \triangleq \dot{x}(t)$ for continuous-time domain and $\delta x(t) \triangleq x(t+1)$ for discrete-time domain, respectively; $x(t) \in \Re^{n_x}$ is the system state; $u(t) \in \Re^{n_u}$ is the control

DOI: 10.1201/9781032695716-1

input; $y(t) \in \Re^{n_y}$ is the system measurement output; $w(t) \in \Re^{n_w}$ is the external disturbance belonging to $L_2[0,\infty)$ for continuous-time domain and belonging to $l_2[0,\infty)$ for discrete-time domain, respectively; and $z(t) \in \Re^{n_z}$ is the regulated output. $f_1(\cdot)$, $f_2(\cdot)$, $f_3(\cdot)$, $g_1(\cdot)$, $g_2(\cdot)$, and $h(\cdot)$ denote smooth nonlinear matrix/vector functions with appropriate dimensions. By using fuzzy modelling methods, for example, the local approximation, the nonlinear system (1.1) can be described by a T-S fuzzy affine model as follows:

Plant Rule $\mathscr{R}^l$: IF $\theta_1(t)$ is $\mathscr{F}_1^l$ and $\cdots$ and $\theta_\varphi(t)$ is $\mathscr{F}_\varphi^l$, THEN

$$\begin{cases} \delta x(t) = (A_l + \Delta A_l)x(t) + a_l + \Delta a_l + (B_l + \Delta B_l)u(t) + D_l w(t) \\ y(t) = C_l x(t) \\ z(t) = L_l x(t) + N_l u(t),\ l \in \mathscr{L} := \{1,2,\ldots,r\}, \end{cases} \tag{1.2}$$

where $\mathscr{R}^l$ is the l-th fuzzy inference rule; r is the number of inference rules; $\mathscr{F}_\phi^l(\phi = 1,\ldots,\varphi)$ represent fuzzy sets; $\theta(t) := \left[\theta_1(t),\ldots,\theta_\varphi(t)\right]$ are the measurable premise variables; and $(A_l, a_l, B_l, C_l, L_l, N_l)$ is the model of the l-th subsystem. ΔA_l, ΔB_l, and Δa_l are the parameter uncertainties satisfying

$$\begin{bmatrix} \Delta A_l & \Delta B_l & \Delta a_l \end{bmatrix} = H_{l1}\Delta_l(t)\begin{bmatrix} H_{l2} & H_{l3} & H_{l4} \end{bmatrix},\ l \in \{1,2,\ldots,r\}, \tag{1.3}$$

where H_{l1}, H_{l2}, H_{l3}, and H_{l4} are known real-valued matrices. $\Delta_l(t) \in \Re^{p_1 \times p_2}$ refers to unknown time-varying matrix functions with Lebesgue measurable elements satisfying

$$\Delta_l^{\mathrm{T}}(t)\Delta_l(t) \leq \mathbf{I},\ l \in \{1,2,\ldots,r\}. \tag{1.4}$$

Remark 1.1.1. Note that the system models shown in (1.2) are in fact local affine models rather than local linear models. Additional offset terms a_l are involved. It has been proved that these models have much improved approximation capabilities compared with the traditional T-S fuzzy systems with local linear models [19, 25, 48].

Remark 1.1.2. To avoid unnecessarily complicated notations, in this monograph, we only consider the parameter uncertainties that appear in the matrices ΔA_l, ΔB_l, and Δa_l. However, the results to be developed in this monograph can be easily extended to the case when the parameter uncertainties also emerge in the matrices D_l, L_l, and N_l, $l \in \{1,2,\ldots,r\}$.

Through a process of a singleton fuzzifier, product inference, and center-average defuzzifier, the global T-S fuzzy affine model can be obtained as follows:

$$\begin{cases} \delta x(t) = \sum\limits_{l=1}^{r} \mu_l[\theta(t)]\{(A_l + \Delta A_l)x(t) + a_l + \Delta a_l + (B_l + \Delta B_l)u(t) \\ \qquad\qquad + D_l w(t)\} \\ y(t) = \sum\limits_{l=1}^{r} \mu_l[\theta(t)]C_l x(t) \\ z(t) = \sum\limits_{l=1}^{r} \mu_l[\theta(t)]\{L_l x(t) + N_l u(t)\} \end{cases} \tag{1.5}$$

where $\mu_l[\theta(t)]$ is defined as the normalized membership function (MF), and

$$\mu_l[\theta(t)] := \frac{\prod_{\phi=1}^{\varphi} \mu_{l\phi}\left[\theta_\phi(t)\right]}{\sum_{v=1}^{r}\prod_{\phi=1}^{\varphi} \mu_{v\phi}\left[\theta_\phi(t)\right]} \geq 0, \ \sum_{l=1}^{r} \mu_l[\theta(t)] = 1 \tag{1.6}$$

and $\mu_{l\phi}\left[\theta_\phi(t)\right]$ represents the grade of membership of $\theta_\phi(t)$ in $\mathscr{F}_\phi^l$.

In view of the stability analysis and controller synthesis of fuzzy systems, one main approach is the direct Lyapunov method. Correspondingly, many stability analysis results have been attained based on common quadratic Lyapunov function (CQLF) [43, 49, 50, 51], piecewise quadratic Lyapunov functions (PQLFs) [19, 25, 52, 53, 54, 55, 56, 57, 58], and fuzzy quadratic Lyapunov functions (FQLFs) [59, 60, 61, 62], respectively. As the CQLF is regarded to be a particular case of the more general PQLFs/FQLFs, the PQLFs/FQLFs-based approaches can tackle the stability analysis, controller, and filtering design problem for a broader class of fuzzy systems over the CQLF-based method. This feature enables the analysis and synthesis results based on PQLFs/FQLFs to be less conservative compared with those based on the CQLF [56, 57, 63, 64]. For instance, [58] proposed a static output feedback controller design method for networked nonlinear systems by T-S fuzzy affine models under a PQLF framework, and two novel controller design approaches are proposed with different degrees of design conservatism and computational burden.

Nowadays, due to the increasing size and complexity of physical processes, many complex nonlinear systems also inevitably encounter multiple time-scale dynamical properties or multi-dimensional (m-D) features, which are known as singular perturbed systems (SPSs) or m-D systems. On one hand, nonlinear SPSs are a class of complex multiple time-scale dynamical systems with inherent ill-conditioned dynamics, which bring much difficulty in controller design. On the other hand, m-D dynamical features extensively exist in massive physical processes or systems, such as circuit analysis, thermal processes, image deblurring, water stream heating, and so on. Among various m-D systems, the two-dimensional (2-D) system has been well-known as a distinct and common type of the m-D systems, whose system states evolve along two independent directions. More recently, based on the powerful approximation competence of T-S fuzzy models, some advanced results on nonlinear SPSs [49, 103, 105] and nonlinear 2-D systems have been obtained [128, 129, 130, 131]. To mention a few, the state feedback and static output feedback problem for T-S fuzzy SPSs was investigated in [49]. The fuzzy modelling and control issue for 2-D systems was studied in [129]. The authors in [131] designed a full-order robust $\mathscr{H}_\infty$ filter for uncertain 2-D T-S fuzzy systems subject to stochastically occurring mixed delays.

1.2 SLIDING MODE CONTROL

In a large number of practical engineering applications such as vehicles, astronautics, aerospace, circuits, and industrial processes, it is usually hard to describe the systems, dynamical features precisely by exact mathematical models. In other words, owing to the variation in system parameters, unmodelled dynamics, or the approximation of complicated system behaviours via a straightforward model, errors inevitably exist between the derived mathematical models and the original systems. In addition, practical systems always suffer from external disturbances in various forms.

As a particular type of variable structure control, sliding mode control (SMC) is recognized to be a powerful nonlinear control method and has been extensively studied due to its high convergence speed, system order reduction, good transient performance, and strong robustness to external disturbances, system parameter variations, and model errors [65, 66, 67]. Loosely speaking, there are two main reasons for these advantages. The first reason is that the system dynamical behaviours can be designed by choosing the sliding surface (or known as switching surface/switching function) specifically. The second reason is that the responses of the sliding motion are generally insensitive to parametric uncertainties and external disturbances. It is also noted that the latter invariance feature of sliding mode control renders the SMC to be an effective and appropriate robust control method. Furthermore, the competence to specify performance is another attractive feature of SMC in the sense of the design point.

In general, SMC is obtained via utilizing a suitable discontinuous control law to force the system states towards a predefined sliding surface containing the origin in finite time, and then the system states can be driven to reach and move along the sliding surface to the origin with some anticipated properties, for example, the stability performance or tracking performance [68, 69, 70, 71]. Recently, in accordance with the great success achieved by the T-S fuzzy-model-based control method, some remarkable results on the SMC problem for nonlinear systems based on T-S fuzzy models have been reported [73, 74, 75, 76, 77, 78, 79, 80, 81, 82, 83, 84, 85]. It is worth mentioning that for many practical engineering situations, the system states are usually not fully measurable. Consequently, some remarkable results on output feedback SMC have been reported; see references [86, 87, 88, 89, 90, 91, 92, 93, 94, 95, 96, 97, 98] for more details on this topic. For instance, the authors in [75] considered the fuzzy SMC problem for stochastic nonlinear systems with time-delay. Some state feedback adaptive sliding mode controller synthesis results for T-S fuzzy systems with matched external disturbance were reported in [79], wherein a memory-based sliding surface was also designed. The authors in [94] proposed some fuzzy adaptive output feedback SMC results for nonlinear systems with mismatched uncertainties, and the bound of external disturbances was estimated by an adaptive law. In [95], an adaptive output feedback SMC design approach was proposed for delta operator fuzzy systems. Note that the above-mentioned fuzzy SMC results in [86, 87, 88, 89, 90, 91, 92, 93, 94, 95] are generally obtained for T-S fuzzy systems with linear local models. In view of the T-S fuzzy affine models having much more enhanced approximation competence, the state feedback SMC and output feedback SMC problems for various types of complex nonlinear systems through T-S fuzzy affine models and piecewise quadratic Lyapunov functions (PQLFs) still need further study.

1.3 USEFUL LEMMAS

In this section, several useful preliminary lemmas are presented for later use.

Lemma A1([25]). Let symmetric matrices $M = M^{\mathrm{T}}$, S, N, and $\Delta(t)$ be real matrices with appropriate dimensions; the inequality $M + S\Delta(t)N + N^{\mathrm{T}}\Delta^{\mathrm{T}}(t)S^{\mathrm{T}} < 0$ holds for all $\Delta^{\mathrm{T}}(t)\Delta(t) < \mathbf{I}$ if and only if for some positive scalar $\varepsilon > 0$ such that $M + \varepsilon SS^{\mathrm{T}} + \varepsilon^{-1}N^{\mathrm{T}}N < 0$.

Lemma A2(Schur Complement [100]). Given matrices $S_1 = S_1^{\mathrm{T}} \in \Re^{n\times n}$, $S_2 \in \Re^{n\times m}$, and $S_3 \in \Re^{m\times m}$, the following conditions are equivalent:

$$\begin{bmatrix} S_1 & S_2 \\ S_2^{\mathrm{T}} & S_3 \end{bmatrix} < 0; \tag{1.7}$$

$$S_1 < 0, S_3 - S_2^{\mathrm{T}}S_1^{-1}S_2 < 0; \tag{1.8}$$

$$S_3 < 0, S_3 - S_2S_3^{-1}S_2^{\mathrm{T}} < 0. \tag{1.9}$$

Schur complement plays an essential role in transforming a nonlinear matrix inequality into an LMI or LMIs as in the 1st item.

Lemma A3(Matrix Inversion Lemma [99]). Given invertible matrices X and Q, and matrices Y, Z with appropriate dimensions, the following condition holds,

$$(X + YQZ)^{-1} = X^{-1} - X^{-1}Y(Q^{-1} + ZX^{-1}Y)^{-1}ZX^{-1}. \tag{1.10}$$

1.4 PREVIEW OF CHAPTERS

This monograph mainly contributes to the sliding mode control problem for complex nonlinear systems by fuzzy model-based methods. Note that the discussed complex nonlinear systems including nonlinear one-dimensional (1-D) systems, nonlinear two-dimensional (2-D) systems, and multiple time-scale dynamical systems (or known as singularly perturbed systems) are extensively used in many practical situations. In this monograph, the sliding mode control issue for these complex systems will be fully studied by means of both state feedback and output feedback methods. Novel stability and robust performance analysis results on the sliding motion are proposed under a universal convex optimization framework, and the sliding surface gain can also be calculated efficiently. Previews of chapters are given as follows.

Chapter 1 shows the research background, motivations, and research problems, which contain model description and stability analysis methods of T-S fuzzy affine systems, and some prevalent fuzzy sliding mode control results on these complex systems.

Chapter 2 investigates the state feedback SMC problem for continuous-time nonlinear 1-D systems through T-S fuzzy affine models. New stability analysis results on the sliding motion are obtained based on CQLF and PQLFs, respectively, and the sliding surface gain can be calculated by solving a set of linear matrix inequalities. Two state feedback SMC design methods are proposed to guarantee that the sliding surface can be reached in finite time.

Chapter 3 proposes some new output feedback SMC design methods for continuous-time nonlinear 1-D systems by T-S fuzzy affine models. Through CQLF and PQLFs, respectively, new stability analysis results on the sliding motion are obtained, and the sliding surface gain can be computed in a convex optimization setup. Then two new output feedback sliding mode controllers are designed to guarantee the local finite-time reachability of the sliding surface.

Chapter 4 studies the state feedback SMC problem for discrete-time nonlinear 1-D systems subject to mismatched parameter uncertainties and external nonlinear disturbance based on T-S fuzzy affine models. A descriptor system is established to characterize the holonomic sliding mode dynamics. The asymptotical stability conditions for the sliding motion are derived under a convex optimization framework based on CQLF and PQLF, respectively. Considering the local models sharing one common control input matrix or not, two-state feedback SMC laws are designed to drive the system states into a neighbourhood of the sliding surface.

Chapter 5 focuses on the fuzzy output feedback SMC for discrete-time nonlinear 1-D systems based on T-S fuzzy affine models. New results on asymptotic stability and robust performance analysis for the sliding motion are proposed under the framework of CQLF and PQLFs, respectively. Considering the upper-bound of the sliding patch being known or not, some new output feedback SMC design approaches are proposed to drive the resulting closed-loop system states into a neighbourhood of the designed sliding surface locally for a prescribed sliding region.

Chapter 6 studies the integral SMC problem for nonlinear SPSs subject to mismatched external disturbances by T-S fuzzy models. Some new stability and robust performance analysis results for sliding motion are proposed under a convex optimization framework. Then a dynamic sliding mode controller is synthesized to drive the system states onto the designed integral sliding surface in finite time.

In Chapter 7, the dynamic SMC design problem for continuous-time nonlinear 2-D systems with mismatched exogenous disturbance is investigated through T-S fuzzy affine models. Associated with the sliding surface, the sliding mode dynamics are embedded into a singular 2-D system. Based on PQLFs, some novel stability and robust performance analysis results on the sliding mode dynamical system are proposed. A robust state feedback dynamic sliding mode controller and an output feedback dynamic dynamic sliding mode controller are designed, respectively, such that the finite-time reachability of the sliding surface can be guaranteed.

Chapter 8 investigates the fuzzy SMC problem for discrete-time nonlinear 2-D systems by T-S fuzzy affine models. By using a singular system-based method and combining it with the sliding surface, the sliding mode dynamical properties are described by a singular piecewise affine 2-D system. Stability and robust performance analysis of the sliding motion are carried out based on the piecewise quadratic Lyapunov functions. Both state feedback and output feedback dynamic SMC laws are designed to guarantee that the system states can converge into a neighbourhood of the sliding surface.

2 State Feedback Fuzzy SMC of Continuous-Time Nonlinear Systems

2.1 INTRODUCTION

This chapter studies the state feedback fuzzy SMC problem of continuous-time nonlinear 1-D systems with exogenous disturbances and mismatched parameter uncertainties through T-S fuzzy affine models. In view of the local models sharing one common control input matrix or not, two state feedback SMC design methods are proposed. Firstly, the stability analysis of the sliding motion will be analyzed based on a CQLF and PQLFs, respectively, and the sliding surface gain can be calculated efficiently by solving a set of linear matrix inequalities (LMIs). Then two sliding mode controllers will be designed to ensure that the sliding surface can be reached in finite time. Note that the dynamic SMC method can also handle the mismatched external disturbance. Finally, simulation studies are given to verify the effectiveness of the proposed methods.

2.2 STATE FEEDBACK SMC DESIGN VIA T-S FUZZY AFFINE MODELS

2.2.1 MODEL DESCRIPTION: CASE I

Consider the first equation of the continuous-time T-S fuzzy affine system described as in (1.2) with a common control input matrix B, which is characterized by r fuzzy **IF-THEN** rules as,

Plant Rule $\mathscr{R}^l$: IF $\theta_1(t)$ is $\mathscr{F}_1^l$ and $\cdots$ and $\theta_\varphi(t)$ is $\mathscr{F}_\varphi^l$, THEN

$$\dot{x}(t) = (A_l + \Delta A_l)x(t) + a_l + \Delta a_l + B\left(u(t) + f(t,x(t))\right),\ l \in \mathscr{L} := \{1,\ldots,r\}, \quad (2.1)$$

where B is the common control input matrix with full-column rank; $f(t,x(t)) \in \Re^{n_u}$ refers to the exogenous disturbance, which exists in a matched form. ΔA_l and Δa_l are the parameter uncertainties in the l-th local model satisfying the conditions in (1.3) as $\begin{bmatrix} \Delta A_l & \Delta a_l \end{bmatrix} = H_{l1}\Delta_l(t)\begin{bmatrix} H_{l2} & H_{l3} \end{bmatrix}, l \in \{1,\ldots,r\}$ with H_{l1}, H_{l2}, and H_{l3} known real-valued matrices and $\|\Delta_l(t)\| \leq \mathbf{I}$. It is assumed that the nonlinear disturbance $f(t,x(t))$ satisfies the following condition as

$$\|f(t,x(t))\| \leq \gamma\|x(t)\| \quad (2.2)$$

with $\gamma > 0$.

DOI: 10.1201/9781032695716-2

Utilizing a singleton fuzzifier, product inference and center-average defuzzifier, one can obtain the following global T-S fuzzy affine model,

$$\dot{x}(t) = (A(\mu) + \Delta A(\mu))x(t) + a(\mu) + \Delta a(\mu) + B(u(t) + f(t,x(t))) \tag{2.3}$$

where

$$\begin{cases} A(\mu) = \sum_{l=1}^{r} \mu_l A_l, \Delta A(\mu) = \sum_{l=1}^{r} \mu_l \Delta A_l, \\ a(\mu) = \sum_{l=1}^{r} \mu_l a_l, \Delta a(\mu) = \sum_{l=1}^{r} \mu_l \Delta a_l. \end{cases} \tag{2.4}$$

2.2.2 SLIDING SURFACE DESIGN

In accordance with the system (2.1) or equivalently (2.3), design the following sliding surface as

$$s(t) = Kx(t) \tag{2.5}$$

with $K \in \Re^{n_u \times n_x}$ being the sliding surface gain to be determined.

As the control input matrix B is full column rank, there always exists a full-row rank matrix $M \in \Re^{(n_x - n_u) \times n_x}$ satisfying $MB = 0$. Consequently, one can construct the following nonsingular transformation matrix T as,

$$T = \begin{bmatrix} M^{\mathrm{T}} & B \end{bmatrix}^{\mathrm{T}}. \tag{2.6}$$

Based on (2.6), it is easy to find that $MT^{-1} = \begin{bmatrix} \mathbf{I}_{n_x - n_u} & \mathbf{0}_{(n_x - n_u) \times n_u} \end{bmatrix}$. Via a state transformation based on the matrix T, one can derive the system (2.3) as

$$T\dot{x}(t) = T\{[A(\mu) + \Delta A(\mu)]x(t) + B[u(t) + f(t,x(t))] + [a(\mu) + \Delta a(\mu)]\}. \tag{2.7}$$

As fuzzy rules and membership functions induce a polyhedral partition of the system state-space, the global model in (2.7) can be viewed as a convex combination of local models in individual regions. Similar to [25], the premise variable space can be decomposed into two kinds of regions: crisp regions and fuzzy regions. The crisp region is the region that possesses only one rule, that is, for some l, $\mu_l[\theta(t)] = 1$ and all other membership functions are zero. The system dynamics in the crisp region are governed by the l-th local model of (2.1). The fuzzy regions denote the regions where $0 < \mu_l[\theta(t)] < 1$ and the system dynamics are characterized by a convex combination of some local models.

Specify the indices of regions $\mathscr{I} = \mathscr{I}_0 \bigcup \mathscr{I}_1$, and $\mathscr{I}_0$ includes the index set of regions covering the origin, and $\mathscr{I}_1$ involves the index set of regions without the origin. Introduce the following set

$$\mathscr{N}(i) := \{n | \mu_n[\theta(t)] > 0,\ \theta(t) \in S_i,\ n \in \mathscr{L},\ i \in \mathscr{I}\} \tag{2.8}$$

to characterize the indices adopted for every region S_i.

Based on (2.8), the system (2.7) can be reformulated into the piecewise-fuzzy affine form as

$$\begin{aligned} T\dot{x}(t) = T(\mathscr{A}_i + \Delta\mathscr{A}_i)x(t) + T(\mathrm{a}_i + \Delta\mathrm{a}_i) + TB(u(t) + f(t,x(t))), \\ \theta(t) \in S_i, i \in \mathscr{I}, \end{aligned} \tag{2.9}$$

where

$$\begin{cases} \mathscr{A}_i := \sum_{n\in\mathscr{N}(i)} \mu_n A_n, \Delta\mathscr{A}_i := \sum_{n\in\mathscr{N}(i)} \mu_n \Delta A_n, \\ \mathrm{a}_i := \sum_{n\in\mathscr{N}(i)} \mu_n a_n, \Delta\mathrm{a}_i := \sum_{n\in\mathscr{N}(i)} \mu_n \Delta a_n \end{cases} \tag{2.10}$$

with $0 < \mu_n[\theta(t)] \le 1$, $\sum_{n\in\mathscr{N}(i)} \mu_n[\theta(t)] = 1$.

Noticing that $MB = 0$, reformulate the system (2.9) as

$$\begin{cases} M\dot{x}(t) = M(\mathscr{A}_i + \Delta\mathscr{A}_i)x(t) + M(\mathrm{a}_i + \Delta\mathrm{a}_i) \\ B^{\mathrm{T}}\dot{x}(t) = B^{\mathrm{T}}(\mathscr{A}_i + \Delta\mathscr{A}_i)x(t) + B^{\mathrm{T}}B(u(t) + f(t,x(t))) + B^{\mathrm{T}}(\mathrm{a}_i + \Delta\mathrm{a}_i), \end{cases} \quad i \in \mathscr{I} \tag{2.11}$$

with $B^{\mathrm{T}}B$ being nonsingular. Notice that the dynamical feature of the sliding motion is characterized via the first equation in (2.11). Additionally, the sliding surface function should satisfy $s(t) = 0$. Hence, based on the above observations, one has the following holonomic sliding mode dynamical equation

$$E\dot{x}(t) = (\mathbf{A}_i + \Delta\mathbf{A}_i)x(t) + \mathbf{a}_i + \Delta\mathbf{a}_i, \ \theta(t) \in S_i, i \in \mathscr{I}, \tag{2.12}$$

where

$$\begin{cases} E = \begin{bmatrix} M \\ 0 \end{bmatrix}, \mathbf{A}_i = \begin{bmatrix} M\mathscr{A}_i \\ K \end{bmatrix}, \mathbf{a}_i = \begin{bmatrix} M\mathrm{a}_i \\ \mathbf{0} \end{bmatrix}, \\ \Delta\mathbf{A}_i = \begin{bmatrix} M\Delta\mathscr{A}_i \\ \mathbf{0} \end{bmatrix}, \Delta\mathbf{a}_i = \begin{bmatrix} M\Delta\mathrm{a}_i \\ \mathbf{0} \end{bmatrix}. \end{cases} \tag{2.13}$$

Conducting a state transformation $z(t) = Tx(t)$ and associating with the fact that $MT^{-1} = \begin{bmatrix} \mathbf{I}_{n_x-n_u} & \mathbf{0}_{(n_x-n_u)\times n_u} \end{bmatrix}$, reformulate the system (2.12) as

$$\tilde{E}\dot{z}(t) = (\tilde{\mathscr{A}}_i + \Delta\tilde{\mathscr{A}}_i)z(t) + \tilde{\mathrm{a}}_i + \Delta\tilde{\mathrm{a}}_i, \ \tilde{\theta}(t) \in \tilde{S}_i, i \in \mathscr{I}, \tag{2.14}$$

where $\tilde{S}_i$ and $\tilde{\theta}(t)$, respectively, denote the new partitioned space and premise variables via the transformation matrix T, and

$$\begin{cases} \tilde{E} = \begin{bmatrix} \mathbf{I}_{(n_x-n_u)} & \mathbf{0} \\ \mathbf{0} & \mathbf{0}_{n_u} \end{bmatrix}, \tilde{\mathrm{a}}_i + \Delta\tilde{\mathrm{a}}_i = \begin{bmatrix} M(\mathrm{a}_i + \Delta\mathrm{a}_i) \\ \mathbf{0} \end{bmatrix}, \\ \tilde{\mathscr{A}}_i + \Delta\tilde{\mathscr{A}}_i = \begin{bmatrix} M(\mathscr{A}_i + \Delta\mathscr{A}_i)T_1 & M(\mathscr{A}_i + \Delta\mathscr{A}_i)T_2 \\ KT_1 & KT_2 \end{bmatrix}, T^{-1} = \begin{bmatrix} T_1 & T_2 \end{bmatrix} \end{cases} \tag{2.15}$$

and $T_1 \in \Re^{n_x\times(n_x-n_u)}$, $T_2 \in \Re^{n_x\times n_u}$ are known constant matrices.

By combining the sliding surface (2.5) with the first equation in system (2.11), the sliding motion can be fully characterized by the singular system (2.14). Therefore, the existence of the proposed sliding surface is ensured by the admissibility of the system (2.14). In the following, new sufficient conditions for the existence of the sliding surface (2.5) will be proposed, and then a state feedback sliding mode controller will be also designed.

2.2.3 STABILITY ANALYSIS OF SLIDING MOTION

In this subsection, new asymptotic stability analysis results on the sliding motion will be proposed through a common quadratic Lyapunov function (CQLF) and piecewise quadratic Lyapunov functions (PQLFs), respectively.

2.2.3.1 Stability Analysis of Sliding Motion through CQLF

It is assumed in this chapter that an ellipsoid $\mathscr{E}_i$ can be used to outer approximate each polyhedral subspace $\tilde{S}_i$ in (2.14) as,

$$\tilde{S}_i \subseteq \mathscr{E}_i,\ \mathscr{E}_i := \{z \| \|Q_i z + q_i\| \leq 1\}. \tag{2.16}$$

The above outer approximation is useful, especially when $\tilde{S}_i$ are slab subspaces. For this case, the parameters Q_i and q_i in (2.16) are ensured to exist, and the approximation is exact; that is to say, $\tilde{S}_i \subseteq \mathscr{E}_i$ and $\mathscr{E}_i \subseteq \tilde{S}_i$. Particularly, when $\tilde{S}_i$ are slabs subject to

$$\tilde{S}_i = \{z | \alpha_i \leq \chi_i^{\mathrm{T}} z \leq \beta_i\},\ i \in \mathscr{I}, \tag{2.17}$$

where $\alpha_i, \beta_i \in \Re$, $\chi_i \in \Re^{n_x}$, and each slab subspace can be expressed via a degenerate ellipsoid as in (2.16) precisely with

$$Q_i = \frac{2\chi_i^{\mathrm{T}}}{\beta_i - \alpha_i},\ q_i = -\frac{\beta_i + \alpha_i}{\beta_i - \alpha_i}. \tag{2.18}$$

Based on (2.16), the following relationship holds for each subspace

$$\begin{bmatrix} 1 \\ z(t) \end{bmatrix}^{\mathrm{T}} \begin{bmatrix} q_i^{\mathrm{T}} q_i - 1 & q_i^{\mathrm{T}} Q_i \\ \star & Q_i^{\mathrm{T}} Q_i \end{bmatrix} \begin{bmatrix} 1 \\ z(t) \end{bmatrix} \leq 0,\ i \in \mathscr{I}, \tag{2.19}$$

where Q_i can be divided as $Q_i = \begin{bmatrix} Q_{i1} & Q_{i2} \end{bmatrix}$, in which $Q_{i1} \in \Re^{1\times(n_x - n_u)}$ and $Q_{i2} \in \Re^{1\times n_u}$.

Then the asymptotic stability analysis results for the sliding motion based on a CQLF are proposed by the following Theorem.

Theorem 2.1. Given a constant α, the system (2.14) is asymptotically stable, if there exist matrices $0 < P_1 = P_1^{\mathrm{T}} \in \Re^{(n_x-n_u)\times(n_x-n_u)}$, $\bar{K} \in \Re^{n_u \times n_x}$, and two sets of scalars $\{\varepsilon_i > 0, i \in \mathscr{I}\}$, $\{\kappa_i < 0, i \in \mathscr{I}_1\}$, such that the following conditions hold,

$$\begin{bmatrix} \Sigma_{11} & \star & \star \\ \Sigma_{21} & \Sigma_{22} & \star \\ H_{n1}^{\mathrm{T}} M^{\mathrm{T}} P_1 & \mathbf{0} & -\varepsilon_i \mathbf{I} \end{bmatrix} < 0, i \in \mathscr{I}_0, n \in \mathscr{N}(i), \tag{2.20}$$

$$\begin{bmatrix} \kappa_i(q_i^{\mathrm{T}} q_i - 1) + \varepsilon_i H_{n3}^{\mathrm{T}} H_{n3} & \star & \star & \star \\ \kappa_i Q_{i1}^{\mathrm{T}} q_i + \Sigma_{10} & \kappa_i Q_{i1}^{\mathrm{T}} Q_{i1} + \Sigma_{11} & \star & \star \\ \kappa_i Q_{i2}^{\mathrm{T}} q_i + \varepsilon_i T_2^{\mathrm{T}} H_{n2}^{\mathrm{T}} H_{n3} & \kappa_i Q_{i2}^{\mathrm{T}} Q_{i1} + \Sigma_{21} & \kappa_i Q_{i2}^{\mathrm{T}} Q_{i2} + \Sigma_{22} & \star \\ \mathbf{0} & H_{n1}^{\mathrm{T}} M^{\mathrm{T}} P_1 & \mathbf{0} & -\varepsilon_i \mathbf{I} \end{bmatrix} < 0,$$
$$i \in \mathscr{I}_1, n \in \mathscr{N}(i), \tag{2.21}$$

where

$$\begin{cases} \Sigma_{11} = \mathtt{Sym}\{P_1 M A_n T_1 - \alpha J^{\mathrm{T}} \bar{K} T_1\} + \varepsilon_i T_1^{\mathrm{T}} H_{n2}^{\mathrm{T}} H_{n2} T_1, \\ \Sigma_{21} = \bar{K} T_1 + T_2^{\mathrm{T}} A_n^{\mathrm{T}} M^{\mathrm{T}} P_1^{\mathrm{T}} - \alpha T_2^{\mathrm{T}} \bar{K}^{\mathrm{T}} J + \varepsilon_i T_2^{\mathrm{T}} H_{n2}^{\mathrm{T}} H_{n2} T_1, \\ \Sigma_{22} = \mathtt{Sym}\{\bar{K} T_2\} + \varepsilon_i T_2^{\mathrm{T}} H_{n2}^{\mathrm{T}} H_{n2} T_2, \\ \Sigma_{10} = P_1 M a_n + \varepsilon_i T_1^{\mathrm{T}} H_{n2}^{\mathrm{T}} H_{n3}, \end{cases} \tag{2.22}$$

and $J \in \Re^{n_u \times (n_x - n_u)}$ represents an arbitrary matrix.

The gain of the sliding surface (2.5) can be attained by

$$K = P_2^{-\mathrm{T}} \bar{K} \tag{2.23}$$

where $P_2 \in \Re^{n_u \times n_u}$ is an arbitrary nonsingular matrix.

Proof. Without loss of generality, only the more complicated case for $i \in \mathscr{I}_1$ will be proved. For the system (2.14), consider the following common quadratic Lyapunov function (CQLF) as,

$$V(t) = z^{\mathrm{T}}(t) \tilde{E}^{\mathrm{T}} \tilde{P} z(t), \tag{2.24}$$

where $\tilde{P} = \begin{bmatrix} P_1 & \mathbf{0} \\ P_3 & P_2 \end{bmatrix}$, $0 < P_1 = P_1^{\mathrm{T}} \in \Re^{(n_x - n_u) \times (n_x - n_u)}$, $P_2 \in \Re^{n_u \times n_u}$ is an arbitrary invertible matrix, and $P_3 \in \Re^{n_u \times (n_x - n_u)}$.

The system (2.14) is asymptotically stable if the following inequality holds,

$$\dot{V}(t) = \mathtt{Sym}\left\{ \left[(\tilde{\mathscr{A}}_i + \Delta \tilde{\mathscr{A}}_i) z(t) + \tilde{\mathrm{a}}_i + \Delta \tilde{\mathrm{a}}_i \right]^{\mathrm{T}} \tilde{P} z(t) \right\} < 0. \tag{2.25}$$

Based on the partition information given in (2.19) and exploiting the S-procedure, then for $\kappa_i < 0$, $i \in \mathscr{I}_1$, inequality (2.26) indicates (2.25),

$$\begin{bmatrix} 1 \\ z(t) \end{bmatrix}^{\mathrm{T}} \left\{ \begin{bmatrix} \mathbf{0} & \star \\ \tilde{P}^{\mathrm{T}} (\tilde{\mathrm{a}}_i + \Delta \tilde{\mathrm{a}}_i) & \mathtt{Sym}\{(\tilde{\mathscr{A}}_i + \Delta \tilde{\mathscr{A}}_i)^{\mathrm{T}} \tilde{P}\} \end{bmatrix} + \kappa_i \begin{bmatrix} q_i^{\mathrm{T}} q_i - 1 & \star \\ Q_i^{\mathrm{T}} q_i & Q_i^{\mathrm{T}} Q_i \end{bmatrix} \right\} \begin{bmatrix} 1 \\ z(t) \end{bmatrix} < 0, i \in \mathscr{I}_1. \tag{2.26}$$

Define $\bar{K} = P_2^{\mathrm{T}} K$. Based on the structure of Q_i and $\tilde{P}$, the following inequality implies (2.26),

$$\begin{bmatrix} \kappa_i (q_i^{\mathrm{T}} q_i - 1) & \star & \star \\ \kappa_i Q_{i1}^{\mathrm{T}} q_i + P_1 M (\mathrm{a}_i + \Delta \mathrm{a}_i) & \kappa_i Q_{i1}^{\mathrm{T}} Q_{i1} + \Theta_{11} & \star \\ \kappa_i Q_{i2}^{\mathrm{T}} q_i & \kappa_i Q_{i2}^{\mathrm{T}} Q_{i1} + \Theta_{21} & \kappa_i Q_{i2}^{\mathrm{T}} Q_{i2} + \mathtt{Sym}\{\bar{K} T_2\} \end{bmatrix} < 0, \quad i \in \mathscr{I}_1, \tag{2.27}$$

where

$$\begin{cases} \Theta_{11} = \mathtt{Sym}\{P_1 M (\mathscr{A}_i + \Delta \mathscr{A}_i) T_1 + P_3^{\mathrm{T}} K T_1\}, \\ \Theta_{21} = \bar{K} T_1 + T_2^{\mathrm{T}} (\mathscr{A}_i^{\mathrm{T}} + \Delta \mathscr{A}_i^{\mathrm{T}}) M^{\mathrm{T}} P_1^{\mathrm{T}} + T_2^{\mathrm{T}} K^{\mathrm{T}} P_3. \end{cases} \tag{2.28}$$

Extract the MFs and use Lemma A1 to tackle the parameter uncertainties in (2.27). Then for $\varepsilon_i > 0$, $i \in \mathscr{I}$, the subsequent inequality implies (2.27),

$$\begin{bmatrix} \kappa_i(q_i^{\mathrm{T}}q_i - 1) & \star & \star \\ P_1MA_n + \kappa_i Q_{i1}^{\mathrm{T}}q_i & \kappa_i Q_{i1}^{\mathrm{T}}Q_{i1} + \Upsilon_{11} & \star \\ \kappa_i Q_{i2}^{\mathrm{T}}q_i & \kappa_i Q_{i2}^{\mathrm{T}}Q_{i1} + \Upsilon_{21} & \kappa_i Q_{i2}^{\mathrm{T}}Q_{i2} + \mathtt{Sym}\{\bar{K}T_2\} \end{bmatrix}$$
$$+\varepsilon_i^{-1}\begin{bmatrix} \mathbf{0} \\ P_1MH_{n1} \\ \mathbf{0} \end{bmatrix}(\star) + \varepsilon_i \begin{bmatrix} H_{n3}^{\mathrm{T}} \\ T_1^{\mathrm{T}}H_{n2}^{\mathrm{T}} \\ T_2^{\mathrm{T}}H_{n2}^{\mathrm{T}} \end{bmatrix}(\star) < 0, i \in \mathscr{I}_1, n \in \mathscr{N}(i), \tag{2.29}$$

where

$$\begin{cases} \Upsilon_{11} = \mathtt{Sym}\{P_1MA_nT_1 + P_3^{\mathrm{T}}KT_1\}, \\ \Upsilon_{21} = \bar{K}T_1 + T_2^{\mathrm{T}}A_n^{\mathrm{T}}M^{\mathrm{T}}P_1^{\mathrm{T}} + T_2^{\mathrm{T}}K^{\mathrm{T}}P_3. \end{cases} \tag{2.30}$$

For numerical tractability, specify P_3 to be $P_3 = -\alpha P_2 J$, where α is a scalar and $J \in \Re^{n_u \times (n_x - n_u)}$ is an arbitrary matrix.

Then by Schur complement, (2.21) implies (2.29). Then the proof is completed. ■

Remark 2.2.1. The results in Theorem 2.1 are derived based on a CQLF, which tends to be conservative. To this end, a novel stability analysis method will be presented based on PQLFs in the following subsection.

2.2.3.2 Stability Analysis of Sliding Motion through PQLFs

Introduce the following notation firstly

$$\bar{z}_1(t) := \begin{bmatrix} 1 \\ z_1(t) \end{bmatrix}, \tag{2.31}$$

where $z_1(t) \in \Re^{n_x - n_u}$ denotes the first $(n_x - n_u)$ rows of $z(t)$.

To ensure the continuity of PQLFs aross the region boundaries, the matrices $\bar{F}_i = \begin{bmatrix} f_i & F_i \end{bmatrix}$, $i \in \mathscr{I}$ are constructed to characterize the boundary across the regions with $f_i = 0$ for $i \in \mathscr{I}_0$ as in [25],

$$\bar{F}_i \begin{bmatrix} 1 \\ z_1(t) \end{bmatrix} = \bar{F}_j \begin{bmatrix} 1 \\ z_1(t) \end{bmatrix}, z_1(t) \in \tilde{S}_i \bigcap \tilde{S}_j, i, j \in \mathscr{I}. \tag{2.32}$$

The S-procedure will be also used to further reduce the conservatism via constructing matrices $\bar{G}_i = \begin{bmatrix} g_i & G_i \end{bmatrix}$, with $g_i = 0$ for $i \in \mathscr{I}_0$ and

$$\bar{G}_i \bar{z}_1(t) \succeq 0, z_1(t) \in \tilde{S}_i, i \in \mathscr{I}, \tag{2.33}$$

where $\succeq$ implies that each entry is nonnegative.

For future use, define

$$\begin{bmatrix} \hat{F}_i & \hat{G}_i \end{bmatrix} = \begin{cases} \begin{bmatrix} F_i & G_i \end{bmatrix}, i \in \mathscr{I}_0, \\ \begin{bmatrix} \bar{F}_i & \bar{G}_i \end{bmatrix}, i \in \mathscr{I}_1. \end{cases} \tag{2.34}$$

The system (2.14) can be reformulated into the following compact form,

$$\hat{E}\dot{\hat{z}}(t) = \left(\hat{\mathscr{A}}_i + \Delta\hat{\mathscr{A}}_i\right)\hat{z}(t),\ i \in \mathscr{I}, \tag{2.35}$$

where

$$\left[\ \hat{E} \mid \hat{z}(t) \mid \hat{\mathscr{A}}_i + \Delta\hat{\mathscr{A}}_i\ \right] = \begin{cases} \left[\ \tilde{E} \mid z(t) \mid \tilde{\mathscr{A}}_i + \Delta\tilde{\mathscr{A}}_i\ \right], i \in \mathscr{I}_0, \\ \left[\ \bar{E} \mid \bar{z}(t) \mid \bar{\mathscr{A}}_i + \Delta\bar{\mathscr{A}}_i\ \right], i \in \mathscr{I}_1 \end{cases} \tag{2.36}$$

with $\tilde{E}$, $\tilde{\mathrm{a}}_i + \Delta\tilde{\mathrm{a}}_i$, and $\tilde{\mathscr{A}}_i + \Delta\tilde{\mathscr{A}}_i$ defined in (2.15) and

$$\begin{cases} \bar{E} = \begin{bmatrix} 1 & \mathbf{0} \\ \mathbf{0} & \tilde{E} \end{bmatrix}, \bar{z}(t) = \begin{bmatrix} 1 \\ z(t) \end{bmatrix}, \\ \bar{\mathscr{A}}_i + \Delta\bar{\mathscr{A}}_i = \begin{bmatrix} 0 & \mathbf{0} \\ \tilde{\mathrm{a}}_i + \Delta\tilde{\mathrm{a}}_i & \tilde{\mathscr{A}}_i + \Delta\tilde{\mathscr{A}}_i \end{bmatrix}. \end{cases} \tag{2.37}$$

Theorem 2.2. For a given constant α, the system (2.35) is asymptotically stable, if matrices $X = X^{\mathrm{T}}$, $U_i = U_i^{\mathrm{T}} \succeq 0$, $W_i = W_i^{\mathrm{T}} \succeq 0$, $\bar{K} \in \Re^{n_u \times n_y}$, and scalars $\varepsilon_i > 0$, $i \in \mathscr{I}$ exist, such that the conditions in (2.38)–(2.39) hold

$$\hat{F}_i^{\mathrm{T}} X \hat{F}_i - \hat{G}_i^{\mathrm{T}} U_i \hat{G}_i > 0, i \in \mathscr{I}, \tag{2.38}$$

$$\begin{bmatrix} \hat{\Lambda}_1 & \star & \star \\ \hat{\Lambda}_2 & \mathtt{Sym}\{\bar{K}T_2\} + \varepsilon_i T_2^{\mathrm{T}} H_{n2}^{\mathrm{T}} H_{n2} T_2 & \star \\ H_{n1}^{\mathrm{T}} M^{\mathrm{T}} \hat{R} \hat{F}_i^{\mathrm{T}} X \hat{F}_i & \mathbf{0} & -\varepsilon_i \mathbf{I} \end{bmatrix} < 0, \quad n \in \mathscr{N}(i), i \in \mathscr{I}, \tag{2.39}$$

where $\hat{F}_i$ and $\hat{G}_i$ are defined in (2.34) and

$$\begin{cases} \hat{\Lambda}_1 = \mathtt{Sym}\left\{\hat{F}_i^{\mathrm{T}} X \hat{F}_i \hat{A}_{n1} - \alpha \hat{J}\bar{K}T_1\hat{R}\right\} + \varepsilon_i \hat{\Lambda}_{1(1)} \hat{\Lambda}_{1(1)}^{\mathrm{T}} + \hat{G}_i^{\mathrm{T}} W_i \hat{G}_i, \\ \hat{\Lambda}_2 = \bar{K}T_1\hat{R} + \hat{A}_{n2}^{\mathrm{T}} \hat{F}_i^{\mathrm{T}} X \hat{F}_i - \alpha T_2^{\mathrm{T}} \bar{K}^{\mathrm{T}} \hat{J} + \varepsilon_i T_2^{\mathrm{T}} H_{n2}^{\mathrm{T}} \hat{\Lambda}_{1(1)}^{\mathrm{T}}, \\ \left.\begin{array}{l} \hat{\Lambda}_{1(1)} = T_1^{\mathrm{T}} H_{n2}^{\mathrm{T}},\ \hat{A}_{n1} = MA_n T_1, \\ \hat{A}_{n2} = MA_n T_2, \hat{J} = \tilde{J}, \hat{R} = \mathbf{I}_{(n_x - n_u)}, \end{array}\right\} \text{if } i \in \mathscr{I}_0, \\ \left.\begin{array}{l} \hat{\Lambda}_{1(1)} = \begin{bmatrix} H_{n3}^{\mathrm{T}} \\ T_1^{\mathrm{T}} H_{n2}^{\mathrm{T}} \end{bmatrix}, \hat{A}_{n1} = \begin{bmatrix} 0 & \mathbf{0} \\ Ma_n & MA_n T_1 \end{bmatrix}, \\ \hat{A}_{n2} = \begin{bmatrix} \mathbf{0} \\ MA_n T_2 \end{bmatrix}, \hat{J} = \bar{J}, \hat{R} = \begin{bmatrix} \mathbf{0} & \mathbf{I}_{(n_x - n_u)} \end{bmatrix} \end{array}\right\} \text{if } i \in \mathscr{I}_1, \end{cases} \tag{2.40}$$

and $\tilde{J} \in \Re^{n_u \times (n_x - n_u)}$, $\bar{J} \in \Re^{n_u \times (1 + n_x - n_u)}$ are arbitrary matrices.

The gain of the sliding surface (2.5) can be attained by

$$K = P_2^{-\mathrm{T}} \bar{K}, \tag{2.41}$$

where $P_2 \in \Re^{n_u \times n_u}$ is an arbitrary invertible matrix.

Proof. For the sliding motion (2.35), construct the following PQLF continuous across the region boundaries as,

$$V(t) = \hat{z}^{\mathrm{T}}(t) \hat{E} \hat{\mathscr{P}}_i \hat{z}(t) = \begin{cases} z^{\mathrm{T}}(t) \tilde{E} \tilde{\mathscr{P}}_i z(t), & i \in \mathscr{I}_0, \\ \bar{z}^{\mathrm{T}}(t) \bar{E} \bar{\mathscr{P}}_i \bar{z}(t), & i \in \mathscr{I}_1, \end{cases} \tag{2.42}$$

where

$$\hat{\mathscr{P}}_i = \begin{bmatrix} \hat{F}_i^{\mathrm{T}} X \hat{F}_i & \mathbf{0} \\ \hat{P}_3 & P_2 \end{bmatrix} = \begin{cases} \tilde{\mathscr{P}}_i = \begin{bmatrix} F_i^{\mathrm{T}} X F_i & \mathbf{0} \\ \tilde{P}_3 & P_2 \end{bmatrix}, i \in \mathscr{I}_0, \\ \bar{\mathscr{P}}_i = \begin{bmatrix} \bar{F}_i^{\mathrm{T}} X \bar{F}_i & \mathbf{0} \\ \bar{P}_3 & P_2 \end{bmatrix}, i \in \mathscr{I}_1, \end{cases} \tag{2.43}$$

and $X = X^{\mathrm{T}}$, $P_2 \in \Re^{n_u \times n_u}$ is an arbitrary nonsingular matrix, and $\tilde{P}_3 \in \Re^{n_u \times (n_x - n_u)}$, $\bar{P}_3 \in \Re^{n_u \times (1 + n_x - n_u)}$.

The inequality in (2.38) guarantees the positive definiteness of PQLF in each subspace. According to the particular structure of $\hat{E}$, we have $\hat{E}\hat{\mathscr{P}}_i = \hat{\mathscr{P}}_i^{\mathrm{T}}\hat{E}$. If the following inequality holds, the system (2.35) is asymptotically stable,

$$\dot{V}(t) = \hat{z}^{\mathrm{T}}(t)\left\{\left(\hat{\mathscr{A}}_i^{\mathrm{T}} + \Delta\hat{\mathscr{A}}_i^{\mathrm{T}}\right)\hat{\mathscr{P}}_i + \hat{\mathscr{P}}_i^{\mathrm{T}}\left(\hat{\mathscr{A}}_i + \Delta\hat{\mathscr{A}}_i\right)\right\}\hat{z}(t) < 0. \tag{2.44}$$

Define $\bar{K} = P_2^{\mathrm{T}} K$. Based on the special structure of $\hat{\mathscr{P}}_i$ shown in (2.43) and using S-procedure, the following inequality implies (2.44),

$$\begin{bmatrix} \mathtt{Sym}\left\{\hat{F}_i^{\mathrm{T}} X \hat{F}_i(\hat{\mathscr{A}}_{i1} + \Delta\hat{\mathscr{A}}_{i1}) + \hat{P}_3^{\mathrm{T}} K T_1\right\} + \hat{G}_i^{\mathrm{T}} W_i \hat{G}_i & \star \\ \bar{K} T_1 \hat{R} + (\hat{\mathscr{A}}_{i2}^{\mathrm{T}} + \Delta\hat{\mathscr{A}}_{i2}^{\mathrm{T}})\hat{F}_i^{\mathrm{T}} X \hat{F}_i + T_2^{\mathrm{T}} K^{\mathrm{T}} \hat{P}_3 & \mathtt{Sym}\{\bar{K} T_2\} \end{bmatrix} < 0, \tag{2.45}$$

where $W_i \succeq 0$ and

$$\begin{cases} \left[\hat{\mathscr{A}}_{i1} \mid \hat{\mathscr{A}}_{i2} \mid \Delta\hat{\mathscr{A}}_{i1} \mid \Delta\hat{\mathscr{A}}_{i2} \mid \hat{R}\right] = \begin{cases} \left[\tilde{\mathscr{A}}_{i1} \mid \tilde{\mathscr{A}}_{i2} \mid \Delta\tilde{\mathscr{A}}_{i1} \mid \Delta\tilde{\mathscr{A}}_{i2} \mid \tilde{R}\right], i \in \mathscr{I}_0, \\ \left[\bar{\mathscr{A}}_{i1} \mid \bar{\mathscr{A}}_{i2} \mid \Delta\bar{\mathscr{A}}_{i1} \mid \Delta\bar{\mathscr{A}}_{i2} \mid \bar{R}\right], i \in \mathscr{I}_1, \end{cases} \\ \tilde{\mathscr{A}}_{i1} = M\mathscr{A}_i T_1, \tilde{\mathscr{A}}_{i2} = M\mathscr{A}_i T_2, \\ \Delta\tilde{\mathscr{A}}_{i1} = M\Delta\mathscr{A}_i T_1, \Delta\tilde{\mathscr{A}}_{i2} = M\Delta\mathscr{A}_i T_2, \tilde{R} = \mathbf{I}_{(n_x - n_u)}, \\ \bar{\mathscr{A}}_{i1} = \begin{bmatrix} 0 & \mathbf{0} \\ M\mathrm{a}_i & \tilde{\mathscr{A}}_{i1} \end{bmatrix}, \bar{\mathscr{A}}_{i2} = \begin{bmatrix} \mathbf{0} \\ \tilde{\mathscr{A}}_{i2} \end{bmatrix}, \Delta\bar{\mathscr{A}}_{i1} = \begin{bmatrix} 0 & \mathbf{0} \\ M\Delta\mathrm{a}_i & \Delta\tilde{\mathscr{A}}_{i1} \end{bmatrix}, \\ \Delta\bar{\mathscr{A}}_{i2} = \begin{bmatrix} \mathbf{0} \\ \Delta\tilde{\mathscr{A}}_{i2} \end{bmatrix}, \bar{R} = \begin{bmatrix} \mathbf{0} & \mathbf{I}_{(n_x - n_u)} \end{bmatrix}. \end{cases} \tag{2.46}$$

Expand the fuzzy MFs and then employ Lemma A1 to tackle the parameter uncertainties in (2.45). For scalars $\varepsilon_i > 0$, $i \in \mathscr{I}$, the subsequent inequality indicates (2.45),

$$\begin{aligned} &\begin{bmatrix} \mathtt{Sym}\left\{\hat{F}_i^{\mathrm{T}} X \hat{F}_i \hat{A}_{n1} + \hat{P}_3 K T_1\right\} + \hat{G}_i^{\mathrm{T}} W_i \hat{G}_i & \star \\ \bar{K} T_1 \hat{R} + \hat{A}_{n2}^{\mathrm{T}} \hat{F}_i^{\mathrm{T}} X \hat{F}_i + T_2^{\mathrm{T}} K^{\mathrm{T}} \hat{P}_3 & \mathtt{Sym}\{\bar{K} T_2\} \end{bmatrix} + \varepsilon_i \begin{bmatrix} \hat{\Lambda}_{1(1)} \\ T_2^{\mathrm{T}} H_{n2}^{\mathrm{T}} \end{bmatrix} (\star) \\ &\qquad + \varepsilon_i^{-1} \begin{bmatrix} \hat{F}_i^{\mathrm{T}} X \hat{F}_i \hat{R}^{\mathrm{T}} M H_{n1} \\ \mathbf{0} \end{bmatrix} (\star) < 0, n \in \mathscr{N}(i), i \in \mathscr{I}, \end{aligned} \tag{2.47}$$

where

$$\begin{cases} \left[\hat{\Lambda}_{1(1)} \mid \hat{A}_{n1} \mid \hat{A}_{n2}\right] = \begin{cases} \left[\tilde{\Lambda}_{1(1)} \mid \tilde{A}_{n1} \mid \tilde{A}_{n2}\right], i \in \mathscr{I}_0, \\ \left[\bar{\Lambda}_{1(1)} \mid \bar{A}_{n1} \mid \bar{A}_{n2}\right], i \in \mathscr{I}_1, \end{cases} \\ \tilde{\Lambda}_{1(1)} = T_1^{\mathrm{T}} H_{n2}^{\mathrm{T}}, \tilde{A}_{n1} = M A_n T_1, \tilde{A}_{n2} = M A_n T_2, \\ \bar{\Lambda}_{1(1)} = \begin{bmatrix} H_{n3}^{\mathrm{T}} \\ T_1^{\mathrm{T}} H_{n2}^{\mathrm{T}} \end{bmatrix}, \bar{A}_{n1} = \begin{bmatrix} 0 & \mathbf{0} \\ M a_n & M A_n T_1 \end{bmatrix}, \bar{A}_{n2} = \begin{bmatrix} \mathbf{0} \\ M A_n T_2 \end{bmatrix}. \end{cases} \tag{2.48}$$

Note that the gain K is not involved within the first row of $\hat{A}_{n1}$ and $\hat{A}_{n2}$. For numerical tractability, specify $\hat{P}_3$ to be

$$\hat{P}_3 = -\alpha P_2 \hat{J} = \begin{cases} -\alpha P_2 \tilde{J}, i \in \mathscr{I}_0, \\ -\alpha P_2 \bar{J}, i \in \mathscr{I}_1, \end{cases} \tag{2.49}$$

where α represents a constant, and $\tilde{J} \in \Re^{n_u \times (n_x - n_u)}$ and $\bar{J} \in \Re^{n_u \times (1+n_x-n_u)}$ are arbitrary matrices.

Then (2.39) implies (2.47) based on Schur complement. This ends the proof. ■

Remark 2.2.2. Note that the transformation matrix T given in (2.6) is not unique. As the input matrix B is of full column rank, singular valuable decomposition can be conducted to matrix B as

$$B = \mathscr{J}_1 \begin{bmatrix} \mathscr{J}_2 \\ \mathbf{0} \end{bmatrix} \mathscr{J}_3^{\mathrm{T}}, \mathscr{J}_1 \mathscr{J}_1^{\mathrm{T}} = \mathbf{I}_{n_x}, \mathscr{J}_3 \mathscr{J}_3^{\mathrm{T}} = \mathbf{I}_{n_u}, \tag{2.50}$$

where $\mathscr{J}_1 \in \Re^{n_x \times n_x}$, $\mathscr{J}_2 \in \Re^{n_u \times n_u}$, $\mathscr{J}_3 \in \Re^{n_u \times n_u}$ are nonsingular matrices. From this perspective, one can further parameterize the matrix M given in (2.6) as

$$M = \Phi \begin{bmatrix} \mathbf{0} & \mathbf{I}_{(n_x - n_u)} \end{bmatrix} \mathscr{J}_1^{\mathrm{T}}, \tag{2.51}$$

where $\Phi \in \Re^{(n_x - n_u) \times (n_x - n_u)}$ is a nonsingular matrix. Owing to the nonsingular feature of Φ, the choice of M is not related to the asymptotic stability of system (2.14).

Remark 2.2.3. Particularly, if the LMIs in conditions (2.20)–(2.21) or (2.38)–(2.39) are feasible, one can obtain the corresponding sliding surface gain by $K = P_2^{-\mathrm{T}} \bar{K}$, and P_2 is an arbitrary invertible matrix. Notice that the sliding surface gain K is generally not unique, and for simplicity, one can select P_2 to be $\mathbf{I}$.

In the following, a state feedback sliding mode controller will be designed to drive the system states onto the sliding surface (2.5) in finite time.

2.3 STATE FEEDBACK SLIDING MODE CONTROLLER DESIGN

In view of the external disturbance existing in a matched form, a state feedback sliding mode controller will be designed in this section.

2.3.1 SLIDING MODE CONTROLLER DESIGN

Theorem 2.3. For the sliding surface (2.5), the following state feedback sliding mode controller can drive the system state trajectories of system (2.1) onto the sliding surface in finite time,

$$u(t) = -(KB)^{-1} \sum_{l=1}^{r} \mu_l \left\{ K(A_l x(t) + a_l) + \zeta(t) \operatorname{sgn}(s(t)) + \eta \operatorname{sgn}(s(t)) \right\}, \tag{2.52}$$

where $\eta > 0$ is a given scalar, and

$$\zeta(t) \geq \sum_{l=1}^{r} \mu_l \left\{ (\|KH_{l1}\| \cdot \|H_{l2}\| + \gamma \|KB\|) \|x(t)\| + \|KH_{l1}\| \cdot \|H_{l3}\| \right\}. \tag{2.53}$$

Proof. Consider the following Lyapunov function for all $t > 0$ as,

$$V(t) = \frac{1}{2}s^{\mathrm{T}}(t)s(t). \tag{2.54}$$

Considering the sliding surface in (2.5), one has

$$\dot{s}(t) = K\dot{x}(t). \tag{2.55}$$

Based on the state feedback sliding mode controller (2.52) and considering the matching condition in (2.2), then one has

$$\begin{aligned}
\dot{V}(t) &= s^{\mathrm{T}}(t)\dot{s}(t)\\
&= s^{\mathrm{T}}(t)\sum_{l=1}^{r}\mu_l K\left[(A_l+\Delta A_l)x(t)+a_l+\Delta a_l+B(u(t)+f(t,x))\right]\\
&= s^{\mathrm{T}}(t)\sum_{l=1}^{r}\mu_l\left\{K(A_l+\Delta A_l)x(t)+K(a_l+\Delta a_l)+KBf(t,x)\right.\\
&\quad \left.-K(A_lx(t)+a_l)\right\}-s^{\mathrm{T}}(t)\left\{\zeta(t)\mathrm{sgn}(s(t))+\eta\,\mathrm{sgn}(s(t))\right\}\\
&\le \sum_{l=1}^{r}\mu_l\left\{(\|K\Delta A_l\|+\gamma\|KB\|)\|x(t)\|+\|\Delta a_l\|\right\}\|s(t)\|\\
&\quad -\zeta(t)\|s(t)\|-\eta\|s(t)\|^2\\
&\le \sum_{l=1}^{r}\mu_l\left\{(\|KH_{l1}\|\cdot\|H_{l2}\|+\gamma\|KB\|)\|x(t)\|+\|KH_{l1}\|\cdot\|H_{l3}\|\right\}\|s(t)\|\\
&\quad -\zeta(t)\|s(t)\|-\eta\|s(t)\|^2.
\end{aligned} \tag{2.56}$$

Notice that

$$\zeta(t) \ge \sum_{l=1}^{r}\mu_l\left\{(\|KH_{l1}\|\cdot\|H_{l2}\|+\gamma\|KB\|)\|x(t)\|+\|KH_{l1}\|\cdot\|H_{l3}\|\right\}, \tag{2.57}$$

and one has

$$\dot{V}(t) \le -\eta\|s(t)\|^2, \tag{2.58}$$

which guarantees the finite-time reachability of the sliding surface (2.5). This ends the proof. ■

2.3.2 ILLUSTRATIVE EXAMPLES

Example 2.1. Consider a nonlinear continuous stirred tank reactor (CSTR) system [40], which is described by the following equation:

$$\begin{cases}
\dot{T} = M_0(T_n - T) + M_1 k_0 \mathrm{e}^{-\frac{M_2}{T}} C_A - M_3 T + M_3\mathbf{u}\\
\dot{C}_A = M_0(C_0 - C_A) - M_0 \mathrm{e}^{-\frac{M_2}{T}} C_A
\end{cases} \tag{2.59}$$

where C_A and T, respectively, represent the concentration and temperature of the reactor. The temperature of the coolant stream is the control input $\mathbf{u}$. Choose $\mathbf{x}_1 =$

Table 2.1
Nominal parameters of CSTR system

Parameter	Nominal Value	Parameter	Nominal Value
k_0	7.2×10^{10} 1/min	T_n	350K
C_0	1mol/L	$-\Delta H$	5×10^4J/mol
C_p	0.239J/gK	E/R	8750K
V	100L	F	100L/min
UA	5×10^4 J/minK	ρ	1000g/L

T and $\mathbf{x}_2 = C_A$ as the system states in this example. The nominal parameters are presented in Table 2.1, where $M_0 = \frac{F}{V}$, $M_1 = \frac{-\Delta H}{\rho C_p}$, $M_2 = \frac{E}{R}$, and $M_3 = \frac{UA}{V\rho C_p}$.

At steady input $\mathbf{u}_o = 300$K, the CSTR system has one equilibrium $\mathbf{x}_o = \begin{bmatrix} 350, & 0.5 \end{bmatrix}^{\mathrm{T}}$. Specially, linearize the nonlinear CSTR system around $\mathbf{x}_{o1} = [324.4 \quad 0.875]^{\mathrm{T}}$, $\mathbf{x}_{o2} = [350 \quad 0.5]^{\mathrm{T}}$, $\mathbf{x}_{o3} = [370.65 \quad 0.2]^{\mathrm{T}}$, and denote $x = \mathbf{x} - \mathbf{x}_{o2}$, $u = \mathbf{u} - \mathbf{u}_o$. Considering the external disturbance, and then system (2.59) can be described by a T-S fuzzy affine model as,

Plant Rule $\mathscr{R}^l$: IF $x_1(t)$ is $\mathscr{F}_1^l$, THEN

$$\dot{x}(t) = (A_l + \Delta A_l)x(t) + a_l + \Delta a_l + B(u(t) + f(t, x(t))),\ l \in \{1,2,3\}, \tag{2.60}$$

where the system matrices are

$$\begin{cases} A_1 = \begin{bmatrix} -0.9757 & 29.0894 \\ -0.0101 & -1.1390 \end{bmatrix}, A_2 = \begin{bmatrix} 4.3790 & 209.1908 \\ -0.0357 & -1.999 \end{bmatrix}, \\ A_3 = \begin{bmatrix} 7.6367 & 842.2443 \\ -0.0513 & -5.0259 \end{bmatrix}, a_1 = \begin{bmatrix} -35.8862 \\ 0.1682 \end{bmatrix}, a_2 = \begin{bmatrix} 0 \\ 0 \end{bmatrix}, \\ a_3 = \begin{bmatrix} 94.9660 \\ -0.4488 \end{bmatrix}, B = \begin{bmatrix} 2.0921 \\ 0 \end{bmatrix}, \end{cases} \tag{2.61}$$

and the parameter uncertainties ΔA_l and Δa_l are given in the form of (1.3) with

$$\begin{cases} H_{l1} = \begin{bmatrix} 0.1 & 0.1 \end{bmatrix}^{\mathrm{T}}, H_{l2} = \begin{bmatrix} 0.05 & 0.05 \end{bmatrix}, \\ H_{l3} = 0.01, l \in \{1,2,3\}. \end{cases} \tag{2.62}$$

Figure 2.1 gives the fuzzy MFs with $d_1 = -50$, $d_2 = -25.6$, $d_3 = 20.65$, and $d_4 = 50$. It can be found from Figure 2.1 that

$$\begin{aligned} S_1 &= \{x \in \Re^{n_x} | d_1 \leq x_1 \leq d_2\}, \\ S_2 &= \{x \in \Re^{n_x} | d_2 \leq x_1 \leq d_3\}, \\ S_3 &= \{x \in \Re^{n_x} | d_3 \leq x_1 \leq d_4\}, \end{aligned} \tag{2.63}$$

where S_2 is a fuzzy region with $\mathscr{N}(2) = \{1,2,3\}$, and S_1 and S_3 are crisp regions with $\mathscr{N}(1) = \{1\}$ and $\mathscr{N}(3) = \{3\}$.

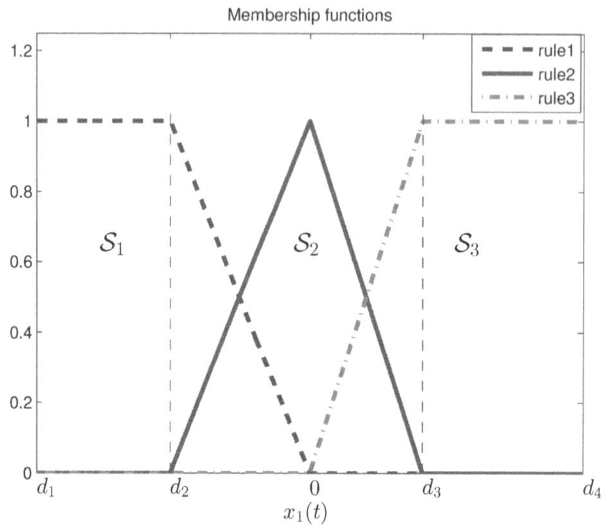

Figure 2.1 Membership functions of Example 2.1.

Based on the method given in this chapter, one can select the transformation matrix T as $T = \begin{bmatrix} 0 & 1 \\ 2.0921 & 0 \end{bmatrix}$, and for this case, $z_2(t) = 2.0921x_1(t)$. One can describe these three regions precisely by degenerate ellipsoids as in (2.16) with

$$\begin{cases} Q_1 = \begin{bmatrix} 0 & \dfrac{2}{d_2 - d_1} \end{bmatrix} \times 2.0921, \\ Q_2 = \begin{bmatrix} 0 & \dfrac{2}{d_3 - d_2} \end{bmatrix} \times 2.0921, \\ Q_3 = \begin{bmatrix} 0 & \dfrac{2}{d_4 - d_3} \end{bmatrix} \times 2.0921, \\ q_1 = \dfrac{d_1 + d_2}{d_1 - d_2}, q_2 = \dfrac{d_3 + d_2}{d_2 - d_3}, q_3 = \dfrac{d_4 + d_3}{d_3 - d_4}. \end{cases} \tag{2.64}$$

In this example, we attempt to design a sliding surface (2.5) and a sliding mode controller (2.52) to drive the state trajectories of the nonlinear CSTR system in (2.60) onto the sliding surface in finite time and guarantee the sliding motion to be asymptotically stable. It is also found that no feasible solution can be acquired by the CQLF-based approach shown in Theorem 2.1. Via employing the PQLFs-based method given in Theorem 2.2 with $\alpha = 1, \tilde{J} = \begin{bmatrix} 50 & 0 \end{bmatrix}, \hat{J} = \begin{bmatrix} 1 & 50 & 0 \end{bmatrix}$, one indeed attains the feasible solutions, and the sliding surface gain is given as $K = \begin{bmatrix} -0.3495 & -2.0425 \end{bmatrix}$ with $P_2 = 1$.

With initial condition $x_0 = \begin{bmatrix} 25 & -0.5 \end{bmatrix}^{\mathrm{T}}$ and nonlinear disturbance function $f(t,x(t)) = \begin{bmatrix} 0.1\sin(t) & 0.3\cos(t) \end{bmatrix} x(t)$, by adopting the sliding mode controller

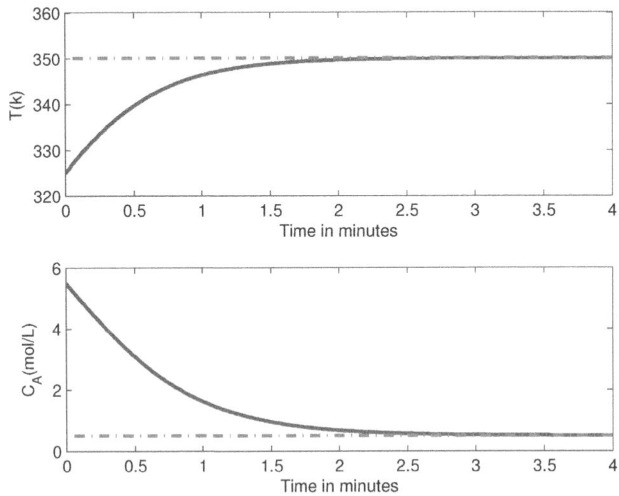

Figure 2.2 The closed-loop system states with sliding mode controller (2.52) in Example 2.1.

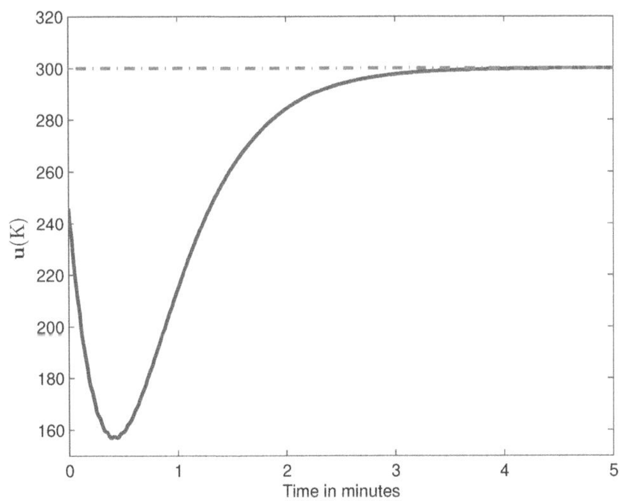

Figure 2.3 The control input with sliding mode controller (2.52) in Example 2.1.

(2.52) with $\gamma = 0.3$ and $\eta = 2.2$, Figure 2.2 and Figure 2.3 show the system states and control input, respectively. The sliding surface is demonstrated in Figure 2.4.

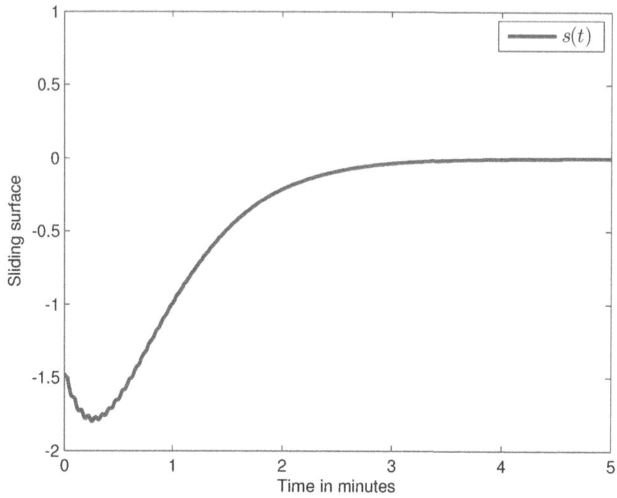

Figure 2.4 The sliding surface with sliding mode controller (2.52) in Example 2.1.

To further show the effectiveness of the proposed state feedback SMC design methods, in the following, consider another classical third-order nonlinear finance system, which is widely studied and has positive enlightening significance for research on some complex economic and financial systems.

Example 2.2. Consider a nonlinear finance system [138, 139], whose dynamics are described by the following equation,

$$\begin{cases} \dot{R}(t) = D(t) + (C(t) - a)R(t) \\ \dot{C}(t) = 1 - bC(t) - R^2(t) \\ \dot{D}(t) = -R(t) - cD(t) \end{cases} \tag{2.65}$$

where $R(t)$ is the interest rate; $C(t)$ is the investment demand; $D(t)$ is the price exponent; $a \geq 0$ denotes the saving amount which means that enterprise unit increases its gross financial; $b \geq 0$ denotes the per-investment cost which is defined as the ratio of original cost less distribution received from target funds; $c \geq 0$ denotes the elasticity of demands of commercials which is a measure of the relationship between a change in the quantity demanded of a particular good and a change in its price.

According to [139], the system (2.65) has a unique equilibrium $(0, 1/b, 0)$ when $c - b - abc \leq 0$. In this example, the parameters a, b, c are chosen as $a = 0.5$, $b = 1.1$, $c = 0.3$. Change the coordinates with $x_1(t) = R(t)$, $x_2(t) = C(t) - 1/b$, $x_3(t) = D(t)$ with $x(t) = \begin{bmatrix} x_1^{\mathrm{T}}(t) & x_2^{\mathrm{T}}(t) & x_3^{\mathrm{T}}(t) \end{bmatrix}^{\mathrm{T}}$, and then the equilibrium has moved to the origin. In order to stabilize the nonlinear finance system (2.65), formulate the system

(2.65) into the following form,

$$\begin{cases} \dot{x}_1(t) = x_3(t) + (x_2(t) + \frac{1}{b} - a)x_1(t) \\ \dot{x}_2(t) = -bx_2(t) - x_1^2(t) + u(t) \\ \dot{x}_3(t) = -x_1(t) - cx_3(t) + u(t) \end{cases} \tag{2.66}$$

where $u(t)$ is the control input.

Linearize the nonlinear finance system (2.66) at three operating points $(0,0,0)$ and $(\pm 0.3,0,0)$, and then the system (2.66) can be described by the following T-S fuzzy affine model with three fuzzy rules as

Plant Rule $\mathscr{R}^l$: IF $x_1(t)$ is $\mathscr{F}_1^l$, THEN

$$\dot{x}(t) = (A_l + \Delta A_l)x(t) + a_l + \Delta a_l + B(u(t) + f(t,x(t))),\ l \in \{1,2,3\}, \tag{2.67}$$

where

$$\begin{cases} A_1 = \begin{bmatrix} 0.4091 & 0 & 1 \\ 0 & -1.1 & 0 \\ -1 & 0 & -0.3 \end{bmatrix}, A_2 = \begin{bmatrix} 0.4091 & -0.3 & 1 \\ 0.6 & -1.1 & 0 \\ -1 & 0 & -0.3 \end{bmatrix}, \\ A_3 = \begin{bmatrix} 0.4091 & 0.3 & 1 \\ -0.6 & -1.1 & 0 \\ -1 & 0 & -0.3 \end{bmatrix}, a_1 = \begin{bmatrix} 0 \\ 0 \\ 0 \end{bmatrix}, a_2 = a_3 = \begin{bmatrix} 0 \\ 0.27 \\ 0 \end{bmatrix}, B = \begin{bmatrix} 0 \\ 1 \\ 1 \end{bmatrix}, \end{cases} \tag{2.68}$$

and the parameter uncertainties ΔA_l and Δa_l are given in the form of (1.3) with

$$\begin{cases} H_{l1} = \begin{bmatrix} 0.01 & 0.01 & 0 \end{bmatrix}^{\mathrm{T}}, H_{l2} = \begin{bmatrix} 0.02 & 0.02 & 0.02 \end{bmatrix}, \\ H_{l3} = 0.01, l \in \{1,2,3\}. \end{cases} \tag{2.69}$$

The fuzzy membership functions are given as in Figure 2.5 with $d_1 = 0.3$ and $d_2 = 1$, and it is found that

$$\begin{aligned} S_1 &= \{x \in \Re^{n_x} | -d_2 \le x_1 \le -d_1\}, \\ S_2 &= \{x \in \Re^{n_x} | -d_1 \le x_1 \le d_1\}, \\ S_3 &= \{x \in \Re^{n_x} | d_1 \le x_1 \le d_2\}, \end{aligned} \tag{2.70}$$

where S_1 and S_3 are fuzzy regions with $\mathscr{N}(1) = \{1,2\}$ and $\mathscr{N}(3) = \{1,3\}$, and S_2 is a crisp region with $\mathscr{N}(2) = \{1\}$. Based on the method given in this chapter, the transformation matrix T can be chosen as $T = \begin{bmatrix} 1 & 0 & 0 \\ 0 & 1 & -1 \\ 0 & 1 & 1 \end{bmatrix}$. These three regions can be precisely described by the degenerate ellipsoids as in (2.16) with

$$\begin{cases} Q_1 = Q_3 = \begin{bmatrix} \frac{2}{d_2 - d_1} & 0 & 0 \end{bmatrix}, Q_2 = \begin{bmatrix} \frac{1}{d_1} & 0 & 0 \end{bmatrix}, \\ q_1 = \frac{d_1 + d_2}{d_2 - d_1}, q_2 = 0, q_3 = \frac{d_1 + d_2}{d_1 - d_2}. \end{cases} \tag{2.71}$$

The purpose is to design a sliding surface (2.5) and a state feedback sliding mode controller (2.52) to drive the state trajectories of the nonlinear finance system in

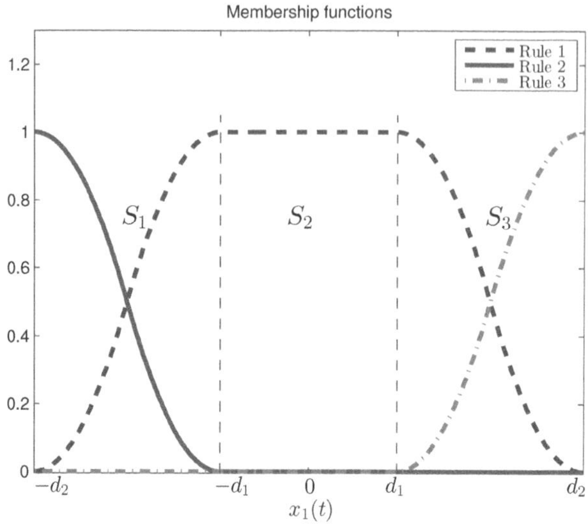

Figure 2.5 Membership functions of Example 2.2–Example 2.4.

(2.67) onto the sliding surface in finite time and guarantee the sliding motion to be asymptotically stable. Utilizing Theorem 3.2 with $\alpha = 1$, $\tilde{J} = \begin{bmatrix} -1 & -1 \end{bmatrix}$, $\hat{J} = \begin{bmatrix} 0 & -1 & -1 \end{bmatrix}$, feasible solutions can be obtained, and the sliding surface gain is given as $K = \begin{bmatrix} -30.1012 & 8.1328 & -22.8836 \end{bmatrix}$ with $P_2 = \mathbf{I}$.

With initial condition $x_0 = \begin{bmatrix} 0.1 & 0 & 0.3 \end{bmatrix}^{\mathrm{T}}$ and nonlinear disturbance function $f(t,x(t)) = \begin{bmatrix} 0.1\sin(t) & 0.2 & 0.3\cos(t) \end{bmatrix} x(t)$, by adopting the sliding mode controller (2.52) with $\gamma = 0.3$ and $\eta = 2.2$, Figure 2.6 and Figure 2.7 show the system states and control input, respectively. The sliding surface is demonstrated in Figure 2.8.

2.4 STATE FEEDBACK DYNAMIC SMC DESIGN VIA T-S FUZZY AFFINE MODELS

Note that the previous state feedback SMC synthesis results are obtained based on an assumption that each local model of T-S fuzzy systems shares one common control input matrix, which tends to be restrictive for many practical situations, for instance, the well-known inverted pendulum system. To relax this assumption, a state feedback dynamic sliding mode controller design method will be proposed in this section.

2.4.1 MODEL DESCRIPTION: CASE II

Consider the first and third equation of the continuous-time T-S fuzzy affine system described as in (1.2), which are characterized by r fuzzy **IF-THEN** rules as,

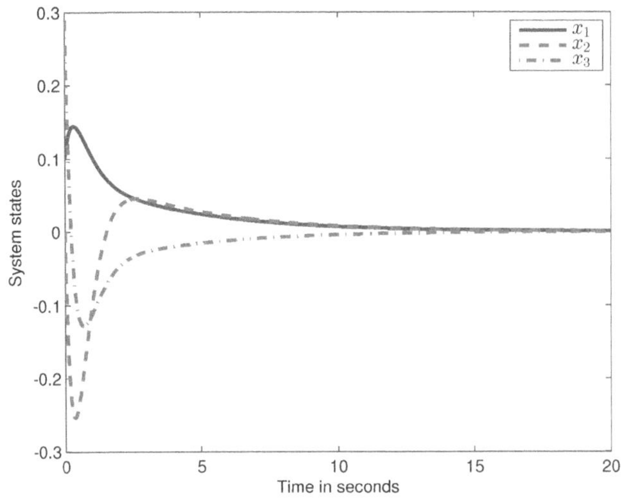

Figure 2.6 The closed-loop system states with sliding mode controller (2.52) in Example 2.2.

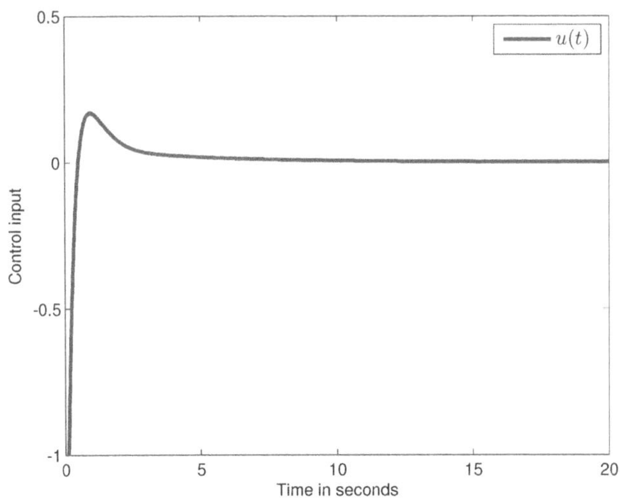

Figure 2.7 The control input with sliding mode controller (2.52) in Example 2.2.

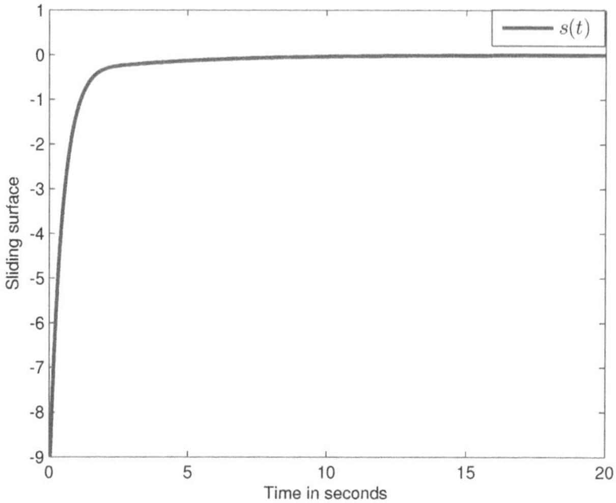

Figure 2.8 The sliding surface with sliding mode controller (2.52) in Example 2.2.

Plant Rule $\mathscr{R}^l$: IF $\theta_1(t)$ is $\mathscr{F}_1^l$ and $\cdots$ and $\theta_\varphi(t)$ is $\mathscr{F}_\varphi^l$, THEN

$$\begin{cases} \dot{x}(t) = (A_l + \Delta A_l)x(t) + a_l + \Delta a_l + (B_l + \Delta B_l)u(t) + D_l w(t) \\ z(t) = L_l x(t) + N_l u(t), \; l \in \mathscr{L} := \{1, \ldots, r\}, \end{cases} \tag{2.72}$$

where the control input matrices are not common and contain parameter uncertainties; $\mathscr{R}^l$ is the l-th fuzzy inference rule; $\mathscr{F}_\phi^l(\phi = 1, \ldots, \varphi)$ represent fuzzy sets; r is the number of inference rules; $\theta(t) := [\theta_1(t), \ldots, \theta_\varphi(t)]$ represent the measurable premise variables; $x(t) \in \Re^{n_x}$ is the system state; $u(t) \in \Re^{n_u}$ is the control input; $w(t) \in \Re^{n_w}$ is the external disturbance belonging to $L_2[0, \infty)$; $z(t) \in \Re^{n_z}$ is the regulated output; ΔA_l, ΔB_l, and Δa_l are the parameter uncertainties satisfying the conditions in (1.3)–(1.4).

Remark 2.4.1 Compared with the fuzzy system model (2.1), each local model does not share one common control input matrix in the T-S fuzzy affine system (2.72), and the external disturbance $w(t)$ is also not in a matched form. When $D_l = B_l = B$ and $\Delta B_l = 0$, the T-S fuzzy affine model (2.72) reduces to the fuzzy model (2.1).

Similarly, the global T-S fuzzy affine model can be obtained as,

$$\begin{cases} \dot{x}(t) = (A(\mu) + \Delta A(\mu))x(t) + a(\mu) + \Delta a(\mu) + (B(\mu) + \Delta B(\mu))u(t) \\ \qquad\quad + D(\mu)w(t) \\ z(t) = L(\mu)x(t) + N(\mu)u(t) \end{cases} \tag{2.73}$$

where

$$\begin{cases} A(\mu) = \sum_{l=1}^{r} \mu_l A_l, \Delta A(\mu) = \sum_{l=1}^{r} \mu_l \Delta A_l, a(\mu) = \sum_{l=1}^{r} \mu_l a_l, \Delta a(\mu) = \sum_{l=1}^{r} \mu_l \Delta a_l, \\ B(\mu) = \sum_{l=1}^{r} \mu_l B_l, \Delta B(\mu) = \sum_{l=1}^{r} \mu_l \Delta B_l, D(\mu) = \sum_{l=1}^{r} \mu_l D_l, L(\mu) = \sum_{l=1}^{r} \mu_l L_l, \\ N(\mu) = \sum_{l=1}^{r} \mu_l N_l. \end{cases} \tag{2.74}$$

Similar to subchapter 2.2, one can divide the premise variable space into fuzzy regions and crisp regions. The fuzzy regions refer to the regions with $0 < \mu_l[\theta(t)] < 1$ and we can use a convex combination of some local models to describe the system dynamics. The crisp region stands for the region containing only one rule.

Denote $\{S_i\}_{i \in \mathscr{I}}$ as the premise variable space partition with $\mathscr{I} = \mathscr{I}_0 \bigcup \mathscr{I}_1$ being the set of region indices, where $\mathscr{I}_0$ represents the index set of regions which contains the origin, and $\mathscr{I}_1$ represents the index set of regions otherwise. The following set is introduced to describe the indices for local models utilized within the interpolation in each region S_i,

$$\mathscr{N}(i) := \{m | \mu_m[\theta(t)] > 0,\ m \in \mathscr{L},\ \theta(t) \in S_i,\ i \in \mathscr{I}\}. \tag{2.75}$$

Based on (2.75), the system (2.73) can be formulated as

$$\begin{cases} \dot{x}(t) = (\mathscr{A}_i + \Delta\mathscr{A}_i)x(t) + \mathrm{a}_i + \Delta\mathrm{a}_i + (\mathscr{B}_i + \Delta\mathscr{B}_i)u(t) + \mathscr{D}_i w(t) \\ z(t) = \mathscr{L}_i x(t) + \mathscr{N}_i u(t), \theta(t) \in S_i, i \in \mathscr{I}, \end{cases} \tag{2.76}$$

where

$$\begin{cases} \mathscr{A}_i := \sum_{m \in \mathscr{N}(i)} \mu_m A_m, \Delta\mathscr{A}_i := \sum_{m \in \mathscr{N}(i)} \mu_m \Delta A_m, \mathrm{a}_i := \sum_{m \in \mathscr{N}(i)} \mu_m a_m, \\ \mathscr{B}_i := \sum_{m \in \mathscr{N}(i)} \mu_m B_m, \Delta\mathscr{B}_i := \sum_{m \in \mathscr{N}(i)} \mu_m \Delta B_m, \Delta\mathrm{a}_i := \sum_{m \in \mathscr{N}(i)} \mu_m \Delta a_m, \\ \mathscr{D}_i := \sum_{m \in \mathscr{N}(i)} \mu_m D_m, \mathscr{L}_i := \sum_{m \in \mathscr{N}(i)} \mu_m L_m, \mathscr{N}_i := \sum_{m \in \mathscr{N}(i)} \mu_m N_m \end{cases} \tag{2.77}$$

with $0 < \mu_m[\theta(t)] \leq 1$, $\sum_{m \in \mathscr{N}(i)} \mu_m[\theta(t)] = 1$.

In the following, a state feedback dynamic sliding mode controller design approach will be proposed for T-S fuzzy affine system (2.72). It will be shown that the finite-time reachability of the sliding surface can be ensured. Some asymptotic stability and robust performance analysis conditions for the sliding motion will be derived through CQLF and PQLFs, respectively. For a prescribed level of disturbance attenuation $\gamma > 0$, the sliding motion is ensured to be asymptotically stable with $w(t) \equiv 0$, and for nonzero $w(t) \in L_2[0, \infty)$, it is satisfied that

$$\int_{t_r}^{\infty} \|z(t)\|^2 dt < \gamma^2 \int_{t_r}^{\infty} \|w(t)\|^2 dt \tag{2.78}$$

under zero initial condition, where t_r represents the time when the system states reach the sliding surface.

2.4.2 SLIDING SURFACE DESIGN

For the fuzzy affine system (2.72) or equivalent (2.76), based on the measurement output and control input, synthesize a new type of sliding surface as

$$s(t) = Kx(t) + u(t) = 0, \tag{2.79}$$

where $K \in \Re^{n_u \times n_x}$ is the sliding surface gain to be determined.

Based on (2.79), one has

$$u(t) = -Kx(t). \tag{2.80}$$

2.4.3 STABILITY AND ROBUST PERFORMANCE ANALYSIS OF SLIDING MOTION

In this section, the asymptotic stability and robust $\mathscr{H}_\infty$ performance analysis of the sliding motion will be conducted through common quadratic Lyapunov function (CQLF) and piecewise quadratic Lyapunov functions (PQLFs), respectively.

2.4.3.1 Stability and Robust Performance Analysis of Sliding Motion through CQLF

Based on (2.80) and the system (2.76), the sliding motion can be derived as

$$\begin{cases} \dot{x}(t) = (\mathbb{A}_i + \Delta\mathbb{A}_i)x(t) + \mathrm{a}_i + \Delta\mathrm{a}_i + \mathscr{D}_i w(t) \\ z(t) = \mathbb{L}_i x(t), i \in \mathscr{I}, \end{cases} \tag{2.81}$$

where $\mathbb{A}_i = \mathscr{A}_i - \mathscr{B}_i K$, $\Delta\mathbb{A}_i = \Delta\mathscr{A}_i - \Delta\mathscr{B}_i K$, $\mathbb{L}_i = \mathscr{L}_i - \mathscr{N}_i K$.

Similar to subchapter 2.2, it is assumed that an ellipsoid $\mathscr{E}_i$ can be used to outer approximate each polyhedral region S_i in (2.76),

$$S_i \subseteq \mathscr{E}_i,\ \mathscr{E}_i := \{x \| \|Q_i x + q_i\| \le 1\}. \tag{2.82}$$

Particularly, when S_i are slabs with $S_i = \{x | \alpha_i \le \chi_i^{\mathrm{T}} x \le \beta_i\}$, $i \in \mathscr{I}$, where $\alpha_i, \beta_i \in \Re$, $\chi_i \in \Re^{n_x}$, and a degenerate ellipsoid as in (2.82) can be adopted to characterize every slab subspace accurately with $Q_i = \dfrac{2\chi_i^{\mathrm{T}}}{\beta_i - \alpha_i}$, $q_i = -\dfrac{\beta_i + \alpha_i}{\beta_i - \alpha_i}$. The following relationship holds for each region,

$$\begin{bmatrix} 1 \\ x(t) \end{bmatrix}^{\mathrm{T}} \begin{bmatrix} q_i^{\mathrm{T}} q_i - 1 & q_i^{\mathrm{T}} Q_i \\ \star & Q_i^{\mathrm{T}} Q_i \end{bmatrix} \begin{bmatrix} 1 \\ x(t) \end{bmatrix} \le 0,\ i \in \mathscr{I}. \tag{2.83}$$

Based on a CQLF, the asymptotic stability and robust $\mathscr{H}_\infty$ performance analysis results for the sliding motion (2.81) will be firstly proposed in Theorem 2.4.

Theorem 2.4. The sliding motion (2.81) is asymptotically stable with robust performance γ, if there exist matrices $0 < \bar{P} = \bar{P}^{\mathrm{T}} \in \Re^{n_x \times n_x}$, $\bar{K} \in \Re^{n_u \times n_x}$, and two sets of scalars $\{\varepsilon_i > 0, i \in \mathscr{I}\}$, $\{\kappa_i < 0, i \in \mathscr{I}_1\}$, such that the following conditions in (2.84)–(2.85) hold,

$$\begin{bmatrix} \mathrm{Sym}\{A_m\bar{P} - B_m\bar{K}\} + \varepsilon_i H_{m1}H_{m1}^{\mathrm{T}} & \star & \star & \star \\ D_m^{\mathrm{T}} & -\gamma^2\mathbf{I} & \star & \star \\ L_m\bar{P} - N_m\bar{K} & \mathbf{0} & -\mathbf{I} & \star \\ H_{m2}\bar{P} - H_{m3}\bar{K} & \mathbf{0} & \mathbf{0} & -\varepsilon_i\mathbf{I} \end{bmatrix} < 0, \quad i \in \mathscr{I}_0, m \in \mathscr{N}(i), \tag{2.84}$$

$$\begin{bmatrix} -\kappa_i(\mathbf{I} - q_iq_i^{\mathrm{T}}) & \star & \star & \star & \star \\ \kappa_i a_m q_i^{\mathrm{T}} + \bar{P}Q_i^{\mathrm{T}} & \Xi_{11} + \kappa_i a_m a_m^{\mathrm{T}} & \star & \star & \star \\ \mathbf{0} & D_m^{\mathrm{T}} & -\gamma^2\mathbf{I} & \star & \star \\ \mathbf{0} & L_m\bar{P} - N_m\bar{K} & \mathbf{0} & -\mathbf{I} & \star \\ \kappa_i H_{m4}q_i^{\mathrm{T}} & \Xi_{41} + \kappa_i H_{m3}a_m^{\mathrm{T}} & \mathbf{0} & \mathbf{0} & -\varepsilon_i\mathbf{I} + \kappa_i H_{m4}H_{m4}^{\mathrm{T}} \end{bmatrix} < 0, \quad i \in \mathscr{I}_1, m \in \mathscr{N}(i), \tag{2.85}$$

where $\Xi_{11} = \mathrm{Sym}\{A_m\bar{P} - B_m\bar{K}\} + \varepsilon_i H_{m1}H_{m1}^{\mathrm{T}}$, $\Xi_{41} = H_{m2}\bar{P} - H_{m3}\bar{K}$.

Then the gain of the sliding surface (2.79) can be attained by

$$K = \bar{K}\bar{P}^{-1}. \tag{2.86}$$

Proof. Only the more complicated case for $i \in \mathscr{I}_1$ will be proved without loss of generality. For the system (2.81), construct the following CQLF as,

$$V(t) = x^{\mathrm{T}}(t)Px(t) \tag{2.87}$$

with $0 < P = P^{\mathrm{T}} \in \Re^{n_x \times n_x}$.

The system (2.81) is asymptotically stable with robust $\mathscr{H}_\infty$ performance γ if the following inequality holds,

$$\dot{V}(t) + z^{\mathrm{T}}(t)z(t) - \gamma^2 w^{\mathrm{T}}(t)w(t) < 0. \tag{2.88}$$

Define $\upsilon(t) = \begin{bmatrix} 1 & x^{\mathrm{T}}(t) & w^{\mathrm{T}}(t) \end{bmatrix}^{\mathrm{T}}$. Along the trajectories of system (2.81), reformulate inequality (2.88) as

$$\upsilon^{\mathrm{T}}(t) \begin{bmatrix} \mathbf{0} & \star & \star \\ P^{\mathrm{T}}(\mathrm{a}_i + \Delta\mathrm{a}_i) & \mathrm{Sym}\{P(\mathbb{A}_i + \Delta\mathbb{A}_i)\} + \mathbb{L}_i^{\mathrm{T}}\mathbb{L}_i & \star \\ \mathbf{0} & \mathscr{D}_i^{\mathrm{T}}P & -\gamma^2\mathbf{I} \end{bmatrix} \upsilon(t) < 0. \tag{2.89}$$

Based on the partition information given in (2.82) and exploiting the S-procedure, then the following inequality implies (2.89) for $\rho_i < 0$, $i \in \mathscr{I}_1$,

$$\begin{bmatrix} \rho_i(q_i^{\mathrm{T}}q_i - 1) & \star & \star \\ P(\mathrm{a}_i + \Delta\mathrm{a}_i) + \rho_i Q_i^{\mathrm{T}}q_i & \mathrm{Sym}\{P(\mathbb{A}_i + \Delta\mathbb{A}_i)\} + \mathbb{L}_i^{\mathrm{T}}\mathbb{L}_i + \rho_i Q_i^{\mathrm{T}}Q_i & \star \\ \mathbf{0} & \mathscr{D}_i^{\mathrm{T}}P & -\gamma^2\mathbf{I} \end{bmatrix} < 0. \tag{2.90}$$

Denote $\bar{P} = P^{-1}$. Multiplying $\text{diag}\{1, \bar{P}, \mathbf{I}_{(n_x+n_z+n_w)}\}$ on both sides of (2.90) and extracting the MFs, then use Lemma A1 to tackle the parameter uncertainties, and one has,

$$\begin{bmatrix} \rho_i(q_i^{\mathrm{T}} q_i - 1) & \star & \star & \star & \star \\ a_m + \rho_i \bar{P} Q_i^{\mathrm{T}} q_i & \mathtt{Sym}\{A_m X - B_m \bar{K}\} + \rho_i \bar{P} Q_i^{\mathrm{T}} Q_i \bar{P} + \varepsilon_i H_{m1}^{\mathrm{T}} H_{m1} & \star & \star & \star \\ \mathbf{0} & D_m^{\mathrm{T}} & -\gamma^2 \mathbf{I} & \star & \star \\ \mathbf{0} & L_m \bar{P} - N_m \bar{K} & \mathbf{0} & -\mathbf{I} & \star \\ H_{m4} & H_{m2} \bar{P} - H_{m3} \bar{K} & \mathbf{0} & \mathbf{0} & -\varepsilon_i \mathbf{I} \end{bmatrix} < 0, \; i \in \mathscr{I}_1, \tag{2.91}$$

where $\bar{K} = K\bar{P}$.

By observing the terms $\rho_i \bar{P} Q_i^{\mathrm{T}} q_i$ and $\rho_i \bar{P} Q_i^{\mathrm{T}} Q_i \bar{P}$ in (2.91), the scalars ρ_i are coupled with the Lyapunov matrix $\bar{P}$. To solve the inequality (2.91) via a linearization procedure efficiently, it will be useful to decouple the Lyapunov matrix $\bar{P}$ from the scalars ρ_i. Based on Schur complement, the following inequality indicates (2.91),

$$\begin{bmatrix} \mathtt{Sym}\{A_m X - B_m \bar{K}\} + \rho_i \bar{P} Q_i^{\mathrm{T}} Q_i \bar{P} + \varepsilon_i H_{m1}^{\mathrm{T}} H_{m1} & \star & \star & \star \\ D_m^{\mathrm{T}} & -\gamma^2 \mathbf{I} & \star & \star \\ L_m \bar{P} - N_m \bar{K} & \mathbf{0} & -\mathbf{I} & \star \\ H_{m2} \bar{P} - H_{m3} \bar{K} & \mathbf{0} & \mathbf{0} & -\varepsilon_i \mathbf{I} \end{bmatrix} + \begin{bmatrix} a_m + \rho_i \bar{P} Q_i^{\mathrm{T}} q_i \\ \mathbf{0} \\ \mathbf{0} \\ H_{m4} \end{bmatrix} (\rho_i(1 - q_i^{\mathrm{T}} q_i))^{-1} (\star) < 0. \tag{2.92}$$

Consider the Matrix Inversion Lemma

$$(\mathbf{X} + \mathbf{YQZ})^{-1} = \mathbf{X}^{-1} - \mathbf{X}^{-1}\mathbf{Y}(\mathbf{Q}^{-1} + \mathbf{ZX}^{-1}\mathbf{Y})^{-1}\mathbf{ZX}^{-1}, \tag{2.93}$$

and yields,

$$\begin{cases} \rho_i^{-1} a_m (1 - q_i^{\mathrm{T}} q_i)^{-1} a_m^{\mathrm{T}} = \rho_i^{-1} a_m a_m^{\mathrm{T}} + \rho_i^{-1} a_m q_i^{\mathrm{T}} (\mathbf{I} - q_i q_i^{\mathrm{T}})^{-1} q_i a_m^{\mathrm{T}}, \\ a_m (1 - q_i^{\mathrm{T}} q_i)^{-1} q_i^{\mathrm{T}} Q_i \bar{P} = a_m q_i^{\mathrm{T}} (\mathbf{I} - q_i q_i^{\mathrm{T}})^{-1} Q_i \bar{P}, \\ \rho_i \bar{P} Q_i^{\mathrm{T}} q_i (1 - q_i^{\mathrm{T}} q_i)^{-1} q_i^{\mathrm{T}} Q_i \bar{P} = \rho_i \bar{P} Q_i^{\mathrm{T}} (\mathbf{I} - q_i q_i^{\mathrm{T}})^{-1} Q_i \bar{P} - \rho_i \bar{P} Q_i^{\mathrm{T}} Q_i \bar{P}, \\ \rho_i^{-1} H_{m4} (1 - q_i^{\mathrm{T}} q_i)^{-1} a_m^{\mathrm{T}} = \rho_i^{-1} H_{m4} a_m^{\mathrm{T}} + \rho_i^{-1} H_{m4} q_i^{\mathrm{T}} (\mathbf{I} - q_i q_i^{\mathrm{T}})^{-1} q_i a_m^{\mathrm{T}}, \\ H_{m4} (1 - q_i^{\mathrm{T}} q_i)^{-1} q_i^{\mathrm{T}} Q_i \bar{P} = H_{m4} q_i^{\mathrm{T}} (\mathbf{I} - q_i q_i^{\mathrm{T}})^{-1} Q_i \bar{P}, \\ \rho_i^{-1} H_{m4} (1 - q_i^{\mathrm{T}} q_i)^{-1} H_{m4}^{\mathrm{T}} = \\ \qquad \rho_i^{-1} H_{m4} H_{m4}^{\mathrm{T}} + \rho_i^{-1} H_{m4} q_i^{\mathrm{T}} (\mathbf{I} - q_i q_i^{\mathrm{T}})^{-1} q_i H_{m4}^{\mathrm{T}}. \end{cases} \tag{2.94}$$

Introduce $\kappa_i = \rho_i^{-1}$ and use Schur complement, condition (2.84) can be obtained. This completes the proof. ■

In Theorem 2.4, the asymptotic stability and robust performance analysis of sliding motion is conducted through a CQLF-based method and the sliding surface gain can be computed simultaneously. Alternatively, in the framework of PQLF, new sufficient conditions for the asymptotic stability and robust $\mathscr{H}_\infty$ performance analysis of the sliding motion will be given in the following.

2.4.3.2 Stability and Robust Performance Analysis of Sliding Motion through PQLFs

To analyze the sliding motion via a PQLFs-based approach, based on (2.80) and system (2.76), the sliding motion can be derived into the following form,

$$\begin{cases} \dot{x}(t) = (\mathscr{A}_i + \Delta\mathscr{A}_i)x(t) + \mathrm{a}_i + \Delta\mathrm{a}_i + (\mathscr{B}_i + \Delta\mathscr{B}_i)u(t) + \mathscr{D}_i w(t) \\ 0 \times \dot{u}(t) = -Kx(t) - u(t) \\ z(t) = \mathscr{L}_i x(t) + \mathscr{N}_i u(t), i \in \mathscr{I}. \end{cases} \tag{2.95}$$

Reformulate the system (2.95) as

$$\begin{cases} E\dot{\bar{x}}(t) = (\bar{\mathscr{A}}_i + \Delta\bar{\mathscr{A}}_i)\bar{x}(t) + \bar{\mathrm{a}}_i + \Delta\bar{\mathrm{a}}_i + \bar{\mathscr{D}}_i w(t) \\ z(t) = \bar{\mathscr{L}}_i \bar{x}(t), i \in \mathscr{I}, \end{cases} \tag{2.96}$$

where

$$\begin{cases} E = \begin{bmatrix} \mathbf{I}_{n_x} & \mathbf{0} \\ \mathbf{0} & \mathbf{0} \end{bmatrix}, \bar{x}(t) = \begin{bmatrix} x(t) \\ u(t) \end{bmatrix}, \bar{\mathscr{A}}_i = \begin{bmatrix} \mathscr{A}_i & \mathscr{B}_i \\ -K & -\mathbf{I} \end{bmatrix}, \Delta\bar{\mathscr{A}}_i = \begin{bmatrix} \Delta\mathscr{A}_i & \Delta\mathscr{B}_i \\ \mathbf{0} & \mathbf{0} \end{bmatrix}, \\ \bar{\mathrm{a}}_i = \begin{bmatrix} \mathrm{a}_i \\ \mathbf{0} \end{bmatrix}, \Delta\bar{\mathrm{a}}_i = \begin{bmatrix} \Delta\mathrm{a}_i \\ \mathbf{0} \end{bmatrix}, \bar{\mathscr{D}}_i = \begin{bmatrix} \mathscr{D}_i \\ \mathbf{0} \end{bmatrix}, \bar{\mathscr{L}}_i = \begin{bmatrix} \mathscr{L}_i & \mathscr{N}_i \end{bmatrix}. \end{cases} \tag{2.97}$$

Via combining the first equation of system (2.76) with the sliding surface (2.79), the sliding mode dynamics can be characterized by the descriptor system (2.96). Therefore, the admissibility of the system (2.96) ensures the existence of the designed sliding surface (2.79). To ensure the PQLFs continuous across boundaries of the subspace, matrices $\tilde{F}_i = \begin{bmatrix} f_i & F_i \end{bmatrix}, i \in \mathscr{I}$, with $f_i = 0$ for $i \in \mathscr{I}_0$ are established to describe the boundary among the regions as in subchapter 2.2,

$$\tilde{F}_i \begin{bmatrix} 1 \\ x(t) \end{bmatrix} = \tilde{F}_j \begin{bmatrix} 1 \\ x(t) \end{bmatrix}, \theta(t) \in S_i \bigcap S_j, \ i, j \in \mathscr{I}. \tag{2.98}$$

To further reduce the conservatism, the S-procedure will be also utilized by constructing matrices $\tilde{G}_i = \begin{bmatrix} g_i & G_i \end{bmatrix}, i \in \mathscr{I}$, with $g_i = 0$ for $i \in \mathscr{I}_0$ and

$$\tilde{G}_i \begin{bmatrix} 1 \\ x(t) \end{bmatrix} \succeq 0, \theta(t) \in S_i, i \in \mathscr{I}, \tag{2.99}$$

where $\succeq$ indicates that each entry is nonnegative.

Theorem 2.5. The sliding mode dynamical system (2.96) is asymptotically stable with robust $\mathscr{H}_\infty$ performance γ, if there exist matrices $X = X^{\mathrm{T}}$, $U_i = U_i^{\mathrm{T}} \succeq 0$, $W_i =$

$W_i^{\mathrm{T}} \succeq 0$, $\bar{K} \in \Re^{n_u \times n_x}$, $P_2 \in \Re^{n_u \times n_u}$, and scalars $\varepsilon_i > 0$, $i \in \mathscr{I}$ such that the LMIs in (2.100)–(2.101) hold,

$$\tilde{F}_i^{\mathrm{T}} X \tilde{F}_i - \tilde{G}_i^{\mathrm{T}} U_i \tilde{G}_i > 0, i \in \mathscr{I}, \tag{2.100}$$

$$\begin{bmatrix} \tilde{\Lambda}_1 & \star & \star & \star \\ \tilde{\Lambda}_2 & \tilde{\Lambda}_3 & \star & \star \\ D_m^{\mathrm{T}} \tilde{R} \tilde{F}_i^{\mathrm{T}} X \tilde{F}_i & \mathbf{0} & -\gamma^2 \mathbf{I} & \star \\ H_{m1}^{\mathrm{T}} \tilde{R} \tilde{F}_i^{\mathrm{T}} X \tilde{F}_i & \mathbf{0} & \mathbf{0} & -\varepsilon_i \mathbf{I} \end{bmatrix} < 0, m \in \mathscr{N}(i), i \in \mathscr{I}, \tag{2.101}$$

where

$$\begin{cases} \tilde{\Lambda}_1 = \mathtt{Sym}\{\tilde{F}_i^{\mathrm{T}} X \tilde{F}_i \tilde{A}_{m1} + \alpha \tilde{J}^{\mathrm{T}} \bar{K} \tilde{R}\} + \tilde{R}^{\mathrm{T}} L_m^{\mathrm{T}} L_m \tilde{R} + \varepsilon_i \tilde{\Lambda}_{1(1)} \tilde{\Lambda}_{1(1)}^{\mathrm{T}} \\ \qquad + \tilde{G}_i^{\mathrm{T}} W_i \tilde{G}_i, \\ \tilde{\Lambda}_2 = -\bar{K} \tilde{R} + B_m^{\mathrm{T}} \tilde{R} \tilde{F}_i^{\mathrm{T}} X \tilde{F}_i + \alpha P_2 \tilde{J} + N_m^{\mathrm{T}} L_m \tilde{R} + \varepsilon_i H_{m3}^{\mathrm{T}} \tilde{\Lambda}_{1(1)}^{\mathrm{T}}, \\ \tilde{\Lambda}_3 = -P_2 - P_2^{\mathrm{T}} + N_m^{\mathrm{T}} N_m + \varepsilon_i H_{m3}^{\mathrm{T}} H_{m3}, \\ \left.\tilde{\Lambda}_{1(1)} = H_{m2}^{\mathrm{T}}, \tilde{A}_{m1} = A_m, \tilde{R} = \mathbf{I}_{n_x}, \tilde{J} \in \Re^{n_u \times n_x}, \right\} \text{if } i \in \mathscr{I}_0, \\ \left.\begin{matrix} \tilde{\Lambda}_{1(1)} = \begin{bmatrix} H_{m4}^{\mathrm{T}} \\ H_{m2}^{\mathrm{T}} \end{bmatrix}, \tilde{A}_{m1} = \begin{bmatrix} 0 & \mathbf{0} \\ a_m & A_m \end{bmatrix}, \tilde{R} = \begin{bmatrix} \mathbf{0} & \mathbf{I}_{n_x} \end{bmatrix}, \\ \tilde{J} \in \Re^{n_u \times (n_x+1)} \end{matrix} \right\} \text{if } i \in \mathscr{I}_1 \end{cases} \tag{2.102}$$

and α is a constant scalar, $\tilde{J} \in \Re^{n_u \times n_x}$, for $i \in \mathscr{I}_0$, and $\tilde{J} \in \Re^{n_u \times (n_x+1)}$, for $i \in \mathscr{I}_1$, are arbitrary matrices.

Moreover, the gain of the sliding surface (2.79) can be obtained via

$$K = P_2^{-\mathrm{T}} \bar{K}. \tag{2.103}$$

Proof. For the sliding mode dynamical system (2.96), consider the following PQLF which is continuous across the region boundaries

$$V(t) = \tilde{x}^{\mathrm{T}}(t) \tilde{E} \tilde{\mathscr{P}}_i \tilde{x}(t), i \in \mathscr{I}, \tag{2.104}$$

where

$$\tilde{\mathscr{P}}_i = \begin{bmatrix} \tilde{F}_i^{\mathrm{T}} X \tilde{F}_i & \mathbf{0} \\ \tilde{P}_3 & P_2 \end{bmatrix} \tag{2.105}$$

and $X = X^{\mathrm{T}}$, $P_2 \in \Re^{n_u \times n_u}$ is an arbitrary nonsingular matrix, and $\tilde{P}_3 \in \Re^{n_u \times n_x}$, for $i \in \mathscr{I}_0$, $\tilde{P}_3 \in \Re^{n_u \times (n_x+1)}$, for $i \in \mathscr{I}_1$.

The inequality in (2.100) ensures the positive definiteness of PQLF in each region. Based on the PQLF shown in (2.104), it suffices to attain the following inequality

$$\dot{V}(t) + z^{\mathrm{T}}(t) z(t) - \gamma^2 w^{\mathrm{T}}(t) w(t) < 0 \tag{2.106}$$

to prove that the sliding mode dynamical system (2.96) is asymptotically stable with robust $\mathscr{H}_\infty$ performance γ.

Based on the structure of $\tilde{\mathscr{P}}_i$ shown in (2.105), one has $\tilde{E}\tilde{\mathscr{P}}_i = \tilde{\mathscr{P}}_i^{\mathrm{T}}\tilde{E}$. Along the state strategies of system (2.96), the inequality (2.106) can be derived as

$$\begin{bmatrix} \tilde{x}(t) \\ w(t) \end{bmatrix}^{\mathrm{T}} \begin{bmatrix} \mathtt{Sym}\{(\tilde{\mathscr{A}}_i + \Delta\tilde{\mathscr{A}}_i)^{\mathrm{T}}\tilde{\mathscr{P}}_i\} + \tilde{\mathscr{L}}_i^{\mathrm{T}}\tilde{\mathscr{L}}_i & \star \\ \tilde{\mathscr{D}}_i^{\mathrm{T}}\tilde{\mathscr{P}}_i & -\gamma^2\mathbf{I} \end{bmatrix} \begin{bmatrix} \tilde{x}(t) \\ w(t) \end{bmatrix} < 0, \tag{2.107}$$

where $\tilde{\mathscr{A}}_i$, $\Delta\tilde{\mathscr{A}}_i$, $\tilde{\mathscr{L}}_i$, $\tilde{\mathscr{D}}_i$ are given as in (2.97).

Define $\bar{K} = P_2^{\mathrm{T}}K$. Using S-procedure, the following inequality implies (2.107),

$$\begin{bmatrix} \tilde{\Upsilon}_1 & \star & \star \\ \tilde{\Upsilon}_2 & -P_2 - P_2^{\mathrm{T}} + \mathscr{N}_i^{\mathrm{T}}\mathscr{N}_i & \star \\ \mathscr{D}_i^{\mathrm{T}}\tilde{R}\tilde{F}_i^{\mathrm{T}}X\tilde{F}_i & \mathbf{0} & -\gamma^2\mathbf{I} \end{bmatrix} < 0, \tag{2.108}$$

where $W_i \succeq 0$ and

$$\begin{cases} \tilde{\Upsilon}_1 = \mathtt{Sym}\{\tilde{F}_i^{\mathrm{T}}X\tilde{F}_i(\tilde{\mathscr{A}}_{i1} + \Delta\tilde{\mathscr{A}}_{i1}) - \tilde{P}_3^{\mathrm{T}}K\tilde{R}\} + \tilde{R}^{\mathrm{T}}\mathscr{L}_i^{\mathrm{T}}\mathscr{L}_i\tilde{R} + \tilde{G}_i^{\mathrm{T}}W_i\tilde{G}_i, \\ \tilde{\Upsilon}_2 = -\bar{K}\tilde{R} + (\mathscr{B}_i^{\mathrm{T}} + \Delta\mathscr{B}_i^{\mathrm{T}})\tilde{R}\tilde{F}_i^{\mathrm{T}}X\tilde{F}_i - \tilde{P}_3 + \mathscr{N}_i^{\mathrm{T}}\mathscr{L}_i\tilde{R}, \\ \tilde{\mathscr{A}}_{i1} = \mathscr{A}_i,\ \Delta\tilde{\mathscr{A}}_{i1} = \Delta\mathscr{A}_i,\ \tilde{R} = \mathbf{I}_{n_x},\ \}\text{if } i \in \mathscr{I}_0, \\ \tilde{\mathscr{A}}_{i1} = \begin{bmatrix} 0 & \mathbf{0} \\ \mathrm{a}_i & \mathscr{A}_i \end{bmatrix}, \Delta\tilde{\mathscr{A}}_{i1} = \begin{bmatrix} 0 & \mathbf{0} \\ \Delta\mathrm{a}_i & \Delta\mathscr{A}_i \end{bmatrix}, \tilde{R} = \begin{bmatrix} \mathbf{0} & \mathbf{I}_{n_x} \end{bmatrix}, \ \}\, i \in \mathscr{I}_1. \end{cases} \tag{2.109}$$

Expand the MFs and use Lemma A1 to handle the parameter uncertainties given as in (1.3). For a set of positive scalars $\varepsilon_i > 0$, $i \in \mathscr{I}$, the following inequality implies (2.108),

$$\begin{bmatrix} \tilde{\Sigma}_1 & \star & \star & \star \\ \tilde{\Sigma}_2 & -P_2 - P_2^{\mathrm{T}} + N_m^{\mathrm{T}}N_m + \varepsilon_i H_{m3}^{\mathrm{T}}H_{m3} & \star & \star \\ D_m^{\mathrm{T}}\tilde{R}\tilde{F}_i^{\mathrm{T}}X\tilde{F}_i & \mathbf{0} & -\gamma^2\mathbf{I} & \star \\ H_{m1}^{\mathrm{T}}\tilde{R}\tilde{F}_i^{\mathrm{T}}X\tilde{F}_i & \mathbf{0} & \mathbf{0} & -\varepsilon_i\mathbf{I} \end{bmatrix} < 0, \quad m \in \mathscr{N}(i), i \in \mathscr{I}, \tag{2.110}$$

where

$$\begin{cases} \tilde{\Sigma}_1 = \mathtt{Sym}\{\tilde{F}_i^{\mathrm{T}}X\tilde{F}_i\tilde{A}_{m1} - \tilde{P}_3^{\mathrm{T}}K\tilde{R}\} + \tilde{R}^{\mathrm{T}}L_m^{\mathrm{T}}L_m\tilde{R} + \tilde{G}_i^{\mathrm{T}}W_i\tilde{G}_i \\ \qquad + \varepsilon_i\tilde{\Lambda}_{1(1)}\tilde{\Lambda}_{1(1)}^{\mathrm{T}}, \\ \tilde{\Sigma}_2 = -\bar{K}\tilde{R} + B_m^{\mathrm{T}}\tilde{R}\tilde{F}_i^{\mathrm{T}}XF_i - P_3 + N_m^{\mathrm{T}}L_m\tilde{R} + \varepsilon_i H_{m3}^{\mathrm{T}}\tilde{\Lambda}_{1(1)}^{\mathrm{T}}, \\ \tilde{\Lambda}_{1(1)} = H_{m2}^{\mathrm{T}}, \tilde{A}_{m1} = A_m,\ \}\text{if } i \in \mathscr{I}_0, \\ \tilde{\Lambda}_{1(1)} = \begin{bmatrix} H_{m4}^{\mathrm{T}} \\ H_{m2}^{\mathrm{T}} \end{bmatrix}, \tilde{A}_{m1} = \begin{bmatrix} 0 & \mathbf{0} \\ a_m & A_m \end{bmatrix}, \ \}\text{if } i \in \mathscr{I}_1. \end{cases} \tag{2.111}$$

For numerical tractability, specify matrix $\tilde{P}_3$ as

$$\tilde{P}_3 = -\alpha P_2\tilde{J}, \tag{2.112}$$

where α is a constant scalar, $\tilde{J} \in \Re^{n_u \times n_x}$, for $i \in \mathscr{I}_0$, and $\tilde{J} \in \Re^{n_u \times (n_x+1)}$, for $i \in \mathscr{I}_1$, are arbitrary matrices.

Based on Schur complement, (2.101) implies (2.110). Thus, the proof is ended. ■

2.5 STATE FEEDBACK DYNAMIC SLIDING MODE CONTROLLER DESIGN

In the following, a state feedback dynamic sliding mode controller will be designed to drive the system states onto the designed sliding surface (2.79) in finite time.

2.5.1 DYNAMIC SLIDING MODE CONTROLLER DESIGN

Based on the $s(t) = Ky(t) + x(t)$ given as in (2.79), the state feedback dynamic sliding mode controller is designed as,

$$\dot{u}(t) = -\sum_{l=1}^{r} \mu_l K\{A_l x(t) + a_l + B_l u(t)\} - (\sigma + \zeta(t))\mathrm{sgn}(s(t)), \tag{2.113}$$

where $\sigma > 0$ is a given scalar, and

$$\begin{aligned}\zeta(t) = &\sum_{l=1}^{r} \mu_l \{\|KH_{l1}\| (\|H_{l2}\| \cdot \|x(t)\| + \|H_{l3}\| \cdot \|u(t)\| + \|H_{l4}\|) \\ &+ \|KD_l\| \varpi(t)\}\end{aligned} \tag{2.114}$$

with $\varpi(t)$ being the uniform upper-bound of the external disturbance $w(t)$.

Theorem 2.6. Given the sliding surface (2.79) and consider the fuzzy affine system (2.72) or equivalently (2.76). The state feedback dynamic sliding mode controller (2.113) can drive the closed-loop system state trajectories onto the sliding surface (2.79) in finite time.

Proof. Consider the Lyapunov function as

$$V(t) = \frac{1}{2} s^{\mathrm{T}}(t) s(t). \tag{2.115}$$

Based on the sliding surface in (2.79), the time-derivative of $s(t)$ can be obtained as,

$$\dot{s}(t) = K\dot{x}(t) + \dot{u}(t). \tag{2.116}$$

Based on the dynamic sliding mode controller (2.113), one has

$$\begin{aligned}\dot{V}(t) &= s^{\mathrm{T}}(t)\dot{s}(t) \\ &= s^{\mathrm{T}}(t) \sum_{l=1}^{r} \mu_l K\{(A_l + \Delta A_l)x(t) + a_l + \Delta a_l + (B_l + \Delta B_l)u(t) \\ &\quad + D_l w(t)\} - s^{\mathrm{T}}(t) \sum_{l=1}^{r} \mu_l K\{A_l x(t) + a_l + B_l u(t)\} - (\sigma + \zeta(t))\|s(t)\| \\ &= s^{\mathrm{T}}(t) \sum_{l=1}^{r} \mu_l K\{\Delta A_l x(t) + \Delta a_l + \Delta B_l u(t) + D_l w(t)\} \\ &\quad - \zeta(t)\|s(t)\| - \sigma\|s(t)\|.\end{aligned} \tag{2.117}$$

Considering the parameter uncertainties described as in (1.3), one has,

$$\begin{aligned}\dot{V}(t) \leq \sum_{l=1}^{r} \mu_l \|s(t)\| \{ &\|KH_{l1}\| \cdot \|H_{l2}\| \cdot \|x(t)\| + \|KH_{l1}\| \cdot \|H_{l4}\| \\ &+ \|KH_{l1}\| \cdot \|H_{l3}\| \cdot \|u(t)\| + \|KD_l\| \varpi(t)\} - \zeta(t)\|s(t)\| - \sigma\|s(t)\|.\end{aligned} \tag{2.118}$$

Noticing that

$$\begin{aligned}\zeta(t) = \sum_{l=1}^{r} \mu_l \{ &\|KH_{l1}\| (\|H_{l2}\| \cdot \|x(t)\| + \|H_{l3}\| \cdot \|u(t)\| + \|H_{l4}\|) \\ &+ \|KD_l\| \varpi(t)\},\end{aligned} \tag{2.119}$$

we have

$$\dot{V}(t) \leq -\sigma\|s(t)\| < 0. \tag{2.120}$$

Thus, it can be concluded that the sliding surface (2.79) can be reached in finite time. Thus, the proof is accomplished. ■

Remark 2.5.1 To reduce the chattering induced by the switching term $\mathtt{sgn}(s(t))$, in the dynamic sliding mode controller (2.113), a smooth continuous term $s(t)/(\|s(t)\| + \delta)$ can be used to replace $\mathtt{sgn}(s(t))$, where δ refers to a small positive scalar.

2.5.2 ILLUSTRATIVE EXAMPLES

In this subsection, two simulation examples will be given to verify the validity of the proposed state feedback dynamic SMC approaches.

Example 2.3. Consider an inverted pendulum system in [77]. The dynamics of the inverted pendulum are described as follows,

$$\begin{cases} \dot{x}_1(t) = x_2(t) \\ \dot{x}_2(t) = \dfrac{g\sin(x_1) - amlx_2^2 \sin(2x_1)/2 - a\cos(x_1)u}{4l/3 - aml\cos^2(x_1)} + w(t) \end{cases} \tag{2.121}$$

where $x_1(t)$ and $x_2(t)$ represent the angle of inverted pendulum from the vertical and the angular velocity, respectively; $w(t)$ is the exogenous disturbance; $g - 9.8\ \mathrm{m/s}^2$ is the gravity constant; $2l = 1\mathrm{m}$ is the length of the pendulum; $a = \dfrac{1}{m+M}$; $M = 8.0$ kg is to the mass of the cart; $m = 2.0$ kg is the mass of pendulum.

Linearize the nonlinear system (2.121) at three operating points $\left(-\frac{\pi}{3}, 0\right)^{\mathrm{T}}$, $(0,0)^{\mathrm{T}}$, and $\left(\frac{\pi}{3}, 0\right)^{\mathrm{T}}$. The regulated output $z(t)$ is given as $z(t) = x_1(t) + x_2(t) + 0.05u(t)$, and then the system (2.121) can be described by a T-S fuzzy affine model with three rules as,

Plant Rule $\mathscr{R}^l$: IF $x_1(t)$ is $\mathscr{F}_1^l$, THEN

$$\begin{cases} \dot{x}(t) = (A_l + \Delta A_l)x(t) + a_l + \Delta a_l + (B_l + \Delta B_l)u(t) + D_l w(t) \\ z(t) = L_l x(t) + N_l u(t),\ l \in \{1,2,3\}, \end{cases} \tag{2.122}$$

where the system matrices are

$$\begin{cases} A_1 = \begin{bmatrix} 0 & 1 \\ 17.2941 & 0 \end{bmatrix}, A_2 = A_3 = \begin{bmatrix} 0 & 1 \\ 13.1321 & 0 \end{bmatrix}, \\ a_1 = \begin{bmatrix} 0 \\ 0 \end{bmatrix}, a_2 = \begin{bmatrix} 0 \\ -7.0992 \end{bmatrix}, a_3 = \begin{bmatrix} 0 \\ 7.0992 \end{bmatrix}, \\ B_1 = \begin{bmatrix} 0 \\ -0.1765 \end{bmatrix}, B_2 = B_3 = \begin{bmatrix} 0 \\ -0.0779 \end{bmatrix}, \\ D_l = \begin{bmatrix} 0 \\ 1 \end{bmatrix}, L_l = \begin{bmatrix} 1 & 1 \end{bmatrix}, N_l = 0.05, l \in \{1,2,3\} \end{cases} \quad (2.123)$$

and the parameter uncertainties ΔA_l, Δa_l and ΔB_l are given in the form of (1.3) with

$$H_{l1} = \begin{bmatrix} 0 \\ 0.01 \end{bmatrix}, H_{l2} = \begin{bmatrix} 0.014 & 0 \end{bmatrix}, H_{l3} = 0.01, H_{l4} = 0.02, l \in \{1,2,3\}. \quad (2.124)$$

The MFs of this example are also shown in Figure 2.5 with $d_1 = \frac{\pi}{12}$ and $d_2 = \frac{\pi}{2}$, which indicates that S_2 is a crisp region with $\mathcal{N}(2) = \{1\}$, and S_1 and S_3 are fuzzy regions with $\mathcal{N}(1) = \{1,2\}$ and $\mathcal{N}(3) = \{1,3\}$. The purpose of this example is to design a sliding surface (2.79) and a state feedback dynamic sliding mode controller (2.113) to ensure that the closed-loop system states can be driven onto the designed sliding surface (2.79) in finite time and the sliding motion is asymptotically stable with a satisfactory robust $\mathcal{H}_\infty$ performance γ. By applying the PQLFs-based method presented in Theorem 2.5 with $\alpha = 450$, $\tilde{J} = \begin{bmatrix} 0 & 1 & 1 \end{bmatrix}$ for $i \in \mathcal{I}_1$, and $\tilde{J} = \begin{bmatrix} 1 & 1 \end{bmatrix}$ for $i \in \mathcal{I}_0$, feasible solutions can be obtained, and the sliding surface gain is given as $K = \begin{bmatrix} -454.8364 & -470.0198 \end{bmatrix}$ with robust $\mathcal{H}_\infty$ performance $\gamma_{\min} = 2.6948$. Note that when using Theorem 2.5 with PQLFs in (2.104), matrices $\tilde{G}_i$ and $\tilde{F}_i$ are given as,

$$\begin{aligned} \left[\begin{array}{c|c|c|c|c|c} F_1 & F_2 & F_3 & f_1 & f_2 & f_3 \end{array}\right] &= \left[\begin{array}{c|c|c|c|c|c} -\rho & \mathbf{0}_{1\times n_x} & \mathbf{0}_{1\times n_x} & -d_1 & 0 & 0 \\ \mathbf{0}_{1\times n_x} & \mathbf{0}_{1\times n_x} & \rho & 0 & 0 & -d_1 \\ \mathbf{I}_{n_x} & \mathbf{I}_{n_x} & \mathbf{I}_{n_x} & \mathbf{0}_{n_x\times 1} & \mathbf{0}_{n_x\times 1} & \mathbf{0}_{n_x\times 1} \end{array}\right], \\ \left[\begin{array}{c|c|c|c|c|c} G_1 & G_2 & G_3 & g_1 & g_2 & g_3 \end{array}\right] &= \left[\begin{array}{c|c|c|c|c|c} \rho & \mathbf{0}_{1\times n_x} & \rho & d_2 & 0 & -d_1 \\ -\rho & \mathbf{0}_{1\times n_x} & -\rho & -d_1 & 0 & d_2 \end{array}\right] \end{aligned} \quad (2.125)$$

with $\rho = \begin{bmatrix} 1 & 0 \end{bmatrix}$ and $n_x = 2$.

To show the effectiveness of the proposed approach, simulations are conducted under different initial conditions. With initial condition $x_0 = \begin{bmatrix} \frac{\pi}{3} & 0 \end{bmatrix}^{\mathrm{T}}$ and exogenous disturbance $w(t) = 30\mathrm{e}^{-0.05t}\sin(0.1t)$, by using the state feedback dynamic sliding mode controller in (2.113) with $\sigma = 2.5$ and $\varpi(t) = 30\mathrm{e}^{-0.05t}$, the closed-loop system states and the control input are shown in Figure 2.9 and Figure 2.10, respectively. Under zero initial condition, the $\mathcal{H}_\infty$ performance is presented in Figure 2.11. The ratio $\frac{\sqrt{\int_0^{t_f} z^{\mathrm{T}}(t)z(t)dt}}{\sqrt{\int_0^{t_f} w^{\mathrm{T}}(t)w(t)dt}}$ is about 0.31, which is lower than the minimum disturbance attenuation level $\gamma_{\min} = 2.6948$.

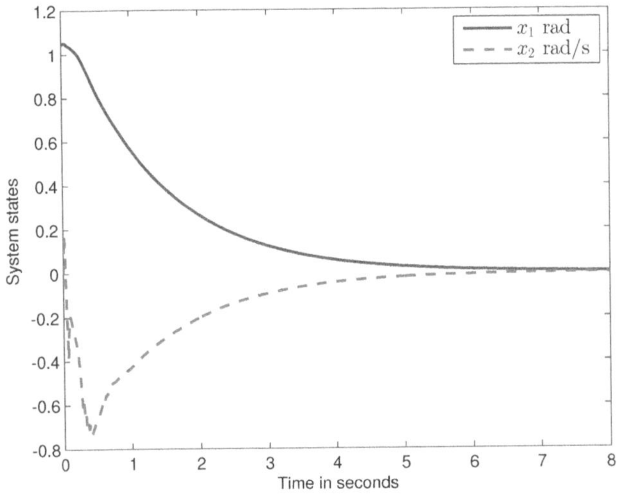

Figure 2.9 The closed-loop system states with dynamic sliding mode controller (2.113) in Example 2.3.

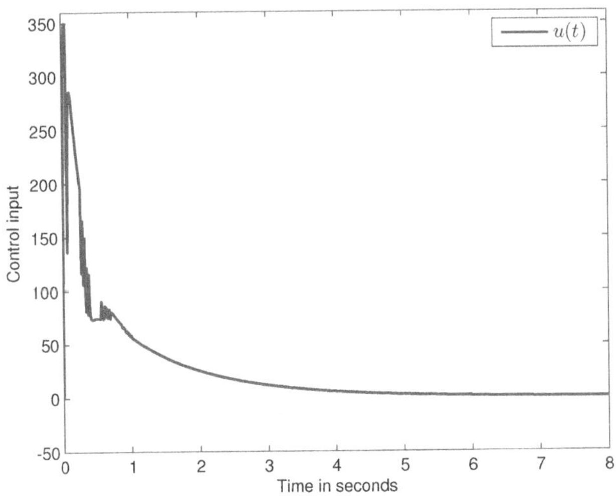

Figure 2.10 The control input with dynamic sliding mode controller (2.113) in Example 2.3.

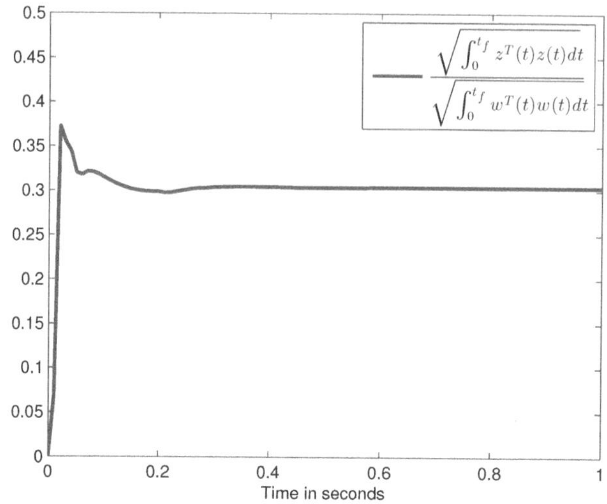

Figure 2.11 Response of the ratio $\frac{\sqrt{\int_0^{t_f} z^T(t)z(t)dt}}{\sqrt{\int_0^{t_f} w^T(t)w(t)dt}}$ with dynamic sliding mode controller (2.113) in Example 2.3.

Nextly, we will consider another classical third-order nonlinear finance system, which is widely studied and has positive enlightening significance for research on some other complex economic and financial systems.

Example 2.4. Consider a nonlinear finance system, whose dynamics have been described as in (2.65). Linearize the nonlinear finance system at three operating points $(0,0,0)$ and $(\pm 0.3,0,0)$. The regulated output $z(t)$ is given as $z(t) = x_1(t) + 0.5x_2(t) + 0.2x_3(t) + 0.8u(t)$. In this example, the control input matrix B is assumed to have parameter uncertainties, and then the system (2.65) can be described by the following T-S fuzzy affine model with three fuzzy rules as

Plant Rule $\mathscr{R}^l$: IF $x_1(t)$ is $\mathscr{F}_1^l$, THEN

$$\begin{cases} \dot{x}(t) = (A_l + \Delta A_l)x(t) + a_l + \Delta a_l + (B_l + \Delta B_l)u(t) + D_l w(t) \\ z(t) = L_l x(t) + N_l u(t),\ l \in \{1,2,3\}, \end{cases} \tag{2.126}$$

where

$$\begin{cases} A_1 = \begin{bmatrix} 0.4091 & 0 & 1 \\ 0 & -1.1 & 0 \\ -1 & 0 & -0.3 \end{bmatrix}, A_2 = \begin{bmatrix} 0.4091 & -0.3 & 1 \\ 0.6 & -1.1 & 0 \\ -1 & 0 & -0.3 \end{bmatrix}, \\ A_3 = \begin{bmatrix} 0.4091 & 0.3 & 1 \\ -0.6 & -1.1 & 0 \\ -1 & 0 & -0.3 \end{bmatrix}, a_1 = \begin{bmatrix} 0 \\ 0 \\ 0 \end{bmatrix}, a_2 = a_3 = \begin{bmatrix} 0 \\ 0.27 \\ 0 \end{bmatrix}, \\ B_l = \begin{bmatrix} 0 \\ 1 \\ 1 \end{bmatrix}, D_l = \begin{bmatrix} 0 \\ 1 \\ 0 \end{bmatrix}, L_l = \begin{bmatrix} 1 & 0.5 & 0.2 \end{bmatrix}, N_l = 0.8, \end{cases} \tag{2.127}$$

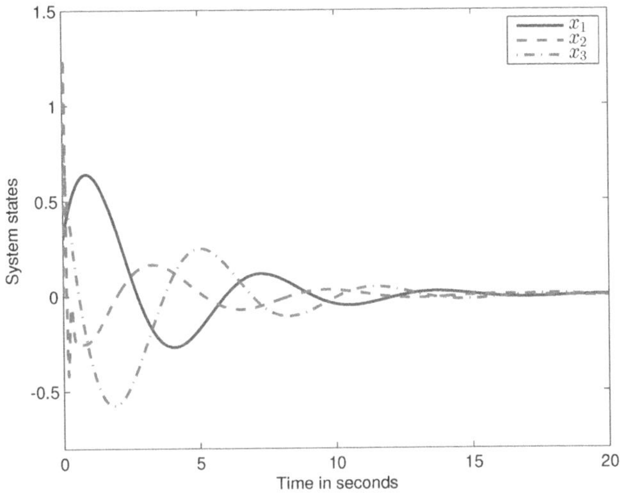

Figure 2.12 The closed-loop system states with dynamic sliding mode controller (2.113) in Example 2.4.

and the parameter uncertainties ΔA_l, Δa_l, and ΔB_l are given in the form of (1.3) with

$$\begin{cases} H_{l1} = \begin{bmatrix} 0.01 & 0.01 & 0 \end{bmatrix}^{\mathrm{T}}, H_{l2} = \begin{bmatrix} 0.02 & 0.02 & 0.02 \end{bmatrix}, \\ H_{l3} = 0, H_{l4} = 0.01, l \in \{1,2,3\}. \end{cases} \tag{2.128}$$

The fuzzy membership functions are also given as in Figure 2.5 with $d_1 = 0.3$ and $d_2 = 1$. By applying the PQLFs-based method presented in Theorem 2.5 with $\alpha = -1$, $\tilde{J} = \begin{bmatrix} 0 & 1 & 1 & 1 \end{bmatrix}$ for $i \in \mathscr{I}_1$, and $\tilde{J} = \begin{bmatrix} 1 & 1 & 1 \end{bmatrix}$ for $i \in \mathscr{I}_0$, feasible solutions can be obtained, and the sliding surface gain is computed as $K = \begin{bmatrix} 1.0061 & 1.0006 & 1.0036 \end{bmatrix}$ with robust $\mathscr{H}_\infty$ performance $\gamma_{\min} = 0.9974$. Note that when using Theorem 2.5 with PQLFs in (2.104), matrices G_i and $\bar{F}_i$ can be computed via (2.125) with $\rho = \begin{bmatrix} 1 & 0 & 0 \end{bmatrix}$ and $n_x = 3$.

To show the effectiveness of the proposed approach, simulations are conducted. Under initial condition $x_0 = \begin{bmatrix} 0.3 & 0 & 0.8 \end{bmatrix}^{\mathrm{T}}$ and external disturbance $w(t) = 60\mathrm{e}^{-0.2t}\cos(0.5t)$, by using the dynamic sliding mode controller (2.113) with $\sigma = 1.5$, the system states and control input are shown in Figures 2.12 and Figure 2.13, respectively. Under zero initial condition, the $\mathscr{H}_\infty$ performance is presented in Figure 2.14. The ratio $\frac{\sqrt{\int_0^{t_f} z^T(t)z(t)dt}}{\sqrt{\int_0^{t_f} w^T(t)w(t)dt}}$ is about 0.142, which is lower than the minimum disturbance attenuation level $\gamma_{\min} = 0.9974$.

The simulation results illustrate that the proposed dynamic SMC design method is effective clearly.

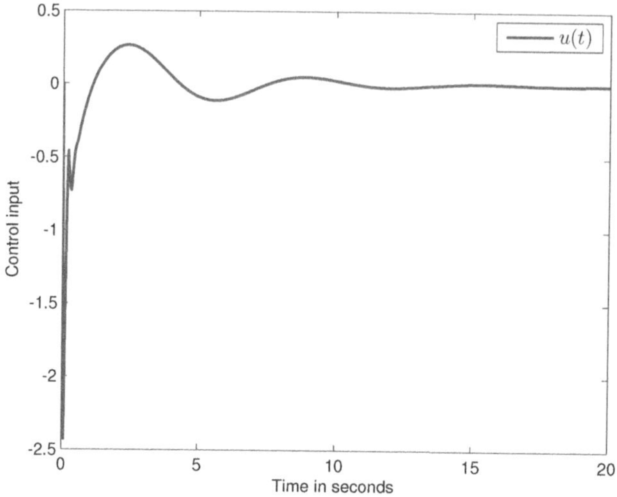

Figure 2.13 The control input with dynamic sliding mode controller (2.113) in Example 2.4.

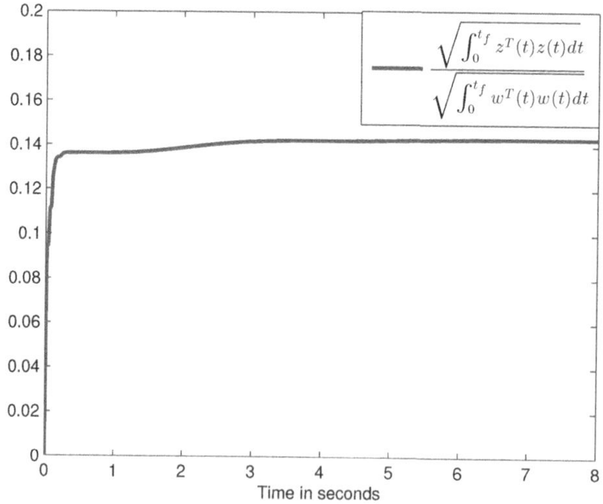

Figure 2.14 Response of the ratio $\frac{\sqrt{\int_0^{t_f} z^T(t)z(t)dt}}{\sqrt{\int_0^{t_f} w^T(t)w(t)dt}}$ with dynamic sliding mode controller (2.113) in Example 2.3.

2.6 CONCLUSION

In this chapter, the state feedback SMC issue for continuous-time T-S fuzzy affine systems has been studied. Via employing PQLFs-based and CQLF-based approaches, respectively, new existence conditions for the sliding surface are derived under a convex optimization setup. Two new sliding mode controller synthesis approaches are proposed to drive the system states onto the sliding surface in finite time. It is also noted that the designed dynamic sliding mode controller can handle the situation that each local model does not share one common input matrix and can deal with mismatched external disturbance. Simulation studies are finally given to show the effectiveness of the developed methods.

3 Output Feedback Fuzzy SMC of Continuous-Time Nonlinear Systems

3.1 INTRODUCTION

In view of some practical constraints, the full system states are generally not measurable. This chapter aims to propose some new output feedback SMC design methods for continuous-time nonlinear 1-D systems subject to exogenous disturbance and mismatched uncertainties based on T-S fuzzy affine models. As the T-S fuzzy affine models have more enhanced approximation competence, the output feedback SMC problem for continuous-time nonlinear systems through T-S fuzzy affine models and piecewise quadratic Lyapunov functions (PQLFs) is a very important and challenging research topic. Associated with the sliding surface, a singular system is adopted to describe the sliding mode dynamical features. Through CQLF and PQLFs, respectively, new stability analysis results on the sliding motion are proposed, and the sliding surface gain can be obtained in a convex optimization setup. Then some new output feedback SMC design methods are proposed to guarantee the finite-time reachability of the sliding surface. Finally, simulation studies are provided to show the effectiveness of the proposed design methods.

3.2 OUTPUT FEEDBACK SMC DESIGN VIA T-S FUZZY AFFINE MODELS

3.2.1 MODEL DESCRIPTION: CASE I

Consider first and second equation of continuous-time T-S fuzzy affine system described as in (1.2) with a common control input matrix B, which is characterized by r fuzzy **IF-THEN** rules as,

Plant Rule $\mathscr{R}^l$: IF $\theta_1(t)$ is $\mathscr{F}_1^l$ and $\cdots$ and $\theta_\varphi(t)$ is $\mathscr{F}_\varphi^l$, THEN

$$\begin{cases} \dot{x}(t) = (A_l + \Delta A_l)x(t) + a_l + \Delta a_l + B\left(u(t) + f(t, x(t))\right) \\ y(t) = Cx(t),\ l \in \mathscr{L} := \{1, \ldots, r\}, \end{cases} \tag{3.1}$$

where $\mathscr{R}^l$ represents the l-th fuzzy inference rule; r is the number of inference rules; $\mathscr{F}_\phi^l(\phi = 1, \ldots, \varphi)$ represent fuzzy sets; $\theta(t) := \left[\theta_1(t), \ldots, \theta_\varphi(t)\right]$ are the measurable premise variables; $x(t) \in \Re^{n_x}$ is the system states; $u(t) \in \Re^{n_u}$ is the control input and B is the common control input matrix with full-column rank; $y(t) \in \Re^{n_y}$ is the system measurement output and $C = C_l$, $l \in \mathscr{L}$ is the common measurement output matrix; $f(t, x(t)) \in \Re^{n_u}$ is the exogenous disturbance. ΔA_l and Δa_l are the

DOI: 10.1201/9781032695716-3

parameter uncertainties in the l-th local model satisfying the condition in (1.3) as $\begin{bmatrix} \Delta A_l & \Delta a_l \end{bmatrix} = H_{l1}\Delta_l(t)\begin{bmatrix} H_{l2} & H_{l3} \end{bmatrix}, l \in \{1,\ldots,r\}$ with H_{l1}, H_{l2}, and H_{l3} known real-valued matrices and $\|\Delta_l(t)\| \leq \mathbf{I}$. It is assumed that the nonlinear disturbance $f(x(t),t)$ satisfies the following matching condition as

$$\|f(t,x(t))\| \leq \lambda(y(t),t) + \gamma\|x(t)\| \tag{3.2}$$

with $\lambda(y(t),t) : [\Re^{n_y} \times \Re^+] \to \Re^+$ referring to a known function and $\gamma > 0$.

Similar to Chapter 2, utilizing a singleton fuzzifier, product inference and center-average defuzzifier, one can obtain the following global T-S fuzzy affine model,

$$\begin{cases} \dot{x}(t) = (A(\mu) + \Delta A(\mu))x(t) + a(\mu) + \Delta a(\mu) + B(u(t) + f(t,x(t))) \\ y(t) = Cx(t) \end{cases} \tag{3.3}$$

where

$$\begin{cases} A(\mu) = \sum_{l=1}^{r} \mu_l A_l, \Delta A(\mu) = \sum_{l=1}^{r} \mu_l \Delta A_l, \\ a(\mu) = \sum_{l=1}^{r} \mu_l a_l, \Delta a(\mu) = \sum_{l=1}^{r} \mu_l \Delta a_l. \end{cases} \tag{3.4}$$

3.2.2 SLIDING SURFACE DESIGN

In accordance with the system (3.1) or equivalently (3.3), based on the system measurement output, design the following sliding surface as

$$s(t) = Ky(t) \tag{3.5}$$

with $K \in \Re^{n_u \times n_y}$ being the sliding surface gain to be determined.

Noticing the fact that the control input matrix B is with the full column rank feature, thus, a full-row rank matrix $M \in \Re^{(n_x - n_u) \times n_x}$ always exists and satisfies $MB = 0$. Consequently, we construct the following nonsingular transformation matrix T as

$$T = \begin{bmatrix} M^{\mathrm{T}} & B \end{bmatrix}^{\mathrm{T}}. \tag{3.6}$$

Based on (3.6), it is easy to find that $MT^{-1} = \begin{bmatrix} \mathbf{I}_{n_x - n_u} & \mathbf{0}_{(n_x - n_u) \times n_u} \end{bmatrix}$. Via a state transformation based on the matrix T, one can derive the system in (3.3) as,

$$T\dot{x}(t) = T\{[A(\mu) + \Delta A(\mu)]x(t) + B[u(t) + f(t,x(t))] + [a(\mu) + \Delta a(\mu)]\}. \tag{3.7}$$

The premise variable space can be decomposed by a similar line in (2.7)–(2.8) in Chapter 2, and to characterize the indices adopted for each region S_i, the following set is introduced,

$$\mathcal{N}(i) := \{n | \mu_n[\theta(t)] > 0,\ \theta(t) \in S_i,\ n \in \mathcal{L},\ i \in \mathcal{I}\}. \tag{3.8}$$

Based on (3.8), the system (3.7) can be reformulated into the following piecewise-fuzzy affine form as,

$$\begin{aligned} T\dot{x}(t) = T(\mathcal{A}_i + \Delta\mathcal{A}_i)x(t) + T(\mathrm{a}_i + \Delta\mathrm{a}_i) + TB(u(t) + f(t,x(t))), \\ \theta(t) \in S_i, i \in \mathcal{I}, \end{aligned} \tag{3.9}$$

where

$$\begin{cases} \mathscr{A}_i := \sum_{n\in\mathscr{N}(i)} \mu_n A_n, \Delta\mathscr{A}_i := \sum_{n\in\mathscr{N}(i)} \mu_n \Delta A_n, \\ \mathrm{a}_i := \sum_{n\in\mathscr{N}(i)} \mu_n a_n, \Delta\mathrm{a}_i := \sum_{n\in\mathscr{N}(i)} \mu_n \Delta a_n \end{cases} \tag{3.10}$$

with $0 < \mu_n[\theta(t)] \leq 1, \sum_{n\in\mathscr{N}(i)} \mu_n[\theta(t)] = 1$.

Noticing that $MB = 0$, one can reformulate system (3.9) as

$$\begin{cases} M\dot{x}(t) = M(\mathscr{A}_i + \Delta\mathscr{A}_i)x(t) + M(\mathrm{a}_i + \Delta\mathrm{a}_i) \\ B^{\mathrm{T}}\dot{x}(t) = B^{\mathrm{T}}(\mathscr{A}_i + \Delta\mathscr{A}_i)x(t) + B^{\mathrm{T}}B(u(t) + f(t,x(t))) + B^{\mathrm{T}}(\mathrm{a}_i + \Delta\mathrm{a}_i), \\ \qquad i \in \mathscr{I} \end{cases} \tag{3.11}$$

with $B^{\mathrm{T}}B$ being nonsingular. Notice that the dynamical features of the sliding motion is characterized via the first equation in (3.11). Additionally, the sliding surface function should satisfy $s(t) = 0$. Hence, with the above observations, we have the following holonomic sliding mode dynamical equation

$$E\dot{x}(t) = (\mathbf{A}_i + \Delta\mathbf{A}_i)x(t) + \mathbf{a}_i + \Delta\mathbf{a}_i, \ \theta(t) \in S_i, i \in \mathscr{I}, \tag{3.12}$$

where

$$\begin{cases} E = \begin{bmatrix} M \\ 0 \end{bmatrix}, \mathbf{A}_i = \begin{bmatrix} M\mathscr{A}_i \\ KC \end{bmatrix}, \mathbf{a}_i = \begin{bmatrix} M\mathrm{a}_i \\ \mathbf{0} \end{bmatrix}, \\ \Delta\mathbf{A}_i = \begin{bmatrix} M\Delta\mathscr{A}_i \\ \mathbf{0} \end{bmatrix}, \Delta\mathbf{a}_i = \begin{bmatrix} M\Delta\mathrm{a}_i \\ \mathbf{0} \end{bmatrix}. \end{cases} \tag{3.13}$$

Through a state transformation $z(t) = Tx(t)$ and associating with the fact that $MT^{-1} = \begin{bmatrix} \mathbf{I}_{n_x-n_u} & \mathbf{0}_{(n_x-n_u)\times n_u} \end{bmatrix}$, then one can reformulate the system (3.12) as

$$\tilde{E}\dot{z}(t) = (\tilde{\mathscr{A}}_i + \Delta\tilde{\mathscr{A}}_i)z(t) + \tilde{\mathrm{a}}_i + \Delta\tilde{\mathrm{a}}_i, \ \tilde{\theta}(t) \in \tilde{S}_i, i \in \mathscr{I}, \tag{3.14}$$

where $\tilde{S}_i$ and $\tilde{\theta}(t)$ denote the new partitioned space and premise variables via the transformation matrix T, respectively, and

$$\begin{cases} \tilde{E} = \begin{bmatrix} \mathbf{I}_{(n_x-n_u)} & \mathbf{0} \\ \mathbf{0} & \mathbf{0}_{n_u} \end{bmatrix}, \tilde{\mathscr{A}}_i + \Delta\tilde{\mathscr{A}}_i = \begin{bmatrix} M(\mathscr{A}_i + \Delta\mathscr{A}_i)T_1 & M(\mathscr{A}_i + \Delta\mathscr{A}_i)T_2 \\ KCT_1 & KCT_2 \end{bmatrix} \\ \tilde{\mathrm{a}}_i + \Delta\tilde{\mathrm{a}}_i = \begin{bmatrix} M(\mathrm{a}_i + \Delta\mathrm{a}_i) \\ \mathbf{0} \end{bmatrix}, T^{-1} = \begin{bmatrix} T_1 & T_2 \end{bmatrix} \end{cases} \tag{3.15}$$

and $T_1 \in \Re^{n_x\times(n_x-n_u)}$, $T_2 \in \Re^{n_x\times n_u}$ are known constant matrices.

By combining the sliding surface (3.5) with the first equation in system (3.11), the sliding motion can be fully characterized by the singular system (3.14). Therefore, the existence of the proposed sliding surface is ensured by the admissibility of the system (3.14). In the following, we will propose sufficient conditions for the existence of the sliding surface (3.5), and synthesize two novel output feedback sliding mode controllers.

3.2.3 STABILITY ANALYSIS OF SLIDING MOTION

In this section, some new results on the asymptotic stability analysis of the sliding motion will be presented through a CQLF and PQLFs, respectively.

3.2.3.1 Stability Analysis of Sliding Motion through CQLF

Along a similar line in Chapter 2, it is assumed that one can use an ellipsoid $\mathscr{E}_i$ to outer approximate every polyhedral subspace $\tilde{S}_i$ in (3.14), and the relationship described in (2.19) holds for each region.

Then we have the following asymptotic stability analysis results for the sliding motion.

Theorem 3.1. Given a constant α, the system (3.14) is asymptotically stable, if there exist matrices $0 < P_1 = P_1^{\mathrm{T}} \in \Re^{(n_x-n_u)\times(n_x-n_u)}$, $\bar{K} \in \Re^{n_u\times n_y}$, and two sets of scalars $\{\varepsilon_i > 0, i \in \mathscr{I}\}$, $\{\kappa_i < 0, i \in \mathscr{I}_1\}$, such that the following conditions in (3.16)–(3.17) hold,

$$\begin{bmatrix} \Sigma_{11} & \star & \star \\ \Sigma_{21} & \Sigma_{22} & \star \\ H_{n1}^{\mathrm{T}}M^{\mathrm{T}}P_1 & \mathbf{0} & -\varepsilon_i\mathbf{I} \end{bmatrix} < 0, i \in \mathscr{I}_0, n \in \mathscr{N}(i), \tag{3.16}$$

$$\begin{bmatrix} \kappa_i(q_i^{\mathrm{T}}q_i-1)+\varepsilon_i H_{n3}^{\mathrm{T}}H_{n3} & \star & \star & \star \\ \kappa_i Q_{i1}^{\mathrm{T}}q_i+\Sigma_{10} & \kappa_i Q_{i1}^{\mathrm{T}}Q_{i1}+\Sigma_{11} & \star & \star \\ \kappa_i Q_{i2}^{\mathrm{T}}q_i+\varepsilon_i T_2^{\mathrm{T}}H_{n2}^{\mathrm{T}}H_{n3} & \kappa_i Q_{i2}^{\mathrm{T}}Q_{i1}+\Sigma_{21} & \kappa_i Q_{i2}^{\mathrm{T}}Q_{i2}+\Sigma_{22} & \star \\ \mathbf{0} & H_{n1}^{\mathrm{T}}M^{\mathrm{T}}P_1 & \mathbf{0} & -\varepsilon_i\mathbf{I} \end{bmatrix} < 0,$$
$$i \in \mathscr{I}_1, n \in \mathscr{N}(i), \tag{3.17}$$

where

$$\begin{cases} \Sigma_{11} = \mathtt{Sym}\{P_1MA_nT_1-\alpha J^{\mathrm{T}}\bar{K}CT_1\}+\varepsilon_i T_1^{\mathrm{T}}H_{n2}^{\mathrm{T}}H_{n2}T_1, \\ \Sigma_{21} = \bar{K}CT_1+T_2^{\mathrm{T}}A_n^{\mathrm{T}}M^{\mathrm{T}}P_1^{\mathrm{T}}-\alpha T_2^{\mathrm{T}}C^{\mathrm{T}}\bar{K}^{\mathrm{T}}J+\varepsilon_i T_2^{\mathrm{T}}H_{n2}^{\mathrm{T}}H_{n2}T_1, \\ \Sigma_{22} = \mathtt{Sym}\{\bar{K}CT_2\}+\varepsilon_i T_2^{\mathrm{T}}H_{n2}^{\mathrm{T}}H_{n2}T_2, \\ \Sigma_{10} = P_1Ma_n+\varepsilon_i T_1^{\mathrm{T}}H_{n2}^{\mathrm{T}}H_{n3} \end{cases} \tag{3.18}$$

and $J \in \Re^{n_u\times(n_x-n_u)}$ represents an arbitrary matrix.

Then the gain of the sliding surface (3.5) can be calculated by

$$K = P_2^{-\mathrm{T}}\bar{K}, \tag{3.19}$$

where $P_2 \in \Re^{n_u\times n_u}$ is an arbitrary nonsingular matrix.

Proof. Only the more complicated case for $i \in \mathscr{I}_1$ will be proved without loss of generality. For system (3.14), construct a common quadratic Lyapunov function (CQLF) as

$$V(t) = z^{\mathrm{T}}(t)\tilde{E}^{\mathrm{T}}\tilde{P}z(t), \tag{3.20}$$

where $\tilde{P} = \begin{bmatrix} P_1 & \mathbf{0} \\ P_3 & P_2 \end{bmatrix}$, $0 < P_1 = P_1^{\mathrm{T}} \in \Re^{(n_x-n_u)\times(n_x-n_u)}$, $P_2 \in \Re^{n_u\times n_u}$ is an arbitrary invertible matrix, and $P_3 \in \Re^{n_u\times(n_x-n_u)}$.

If the following inequality holds, the system (3.14) is asymptotically stable

$$\dot{V}(t) = \mathtt{Sym}\left\{\left[(\tilde{\mathscr{A}}_i + \Delta\tilde{\mathscr{A}}_i)z(t) + \tilde{\mathrm{a}}_i + \Delta\tilde{\mathrm{a}}_i\right]^{\mathrm{T}} \tilde{P}z(t)\right\} < 0. \tag{3.21}$$

Based on the partition information described as in (2.19) and exploiting the S-procedure, then inequality (3.22) indicates (3.21) with $\kappa_i < 0$, $i \in \mathscr{I}_1$,

$$\begin{bmatrix} 1 \\ z(t) \end{bmatrix}^{\mathrm{T}} \left\{ \begin{bmatrix} \mathbf{0} & \star \\ \tilde{P}^{\mathrm{T}}(\tilde{\mathrm{a}}_i + \Delta\tilde{\mathrm{a}}_i) & \mathtt{Sym}\{(\tilde{\mathscr{A}}_i + \Delta\tilde{\mathscr{A}}_i)^{\mathrm{T}}\tilde{P}\} \end{bmatrix} + \kappa_i \begin{bmatrix} q_i^{\mathrm{T}} q_i - 1 & \star \\ Q_i^{\mathrm{T}} q_i & Q_i^{\mathrm{T}} Q_i \end{bmatrix} \right\} \begin{bmatrix} 1 \\ z(t) \end{bmatrix} < 0, i \in \mathscr{I}_1. \tag{3.22}$$

Define $\bar{K} = P_2^{\mathrm{T}} K$. Based on the structure of Q_i and $\tilde{P}$, the subsequent inequality implies (3.22),

$$\begin{bmatrix} \kappa_i(q_i^{\mathrm{T}} q_i - 1) & \star & \star \\ \kappa_i Q_{i1}^{\mathrm{T}} q_i + P_1 M(\mathrm{a}_i + \Delta\mathrm{a}_i) & \kappa_i Q_{i1}^{\mathrm{T}} Q_{i1} + \Theta_{11} & \star \\ \kappa_i Q_{i2}^{\mathrm{T}} q_i & \kappa_i Q_{i2}^{\mathrm{T}} Q_{i1} + \Theta_{21} & \kappa_i Q_{i2}^{\mathrm{T}} Q_{i2} + \mathtt{Sym}\{\bar{K}CT_2\} \end{bmatrix} < 0, \quad i \in \mathscr{I}_1, \tag{3.23}$$

where

$$\begin{cases} \Theta_{11} = \mathtt{Sym}\{P_1 M(\mathscr{A}_i + \Delta\mathscr{A}_i)T_1 + P_3^{\mathrm{T}} KCT_1\}, \\ \Theta_{21} = \bar{K}CT_1 + T_2^{\mathrm{T}}(\mathscr{A}_i^{\mathrm{T}} + \Delta\mathscr{A}_i^{\mathrm{T}})M^{\mathrm{T}} P_1^{\mathrm{T}} + T_2^{\mathrm{T}} C^{\mathrm{T}} K^{\mathrm{T}} P_3. \end{cases} \tag{3.24}$$

Extract the MFs and use Lemma A1 to tackle the parameter uncertainties. Then for $\varepsilon_i > 0$, $i \in \mathscr{I}$, the following inequality implies (3.23),

$$\begin{bmatrix} \kappa_i(q_i^{\mathrm{T}} q_i - 1) & \star & \star \\ P_1 M a_n + \kappa_i Q_{i1}^{\mathrm{T}} q_i & \kappa_i Q_{i1}^{\mathrm{T}} Q_{i1} + \Upsilon_{11} & \star \\ \kappa_i Q_{i2}^{\mathrm{T}} q_i & \kappa_i Q_{i2}^{\mathrm{T}} Q_{i1} + \Upsilon_{21} & \kappa_i Q_{i2}^{\mathrm{T}} Q_{i2} + \mathtt{Sym}\{\bar{K}CT_2\} \end{bmatrix} + \varepsilon_i^{-1} \begin{bmatrix} \mathbf{0} \\ P_1 M H_{n1} \\ \mathbf{0} \end{bmatrix} (\star) + \varepsilon_i \begin{bmatrix} H_{n3}^{\mathrm{T}} \\ T_1^{\mathrm{T}} H_{n2}^{\mathrm{T}} \\ T_2^{\mathrm{T}} H_{n2}^{\mathrm{T}} \end{bmatrix} (\star) < 0, \; i \in \mathscr{I}_1, n \in \mathscr{N}(i), \tag{3.25}$$

where

$$\begin{cases} \Upsilon_{11} = \mathtt{Sym}\{P_1 M A_n T_1 + P_3^{\mathrm{T}} KCT_1\}, \\ \Upsilon_{21} = \bar{K}CT_1 + T_2^{\mathrm{T}} A_n^{\mathrm{T}} M^{\mathrm{T}} P_1^{\mathrm{T}} + T_2^{\mathrm{T}} C^{\mathrm{T}} K^{\mathrm{T}} P_3. \end{cases} \tag{3.26}$$

For numerical tractability, specify P_3 to be $P_3 = -\alpha P_2 J$, where α is a scalar and $J \in \Re^{n_u\times(n_x-n_u)}$ is an arbitrary matrix.

Then by Schur complement, (3.17) implies (3.25). Then the proof is completed. ■

The stability analysis results on the sliding motion given as in Theorem 3.1 are attained relying on a CQLF. To further reduce the conservatism, a novel stability analysis method will be presented based on PQLFs in the following.

3.2.3.2 Stability Analysis of Sliding Motion through PQLFs

Introduce the following notation firstly

$$\bar{z}_1(t) := \begin{bmatrix} 1 \\ z_1(t) \end{bmatrix}, \tag{3.27}$$

where $z_1(t) \in \Re^{n_x - n_u}$ denotes the first $(n_x - n_u)$ rows of $z(t)$.

To ensure the continuity of PQLFs among the region boundaries, as in subchapter 2.2, the matrices $\bar{F}_i = \begin{bmatrix} f_i & F_i \end{bmatrix}, i \in \mathscr{I}$ are constructed to characterize the boundary across the regions with $f_i = 0$ for $i \in \mathscr{I}_0$,

$$\bar{F}_i \begin{bmatrix} 1 \\ z_1(t) \end{bmatrix} = \bar{F}_j \begin{bmatrix} 1 \\ z_1(t) \end{bmatrix}, z_1(t) \in \tilde{S}_i \bigcap \tilde{S}_j, i, j \in \mathscr{I}. \tag{3.28}$$

The S-procedure will be also utilized to further reduce the conservatism via constructing matrices $\bar{G}_i = \begin{bmatrix} g_i & G_i \end{bmatrix}$, with $g_i = 0$ for $i \in \mathscr{I}_0$ and

$$\bar{G}_i \bar{z}_1(t) \succeq 0, z_1(t) \in \tilde{S}_i, i \in \mathscr{I} \tag{3.29}$$

with $\succeq$ implying that every entry is nonnegative.

For future use, we define

$$\begin{bmatrix} \hat{F}_i & \hat{G}_i \end{bmatrix} = \begin{cases} \begin{bmatrix} F_i & G_i \end{bmatrix}, i \in \mathscr{I}_0, \\ \begin{bmatrix} \bar{F}_i & \bar{G}_i \end{bmatrix}, i \in \mathscr{I}_1. \end{cases} \tag{3.30}$$

The system (3.14) can be reformulated into the following compact form,

$$\hat{E}\dot{\hat{z}}(t) = \left(\hat{\mathscr{A}}_i + \Delta\hat{\mathscr{A}}_i\right)\hat{z}(t), \ i \in \mathscr{I}, \tag{3.31}$$

where

$$\begin{bmatrix} \hat{E} \mid \hat{z}(t) \mid \hat{\mathscr{A}}_i + \Delta\hat{\mathscr{A}}_i \end{bmatrix} = \begin{cases} \begin{bmatrix} \tilde{E} & z(t) & \tilde{\mathscr{A}}_i + \Delta\tilde{\mathscr{A}}_i \end{bmatrix}, i \in \mathscr{I}_0, \\ \begin{bmatrix} \bar{E} & \bar{z}(t) & \bar{\mathscr{A}}_i + \Delta\bar{\mathscr{A}}_i \end{bmatrix}, \imath \in \mathscr{I}_1 \end{cases} \tag{3.32}$$

with $\tilde{E}$, $\tilde{\text{a}}_i + \Delta\tilde{\text{a}}_i$, and $\tilde{\mathscr{A}}_i + \Delta\tilde{\mathscr{A}}_i$ defined in (3.15) and

$$\begin{cases} \bar{E} = \begin{bmatrix} 1 & \mathbf{0} \\ \mathbf{0} & \tilde{E} \end{bmatrix}, \bar{z}(t) = \begin{bmatrix} 1 \\ z(t) \end{bmatrix}, \\ \bar{\mathscr{A}}_i + \Delta\bar{\mathscr{A}}_i = \begin{bmatrix} 0 & \mathbf{0} \\ \tilde{\text{a}}_i + \Delta\tilde{\text{a}}_i & \tilde{\mathscr{A}}_i + \Delta\tilde{\mathscr{A}}_i \end{bmatrix}. \end{cases} \tag{3.33}$$

Theorem 3.2. Given a constant α, the system (3.31) is asymptotically stable, if matrices $X = X^{\mathrm{T}}$, $U_i = U_i^{\mathrm{T}} \succeq 0$, $W_i = W_i^{\mathrm{T}} \succeq 0$, $\bar{K} \in \Re^{n_u \times n_y}$, and scalars $\varepsilon_i > 0, i \in \mathscr{I}$ exist, such that the conditions in (3.34)-(3.35) hold

$$\hat{F}_i^{\mathrm{T}} X \hat{F}_i - \hat{G}_i^{\mathrm{T}} U_i \hat{G}_i > 0, i \in \mathscr{I}, \tag{3.34}$$

$$\begin{bmatrix} \hat{\Lambda}_1 & \star & \star \\ \hat{\Lambda}_2 & \mathrm{Sym}\{\bar{K}CT_2\}+\varepsilon_i T_2^{\mathrm{T}}H_{n2}^{\mathrm{T}}H_{n2}T_2 & \star \\ H_{n1}^{\mathrm{T}}M^{\mathrm{T}}\hat{R}\hat{F}_i^{\mathrm{T}}X\hat{F}_i & \mathbf{0} & -\varepsilon_i\mathbf{I} \end{bmatrix} < 0, \quad s\in\mathscr{N}(i), i\in\mathscr{I}, \tag{3.35}$$

where $\hat{F}_i$, $\hat{G}_i$ are defined in (3.30) and

$$\begin{cases} \hat{\Lambda}_1 = \mathrm{Sym}\left\{\hat{F}_i^{\mathrm{T}}X\hat{F}_i\hat{A}_{n1} - \alpha\hat{J}\bar{K}CT_1\hat{R}\right\} + \varepsilon_i\hat{\Lambda}_{1(1)}\hat{\Lambda}_{1(1)}^{\mathrm{T}} + \hat{G}_i^{\mathrm{T}}W_i\hat{G}_i, \\ \hat{\Lambda}_2 = \bar{K}CT_1\hat{R} + \hat{A}_{n2}^{\mathrm{T}}\hat{F}_i^{\mathrm{T}}X\hat{F}_i - \alpha T_2^{\mathrm{T}}C^{\mathrm{T}}\bar{K}^{\mathrm{T}}\hat{J} + \varepsilon_i T_2^{\mathrm{T}}H_{n2}^{\mathrm{T}}\hat{\Lambda}_{1(1)}^{\mathrm{T}}, \\ \left.\begin{array}{l}\hat{\Lambda}_{1(1)} = T_1^{\mathrm{T}}H_{n2}^{\mathrm{T}},\ \hat{A}_{n1} = MA_nT_1, \\ \hat{A}_{n2} = MA_nT_2, \hat{J} = \tilde{J}, \hat{R} = \mathbf{I}_{(n_x-n_u)},\end{array}\right\}\text{if } i\in\mathscr{I}_0, \\ \left.\begin{array}{l}\hat{\Lambda}_{1(1)} = \begin{bmatrix} H_{n3}^{\mathrm{T}} \\ T_1^{\mathrm{T}}H_{n2}^{\mathrm{T}} \end{bmatrix}, \hat{A}_{n1} = \begin{bmatrix} 0 & \mathbf{0} \\ Ma_n & MA_nT_1 \end{bmatrix}, \\ \hat{A}_{n2} = \begin{bmatrix} \mathbf{0} \\ MA_nT_2 \end{bmatrix}, \hat{J} = \bar{J}, \hat{R} = \begin{bmatrix} \mathbf{0} & \mathbf{I}_{(n_x-n_u)} \end{bmatrix}\end{array}\right\}\text{if } i\in\mathscr{I}_1 \end{cases} \tag{3.36}$$

and $\tilde{J}\in\Re^{n_u\times(n_x-n_u)}$, $\bar{J}\in\Re^{n_u\times(1+n_x-n_u)}$ are arbitrary matrices.

The gain of the sliding surface (3.5) can be computed by

$$K = P_2^{-\mathrm{T}}\bar{K}, \tag{3.37}$$

where $P_2\in\Re^{n_u\times n_u}$ is an arbitrary invertible matrix.

Proof. For the sliding mode dynamical system (3.31), construct the following PQLF continuous among the region boundaries as,

$$V(t) = \hat{z}^{\mathrm{T}}(t)\hat{E}\hat{\mathscr{P}}_i\hat{z}(t) = \begin{cases} z^{\mathrm{T}}(t)\tilde{E}\tilde{\mathscr{P}}_iz(t), & i\in\mathscr{I}_0, \\ \bar{z}^{\mathrm{T}}(t)\bar{E}\bar{\mathscr{P}}_i\bar{z}(t), & i\in\mathscr{I}_1, \end{cases} \tag{3.38}$$

where

$$\hat{\mathscr{P}}_i = \begin{bmatrix} \hat{F}_i^{\mathrm{T}}X\hat{F}_i & \mathbf{0} \\ \hat{P}_3 & P_2 \end{bmatrix} = \begin{cases} \tilde{\mathscr{P}}_i = \begin{bmatrix} F_i^{\mathrm{T}}XF_i & \mathbf{0} \\ \tilde{P}_3 & P_2 \end{bmatrix}, i\in\mathscr{I}_0, \\ \bar{\mathscr{P}}_i = \begin{bmatrix} \bar{F}_i^{\mathrm{T}}X\bar{F}_i & \mathbf{0} \\ \bar{P}_3 & P_2 \end{bmatrix}, i\in\mathscr{I}_1 \end{cases} \tag{3.39}$$

and $X = X^{\mathrm{T}}$, $P_2\in\Re^{n_u\times n_u}$ is an arbitrary nonsingular matrix, and $\tilde{P}_3\in\Re^{n_u\times(n_x-n_u)}$, $\bar{P}_3\in\Re^{n_u\times(1+n_x-n_u)}$.

The inequality in (3.34) guarantees the positive definiteness of PQLF within each subspace. According to the particular structure of $\hat{E}$, we have $\hat{E}\hat{\mathscr{P}}_i = \hat{\mathscr{P}}_i^{\mathrm{T}}\hat{E}$. If the following inequality holds, the system (3.31) is asymptotically stable,

$$\dot{V}(t) = \hat{z}^{\mathrm{T}}(t)\left\{\left(\hat{\mathscr{A}}_i^{\mathrm{T}} + \Delta\hat{\mathscr{A}}_i^{\mathrm{T}}\right)\hat{\mathscr{P}}_i + \hat{\mathscr{P}}_i^{\mathrm{T}}\left(\hat{\mathscr{A}}_i + \Delta\hat{\mathscr{A}}_i\right)\right\}\hat{z}(t) < 0. \tag{3.40}$$

Define $\bar{K} = P_2^{\mathrm{T}}K$. By the special structure of $\hat{\mathscr{P}}_i$ shown in (3.39) and using S-procedure, the following inequality implies (3.40),

$$\begin{bmatrix} \mathrm{Sym}\left\{\hat{F}_i^{\mathrm{T}}X\hat{F}_i(\hat{\mathscr{A}}_{i1} + \Delta\hat{\mathscr{A}}_{i1}) + \hat{P}_3^{\mathrm{T}}KCT_1\right\} + \hat{G}_i^{\mathrm{T}}W_i\hat{G}_i & \star \\ \bar{K}CT_1\hat{R} + (\hat{\mathscr{A}}_{i2}^{\mathrm{T}} + \Delta\hat{\mathscr{A}}_{i2}^{\mathrm{T}})\hat{F}_i^{\mathrm{T}}X\hat{F}_i + T_2^{\mathrm{T}}C^{\mathrm{T}}K^{\mathrm{T}}\hat{P}_3 & \mathrm{Sym}\{\bar{K}CT_2\} \end{bmatrix} < 0, \tag{3.41}$$

where $W_i \succeq 0$ and

$$\begin{cases} \left[\hat{\mathscr{A}}_{i1} \mid \hat{\mathscr{A}}_{i2} \mid \Delta\hat{\mathscr{A}}_{i1} \mid \Delta\hat{\mathscr{A}}_{i2} \mid \hat{R}\right] = \begin{cases} \left[\tilde{\mathscr{A}}_{i1} \mid \tilde{\mathscr{A}}_{i2} \mid \Delta\tilde{\mathscr{A}}_{i1} \mid \Delta\tilde{\mathscr{A}}_{i2} \mid \tilde{R}\right], i \in \mathscr{I}_0, \\ \left[\bar{\mathscr{A}}_{i1} \mid \bar{\mathscr{A}}_{i2} \mid \Delta\bar{\mathscr{A}}_{i1} \mid \Delta\bar{\mathscr{A}}_{i2} \mid \bar{R}\right], i \in \mathscr{I}_1, \end{cases} \\ \tilde{\mathscr{A}}_{i1} = M\mathscr{A}_i T_1, \tilde{\mathscr{A}}_{i2} = M\mathscr{A}_i T_2, \\ \Delta\tilde{\mathscr{A}}_{i1} = M\Delta\mathscr{A}_i T_1, \Delta\tilde{\mathscr{A}}_{i2} = M\Delta\mathscr{A}_i T_2, \tilde{R} = \mathbf{I}_{(n_x-n_u)}, \\ \bar{\mathscr{A}}_{i1} = \begin{bmatrix} 0 & \mathbf{0} \\ M\mathrm{a}_i & \tilde{\mathscr{A}}_{i1} \end{bmatrix}, \bar{\mathscr{A}}_{i2} = \begin{bmatrix} \mathbf{0} \\ \tilde{\mathscr{A}}_{i2} \end{bmatrix}, \Delta\bar{\mathscr{A}}_{i1} = \begin{bmatrix} 0 & \mathbf{0} \\ M\Delta\mathrm{a}_i & \Delta\tilde{\mathscr{A}}_{i1} \end{bmatrix}, \\ \Delta\bar{\mathscr{A}}_{i2} = \begin{bmatrix} \mathbf{0} \\ \Delta\tilde{\mathscr{A}}_{i2} \end{bmatrix}, \bar{R} = \begin{bmatrix} \mathbf{0} & \mathbf{I}_{(n_x-n_u)} \end{bmatrix}. \end{cases} \tag{3.42}$$

Expand the fuzzy MFs and then employ Lemma A1 to tackle the uncertainty terms presented as in (1.3). For scalars $\varepsilon_i > 0$, $i \in \mathscr{I}$, the subsequent inequality indicates (3.41),

$$\begin{bmatrix} \mathrm{Sym}\{\hat{F}_i^{\mathrm{T}} X \hat{F}_i \hat{A}_{n1} + \hat{P}_3 K C T_1\} + \hat{G}_i^{\mathrm{T}} W_i \hat{G}_i & \star \\ \bar{K} C T_1 \hat{R} + \hat{A}_{n2}^{\mathrm{T}} \hat{F}_i^{\mathrm{T}} X \hat{F}_i + T_2^{\mathrm{T}} C^{\mathrm{T}} K^{\mathrm{T}} \hat{P}_3 & \mathrm{Sym}\{\bar{K} C T_2\} \end{bmatrix}$$
$$+ \varepsilon_i \begin{bmatrix} \hat{\Lambda}_{1(1)} \\ T_2^{\mathrm{T}} H_{n2}^{\mathrm{T}} \end{bmatrix} (\star) + \varepsilon_i^{-1} \begin{bmatrix} \hat{F}_i^{\mathrm{T}} X \hat{F}_i \hat{R}^{\mathrm{T}} M H_{n1} \\ \mathbf{0} \end{bmatrix} (\star) < 0, \; n \in \mathscr{N}(i), i \in \mathscr{I}, \tag{3.43}$$

where

$$\begin{cases} \left[\hat{\Lambda}_{1(1)} \mid \hat{A}_{n1} \mid \hat{A}_{n2}\right] = \begin{cases} \left[\tilde{\Lambda}_{1(1)} \mid \tilde{A}_{n1} \mid \tilde{A}_{n2}\right], i \in \mathscr{I}_0, \\ \left[\bar{\Lambda}_{1(1)} \mid \bar{A}_{n1} \mid \bar{A}_{n2}\right], i \in \mathscr{I}_1, \end{cases} \\ \tilde{\Lambda}_{1(1)} = T_1^{\mathrm{T}} H_{n2}^{\mathrm{T}}, \tilde{A}_{n1} = M A_n T_1, \tilde{A}_{n2} = M A_n T_2, \\ \bar{\Lambda}_{1(1)} = \begin{bmatrix} H_{n3}^{\mathrm{T}} \\ T_1^{\mathrm{T}} H_{n2}^{\mathrm{T}} \end{bmatrix}, \bar{A}_{n1} = \begin{bmatrix} 0 & \mathbf{0} \\ M a_n & M A_n T_1 \end{bmatrix}, \bar{A}_{n2} = \begin{bmatrix} \mathbf{0} \\ M A_n T_2 \end{bmatrix}. \end{cases} \tag{3.44}$$

Note that the gain K is not involved within the first row of $\hat{A}_{n1}$ and $\hat{A}_{n2}$. For numerical tractability, specify $\hat{P}_3$ to be

$$\hat{P}_3 = -\alpha P_2 \hat{J} = \begin{cases} -\alpha P_2 \tilde{J}, i \in \mathscr{I}_0, \\ -\alpha P_2 \bar{J}, i \in \mathscr{I}_1, \end{cases} \tag{3.45}$$

where α represents a constant, and $\tilde{J} \in \Re^{n_u \times (n_x - n_u)}$ and $\bar{J} \in \Re^{n_u \times (1 + n_x - n_u)}$ are arbitrary matrices.

Then (3.35) implies (3.43) on the basis of Schur complement. Then the proof is completed. ■

Particularly, if the LMIs in (3.16)–(3.17) or (3.34)–(3.35) are feasible, one can obtain the corresponding sliding surface gain by $K = P_2^{-\mathrm{T}} \bar{K}$, and P_2 is an arbitrary invertible matrix.

In following section, we will design two output feedback sliding mode controllers to drive the system states onto the sliding surface (3.5) in finite time locally.

3.3 OUTPUT FEEDBACK SLIDING MODE CONTROLLER DESIGN

Firstly, along a similar process in [72], a sliding patch Ω is introduced as,

$$\Omega = \{x(t) \in \Re^{n_x} : \|x(t)\| \le \sigma\}. \tag{3.46}$$

3.3.1 SLIDING MODE CONTROLLER DESIGN

Theorem 3.3. For the sliding surface (3.5) and the region Ω, the following output feedback sliding mode controller can drive the state trajectories of system (3.1) to the sliding surface in finite time locally,

$$u(t) = -(KCB)^{-1}\{[\zeta(t) + \|KCB\| \cdot \lambda(y(t),t)]\,\mathrm{sgn}(s(t)) + \eta\,\mathrm{sgn}(s(t))\}, \quad (3.47)$$

where

$$\begin{aligned}\zeta(t) \geq & \sum_{l=1}^{r} \mu_l \{\|KCA_l\| + \|KCH_{l1}\| \cdot \|H_{l2}\| + \gamma\|KCB\|\}\,\sigma \\ & + \sum_{l=1}^{r} \mu_l \{\|KCa_l\| + \|KCH_{l1}\| \cdot \|H_{l3}\|\}\end{aligned} \quad (3.48)$$

with $\eta > 0$ and $\sigma > 0$ being given scalars.

Proof. Consider the following Lyapunov function for all $t > 0$ as follows,

$$V(t) = \frac{1}{2}s^{\mathrm{T}}(t)s(t). \quad (3.49)$$

Considering the sliding surface in (3.5), provides

$$\dot{s}(t) = KC\dot{x}(t). \quad (3.50)$$

Based on the output feedback sliding mode controller (3.47) and considering the matching condition in (3.2) and the sliding patch introduced by (3.46), then one has

$$\begin{aligned}\dot{V}(t) &= s^{\mathrm{T}}(t)\dot{s}(t) \\ &= s^{\mathrm{T}}(t)\sum_{l=1}^{r} \mu_l KC[(A_l + \Delta A_l)x(t) + a_l + \Delta a_l + B(u(t) + f(t,x))] \\ &= s^{\mathrm{T}}(t)\sum_{l=1}^{r} \mu_l \{KC(A_l + \Delta A_l)x(t) + KC(a_l + \Delta a_l) + KCBf(t,x) \\ &\qquad -\eta\,\mathrm{sgn}(s(t)) - [\zeta(t) + \|KCB\| \cdot \lambda(y,t)]\,\mathrm{sgn}(s(t))\} \\ &\leq \sum_{l=1}^{r} \mu_l \{\|KCA_l\| + \|KC\Delta A_l\| + \gamma\|KCB\|\} \cdot \sigma\|s(t)\| \\ &\quad + \sum_{l=1}^{r} \mu_l \{\|KCa_l\| + \|KC\Delta a_l\|\}\,\|s(t)\| - \eta\|s(t)\| - \zeta(t)\|s(t)\| \\ &\leq \sum_{l=1}^{r} \mu_l \{\|KCA_l\| + \|KCH_{l1}\| \cdot \|H_{l2}\| + \gamma\|KCB\|\} \cdot \sigma\|s(t)\| \\ &\quad + \sum_{l=1}^{r} \mu_l \{\|KCa_l\| + \|KCH_{l1}\| \cdot \|H_{l3}\|\}\,\|s(t)\| - \zeta(t)\|s(t)\| - \eta\|s(t)\|.\end{aligned} \quad (3.51)$$

Notice that

$$\begin{aligned}\zeta(t) \geq & \sum_{l=1}^{r} \mu_l \{\|KCA_l\| + \|KCH_{l1}\| \cdot \|H_{l2}\| + \gamma\|KCB\|\}\,\sigma \\ & + \sum_{l=1}^{r} \mu_l \{\|KCa_l\| + \|KCH_{l1}\| \cdot \|H_{l3}\|\},\end{aligned} \quad (3.52)$$

and one has

$$\dot{V}(t) \leq -\eta \|s(t)\|, \tag{3.53}$$

which ensures the finite-time reachability of the sliding surface (3.5). The proof is completed. ■

Remark 3.3.1. As in the sliding mode controller (3.47), an assumption that $\|x(t)\| \leq \sigma$ is introduced. In other words, only the finite-time reachability in the region Ω to the sliding surface (3.5) can be ensured by the output feedback sliding mode controller (3.47). To relax the above-mentioned assumption, we will develop an adaptive output feedback sliding mode controller for the case when σ is not available.

3.3.2 ADAPTIVE SLIDING MODE CONTROLLER DESIGN

Theorem 3.4. For the sliding surface (3.5) and the region Ω, the following adaptive output feedback sliding mode controller can force the state trajectories of system (3.1) onto the sliding surface in finite time,

$$u(t) = -(KCB)^{-1}\left\{\left[\check{\zeta}(t) + \|KCB\| \cdot \lambda(y(t),t)\right] \operatorname{sgn}(s(t)) + \eta \operatorname{sgn}(s(t))\right\}, \tag{3.54}$$

where

$$\begin{aligned} \check{\zeta}(t) \geq & \sum_{l=1}^{r} \mu_l \left\{\|KCA_l\| + \|KCH_{l1}\| \cdot \|H_{l2}\| + \gamma\|KCB\|\right\} \check{\sigma}(t) \\ & + \sum_{l=1}^{r} \mu_l \left\{\|KCa_l\| + \|KCH_{l1}\| \cdot \|H_{l3}\|\right\}, \end{aligned} \tag{3.55}$$

and the following adaptive law generates $\check{\sigma}(t)$,

$$\dot{\check{\sigma}}(t) = \varpi \sum_{l=1}^{r} \mu_l \left\{\|KCA_l\| + \gamma\|KCB\| + \|KCH_{l1}\| \cdot \|H_{l2}\|\right\} \|s(t)\|, \tag{3.56}$$

where $\check{\sigma}(t)$ is the estimation of σ, and ϖ and η are two given positive parameters.

Proof. Consider the following Lyapunov function for all $t > 0$ as follows,

$$V(t) = \frac{1}{2}\left[\frac{1}{\varpi}\sigma_e^2(t) + s^{\mathrm{T}}(t)s(t)\right], \tag{3.57}$$

where $\sigma_e(t) = \sigma - \check{\sigma}(t)$.

Based on the adaptive sliding mode controller (3.54) and the matching condition in (3.2), one has

$$\begin{aligned}
\dot{V}(t) &= s^{\mathrm{T}}(t)\dot{s}(t) - \frac{1}{\varpi}\sigma_e(t)\dot{\check{\sigma}}(t) \\
&= s^{\mathrm{T}}(t)\sum_{l=1}^{r}\mu_l KC\{(A_l+\Delta A_l)x(t)+a_l+\Delta a_l+B(u(t)+f(t,x))\} \\
&\quad - \frac{1}{\varpi}\sigma_e(t)\dot{\check{\sigma}}(t) \\
&= s^{\mathrm{T}}(t)\sum_{l=1}^{r}\mu_l\{KC(A_l+\Delta A_l)x(t)+KC(a_l+\Delta a_l)+KCBf(t,x) \\
&\quad -\eta\,\mathrm{sgn}(s(t)) - \left[\|KCB\|\cdot\lambda(y,t)+\check{\zeta}(t)\right]\mathrm{sgn}(s(t))\Big\} - \frac{1}{\varpi}\sigma_e(t)\dot{\check{\sigma}}(t) \\
&\le \sum_{l=1}^{r}\mu_l\{\|KCA_l\|+\|KCH_{l1}\|\cdot\|H_{l2}\|+\gamma\|KCB\|\}\cdot\sigma\|s(t)\| \\
&\quad + \sum_{l=1}^{r}\mu_l\{\|KCa_l\|+\|KCH_{l1}\|\cdot\|H_{l3}\|\}\|s(t)\| \\
&\quad -(\sigma-\check{\sigma}(t))\sum_{l=1}^{r}\mu_l\{\gamma\|KCB\|+\|KCA_l\|+\|KCH_{l1}\|\cdot\|H_{l2}\|\}\|s(t)\| \\
&\quad -\check{\zeta}(t)\|s(t)\|-\eta\|s(t)\|.
\end{aligned} \tag{3.58}$$

Based on (3.55)–(3.56), one has

$$\dot{V}(t) \le -\eta\|s(t)\|. \tag{3.59}$$

Thus, the adaptive sliding mode controller (3.54) can drive the state trajectories of the fuzzy affine system (3.1) towards the sliding surface within the region Ω. The proof is consequently completed. ■

Remark 3.3.2. Detailed controller design procedures are given as follows.

Step 1. Based on the normalized membership functions and space partitions of the system (3.1), calculate the parameters Q_i and q_i for Theorem 3.1 via (2.18), or the matrices matrices $\hat{G}_i$ and $\hat{F}_i$ for Theorem 3.2 via (3.28)–(3.29).

Step 2. For given parameter α and matrix J, together with those parameters computed in Step 1, solve the LMI problems given in Theorem 3.1 or Theorem 3.2 over the variables matrices P_2, $\bar{K}$, and scalars $\{\varepsilon_i > 0, i \in \mathscr{I}\}$.

Step 3. Based on the obtained matrices P_2 and $\bar{K}$ in Step 2, the sliding surface gain can be then computed via $K = P_2^{-\mathrm{T}}\bar{K}$.

Step 4. Design the sliding mode controllers in (3.47) or adaptive sliding mode controllers (3.54), respectively, for situations that the upper-bound σ is given or not. Note that one can design the controller parameters $\zeta(t)$ via (3.48), or design $\check{\zeta}(t)$ and $\check{\sigma}(t)$ via (3.55)–(3.56).

3.3.3 ILLUSTRATIVE EXAMPLES

Example 3.1. Consider a nonlinear continuous stirred tank reactor (CSTR) system [40], which is described by the following equation

$$\begin{cases} \dot{T} = M_0(T_n - T) + M_1 k_0 e^{-\frac{M_2}{T}} C_A - M_3 T + M_3 \mathbf{u} \\ \dot{C}_A = M_0(C_0 - C_A) - M_0 e^{-\frac{M_2}{T}} C_A \end{cases} \tag{3.60}$$

and the parameters are given as in Example 2.1. Following a same fuzzy modelling process given as in Example 2.1 and considering the exogenous disturbance, then the CSTR system can be characterized by a T-S fuzzy affine model as,

Plant Rule $\mathscr{R}^l$: IF $x_1(t)$ is $\mathscr{F}_1^l$, THEN

$$\begin{cases} \dot{x}(t) = (A_l + \Delta A_l)x(t) + a_l + \Delta a_l + B(u(t) + f(t, x(t))) \\ y(t) = Cx(t),\ l \in \{1,2,3\}, \end{cases} \tag{3.61}$$

where the system matrices are given as in (2.61), and $y(t) = x_1(t)$ is the measurement output with $C = \begin{bmatrix} 1 & 0 \end{bmatrix}$, and the parameter uncertainties ΔA_l and Δa_l are given in the form of (1.3) with matrices H_{l1}, H_{l2}, and H_{l3} given as in (2.62).

Figure 3.1 gives the fuzzy membership functions with $d_1 = -50$, $d_2 = -25.6$, $d_3 = 20.65$ and $d_4 = 50$. By observing Figure 3.1, the regions can be described as in (2.63), where S_2 is a fuzzy region with $\mathscr{N}(2) = \{1,2,3\}$, and S_1 and S_3 are crisp regions with $\mathscr{N}(1) = \{1\}$ and $\mathscr{N}(3) = \{3\}$.

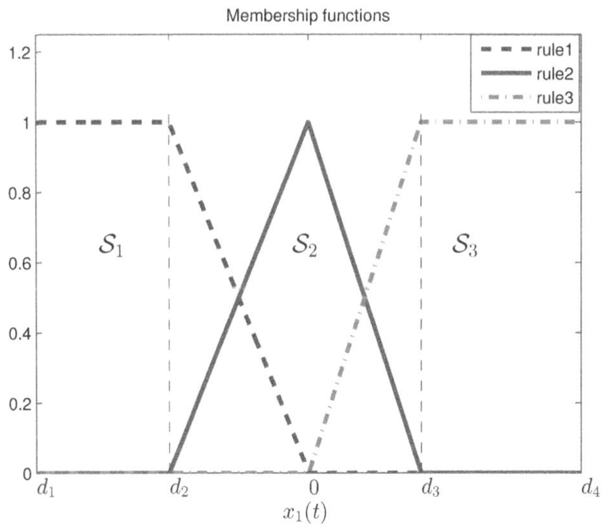

Figure 3.1 Membership functions of Example 3.1.

Based on the method given in this chapter, one can select the transformation matrix T as $T = \begin{bmatrix} 0 & 1 \\ 2.0921 & 0 \end{bmatrix}$, and for this case, $z_2(t) = 2.0921x_1(t)$. One can describe these three regions precisely by degenerate ellipsoids as in (2.16) with Q_i and q_i computed by (2.64).

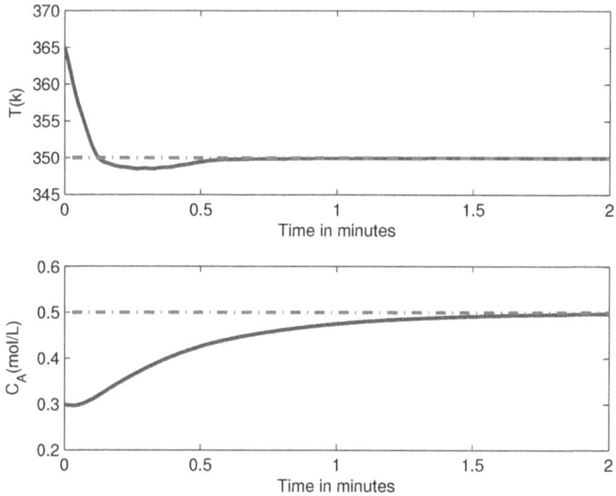

Figure 3.2 The closed-loop system states with sliding mode controller (3.47) in Example 3.1.

We attempt to synthesize a sliding surface (3.5) and a sliding mode controller (3.47) or (3.54) to drive the state trajectories of the nonlinear CSTR system in (3.60) to the sliding surface and guarantee the sliding motion to be asymptotically stable. *The full system states are assumed to be not available in this example.* It is also found that no feasible solution can be acquired by the CQLF-based approach shown in Theorem 3.1. However, via employing the PQLFs-based method given in Theorem 3.2 with $\alpha = 1$, $\tilde{J} = \begin{bmatrix} 60 & 0 \end{bmatrix}$, $\hat{J} = \begin{bmatrix} 1 & 60 & 0 \end{bmatrix}$, one indeed attains the feasible solutions, and the sliding surface gain is given as $K = -1.1076$ with $P_2 = \mathbf{I}$.

With initial condition $x_0 = \begin{bmatrix} 20 & -0.25 \end{bmatrix}^{\mathrm{T}}$ and nonlinear disturbance function $f(t,x(t)) = \cos(t)y(t) + \begin{bmatrix} 0.1 & 0.2 \end{bmatrix} x(t)$, by adopting the sliding mode controller (3.47) with $\gamma = 0.5$, $\eta = 1.2$, and $\sigma = 0.2$, Figure 3.2 and Figure 3.3, respectively, show the system states and control input. The sliding surface is demonstrated in Figure 3.4.

When the parameter σ is not accessible, the output feedback SMC approach given in [72] and the sliding mode controller (3.47) are not applicable. In addition, due to the existence of $a_l + \Delta a_l$, much difficulty is also introduced in the stability analysis of the sliding motion. For this case, the adaptive output feedback SMC in (3.54) can be

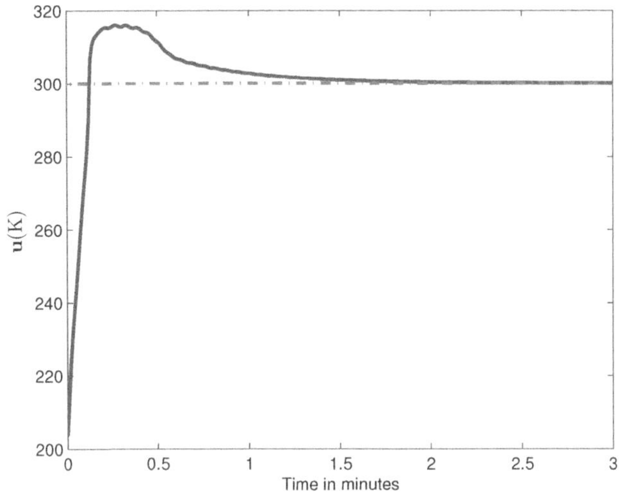

Figure 3.3 The control input with sliding mode controller (3.47) in Example 3.1.

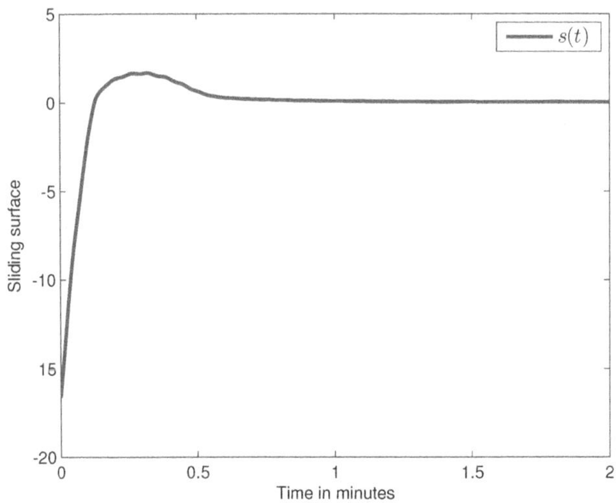

Figure 3.4 The sliding surface with sliding mode controller (3.47) in Example 3.1.

utilized instead. With the same nonlinear disturbance $f(t,x(t))$ and initial condition, exploiting the adaptive sliding mode controller (3.54) with $\gamma = 0.5$, $\eta = 1.2$, and

$\varpi = 0.0001$, Figure 3.5 and Figure 3.6 show the system states and the control input, respectively. Figure 3.7 presents the sliding surface.

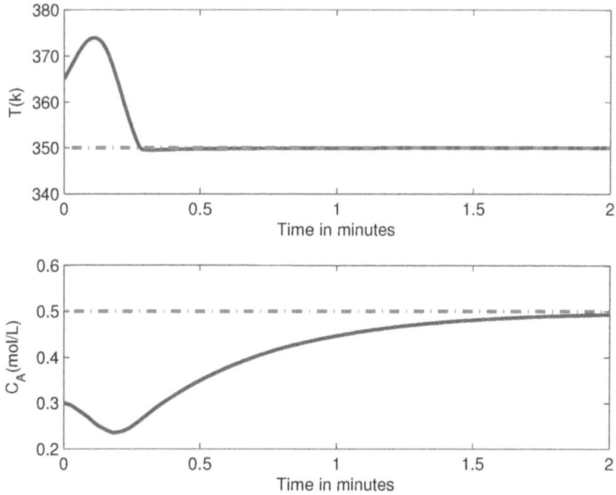

Figure 3.5 The closed-loop system states with adaptive sliding mode controller (3.54) in Example 3.1.

To further show the effectiveness of the proposed output feedback fuzzy SMC methods, in the following, consider an autonomous land vehicle system borrowed form [25], which is widely used as a servo system in many practical situations, for example, the automated guided vehicles (AGVs).

Example 3.2. Consider an autonomous land vehicle system shown in Figure 3.8, and the aim is to synthesize a controller to force a cart to follow the straight line $\mathbf{y} = 0$ with a constant velocity $v_0 = 1$ m/s. The torque u is used to control the path of the cart, which is characterized by the following motion equation,

$$\begin{bmatrix} \dot{\theta}(t) \\ \dot{\psi}(t) \\ \dot{\mathbf{y}}(t) \end{bmatrix} = \begin{bmatrix} 0 & 1 & 0 \\ 0 & -\frac{k}{I} & 0 \\ 0 & 0 & 0 \end{bmatrix} \begin{bmatrix} \theta(t) \\ \psi(t) \\ \mathbf{y}(t) \end{bmatrix} + \begin{bmatrix} 0 \\ 0 \\ v_0 \sin(\theta(t)) \end{bmatrix} + \begin{bmatrix} 0 \\ \frac{1}{I} \\ 0 \end{bmatrix} (u(t) + f(t,x(t))) \tag{3.62}$$

where $\theta(t)$ denotes the heading angle with time derivative $\psi(t)$; $I = 1$ kg $\cdot$ m^2 denotes the moment of inertia of the cart regarding to the center of mass; the damping coefficient is set as $k = 0.01$; $u(t)$ denotes the control torque; $f(t,x(t))$ represents the nonlinear disturbance of the cart.

Denote the state variables as $x_1(t) = \theta(t)$, $x_2(t) = \psi(t)$, and $x_3(t) = \mathbf{y}(t)$. Linearize the system (3.62) at three operating points $(0,0,0)^{\mathrm{T}}$, $\left(-\frac{2\pi}{5},0,0\right)^{\mathrm{T}}$, and $\left(\frac{2\pi}{5},0,0\right)^{\mathrm{T}}$, and then the system (3.62) can be characterized by a T-S fuzzy affine model as,

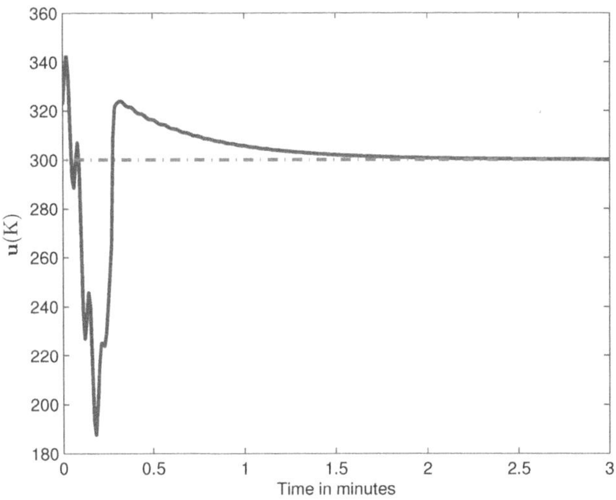

Figure 3.6 The control input with adaptive sliding mode controller (3.54) in Example 3.1.

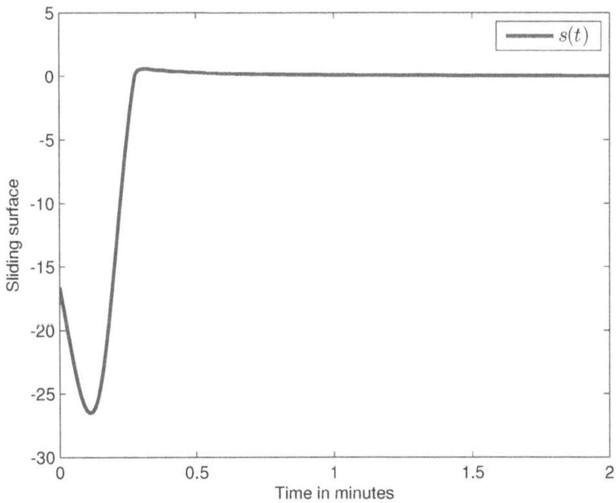

Figure 3.7 The sliding surface with adaptive sliding mode controller (3.54) in Example 3.1.

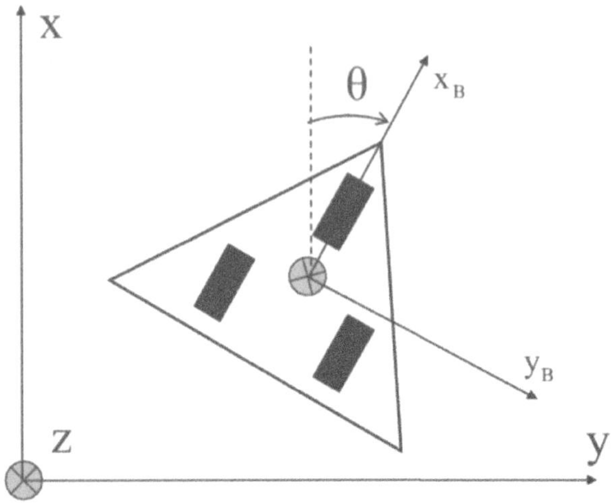

Figure 3.8 Autonomous land vehicle in Example 3.2.

Plant Rule $\mathscr{R}^l$: IF $x_1(t)$ is $\mathscr{F}_1^l$, THEN

$$\begin{cases} \dot{x}(t) = (A_l + \Delta A_l)x(t) + a_l + \Delta a_l + B(u(t) + f(t,x(t))), \\ y(t) = Cx(t),\ l \in \{1,2,3\}, \end{cases} \tag{3.63}$$

where the system matrices are

$$\begin{cases} A_1 = \begin{bmatrix} 0 & 1 & 0 \\ 0 & -0.01 & 0 \\ 1 & 0 & 0 \end{bmatrix}, A_2 = A_3 = \begin{bmatrix} 0 & 1 & 0 \\ 0 & -0.01 & 0 \\ 0.3090 & 0 & 0 \end{bmatrix}, \\ a_1 = \begin{bmatrix} 0 \\ 0 \\ 0 \end{bmatrix}, a_2 = \begin{bmatrix} 0 \\ 0 \\ -0.5627 \end{bmatrix}, a_3 = \begin{bmatrix} 0 \\ 0 \\ 0.5627 \end{bmatrix}, B = \begin{bmatrix} 0 \\ 1 \\ 0 \end{bmatrix}, \\ C = \begin{bmatrix} 1.5 & 1 & 0.5 \end{bmatrix} \end{cases} \tag{3.64}$$

and the parameter uncertainties ΔA_l and Δa_l are given in the form of (1.3) with

$$\begin{cases} H_{l1} = \begin{bmatrix} 0 & 0 & 0.01 \end{bmatrix}^{\mathrm{T}}, H_{l2} = \begin{bmatrix} 0.02 & 0 & 0 \end{bmatrix}, \\ H_{l3} = 0.01, l \in \{1,2,3\}. \end{cases} \tag{3.65}$$

Figure 3.9 gives the MFs in this example with $d_1 = \frac{\pi}{15}$ and $d_2 = \frac{3\pi}{5}$. It can be found from Figure 3.9 that

$$\begin{aligned} S_1 &= \{x \in \Re^{n_x} \mid -d_2 \le x_1 \le -d_1\}, \\ S_2 &= \{x \in \Re^{n_x} \mid -d_1 \le x_1 \le d_1\}, \\ S_3 &= \{x \in \Re^{n_x} \mid d_1 \le x_1 \le d_2\}. \end{aligned} \tag{3.66}$$

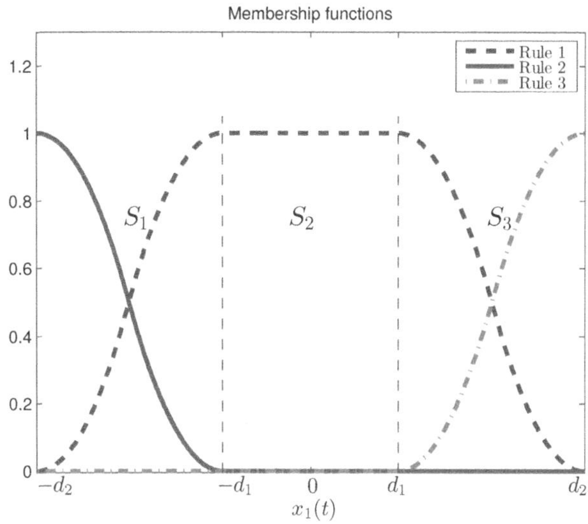

Figure 3.9 Membership functions of Example 3.2–Example 3.5.

Based on the method given in this chapter, one can select the transformation matrix T as $T = \begin{bmatrix} 1 & 0 & 0 \\ 0 & 0 & 1 \\ 0 & 1 & 0 \end{bmatrix}$. One can describe these three regions precisely by degenerate ellipsoids as in (2.16) with

$$\begin{cases} Q_1 = Q_3 = \begin{bmatrix} \dfrac{2}{d_2 - d_1} & 0 & 0 \end{bmatrix}, Q_2 = \begin{bmatrix} \dfrac{1}{d_1} & 0 & 0 \end{bmatrix}, \\ q_1 = \dfrac{d_1 + d_2}{d_2 - d_1}, q_2 = 0, q_3 = \dfrac{d_1 + d_2}{d_1 - d_2}. \end{cases} \tag{3.67}$$

The purpose is to synthesize a sliding surface (3.5) and a sliding mode controller (3.47) or an adaptive sliding mode controller (3.54) to drive the state trajectories of the autonomous land vehicle in (3.62) to the sliding surface in finite time and guarantee the sliding motion to be asymptotically stable. It is also found that no feasible solution can be acquired by the CQLF-based approach shown in Theorem 3.1. However, via employing the PQLFs-based method given in Theorem 3.2 with $\alpha = 1$, $\tilde{J} = \begin{bmatrix} -1 & -1 \end{bmatrix}$, $\hat{J} = \begin{bmatrix} 0 & -1 & -1 \end{bmatrix}$, one indeed attains the feasible solutions, and the sliding surface gain is obtained as $K = -1.5948$ with $P_2 = \mathbf{I}$. Note that, when utilizing Theorem 3.2 with PQLFs in (3.38), the characterizing matrices $\hat{G}_i$ and $\hat{F}_i$

are given as,

$$[F_1 \mid F_2 \mid F_3 \mid f_1 \mid f_2 \mid f_3] = \left[\begin{array}{c|c|c|c|c|c} -\rho & \mathbf{0}_{1\times n_x} & \mathbf{0}_{1\times n_x} & -d_1 & 0 & 0 \\ \mathbf{0}_{1\times n_x} & \mathbf{0}_{1\times n_x} & \rho & 0 & 0 & -d_1 \\ \mathbf{I}_{n_x} & \mathbf{I}_{n_x} & \mathbf{I}_{n_x} & \mathbf{0}_{n_x\times 1} & \mathbf{0}_{n_x\times 1} & \mathbf{0}_{n_x\times 1} \end{array}\right], \quad (3.68)$$
$$[G_1 \mid G_2 \mid G_3 \mid g_1 \mid g_2 \mid g_3] = \left[\begin{array}{c|c|c|c|c|c} \rho & \mathbf{0}_{1\times n_x} & \rho & d_2 & 0 & -d_1 \\ -\rho & \mathbf{0}_{1\times n_x} & -\rho & -d_1 & 0 & d_2 \end{array}\right]$$

with $\rho = \begin{bmatrix} 1 & 0 \end{bmatrix}$ and $n_x = 2$.

With initial condition $x_0 = \begin{bmatrix} 1.2 & 0.5 & 0.3 \end{bmatrix}^{\mathrm{T}}$ and nonlinear disturbance function $f(t,x(t)) = \cos(t)y(t) + \begin{bmatrix} 0.1 & 0.2 & 0.3 \end{bmatrix} x(t)$, by adopting the sliding mode controller in (3.47) with $\gamma = 0.5$, $\eta = 1.2$, and $\sigma = 1.3$, Figure 3.10 and Figure 3.11, respectively, show the system states and control input. The sliding surface is demonstrated in Figure 3.12.

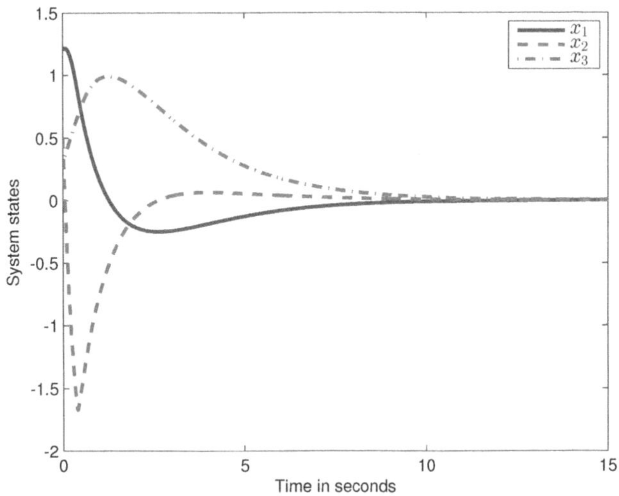

Figure 3.10 The closed-loop system states with sliding mode controller (3.47) in Example 3.2.

When the parameter σ is not accessible, the adaptive output feedback SMC in (3.54) can be utilized instead. With the same disturbance $f(t,x(t))$ and initial condition, utilizing the adaptive sliding mode controller in (3.54) with $\gamma = 0.5$, $\eta = 1.2$, and $\varpi = 0.01$, Figure 3.13 and Figure 3.14 show the system states and the control input, respectively. Figure 3.15 gives the sliding surface.

It is worth mentioning that the proposed output feedback fuzzy SMC methods in this chapter can be used not only in some servo systems or industrial processes but also in some society phenomenna or finance systems analysis and control. In the following, another classical third-order nonlinear finance system will be considered,

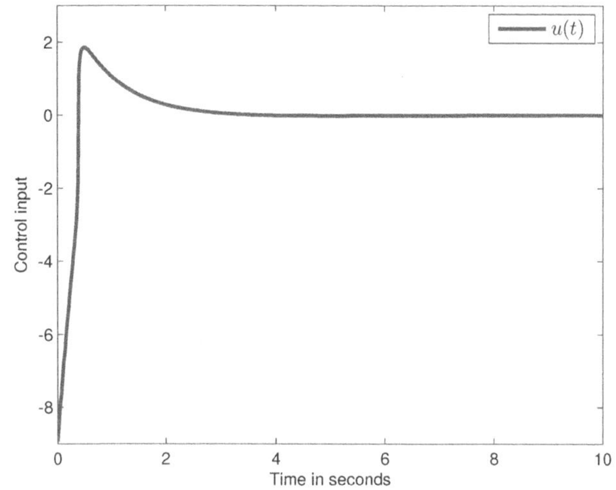

Figure 3.11 The control input with sliding mode controller (3.47) in Example 3.2.

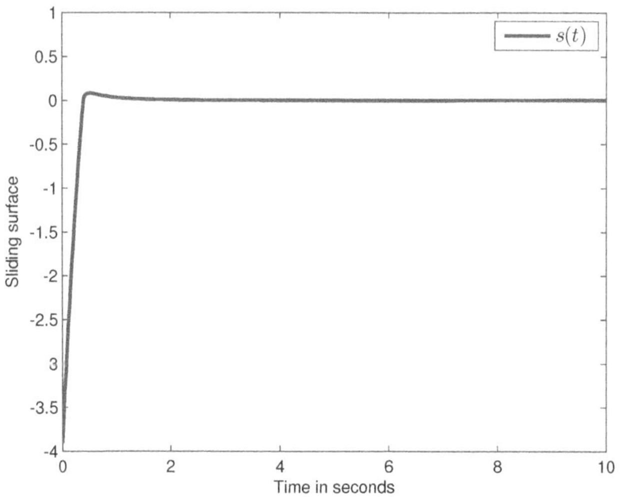

Figure 3.12 The sliding surface with sliding mode controller (3.47) in Example 3.2.

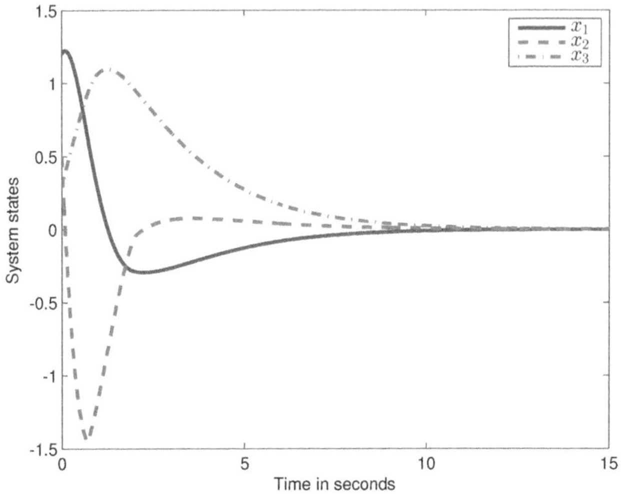

Figure 3.13 The closed-loop system states with adaptive sliding mode controller (3.54) in Example 3.2.

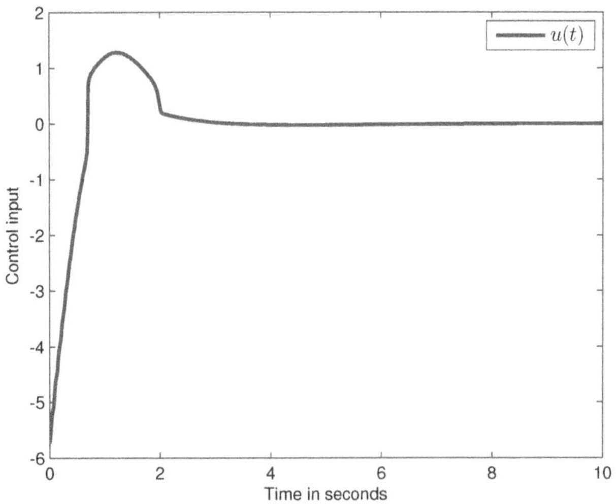

Figure 3.14 The control input with adaptive sliding mode controller (3.54) in Example 3.2.

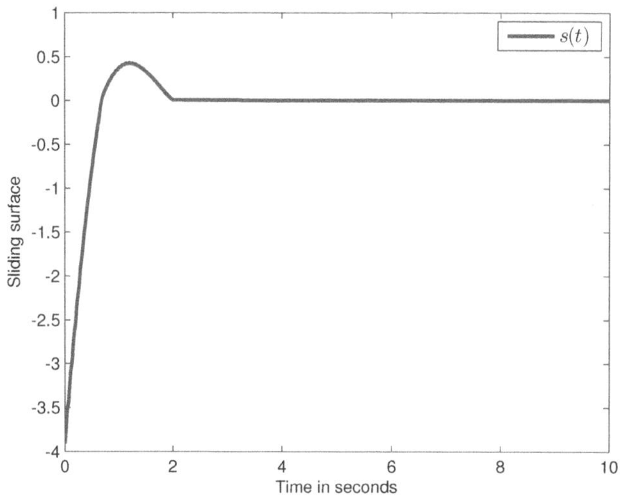

Figure 3.15 The sliding surface with adaptive sliding mode controller (3.54) in Example 3.2.

which is widely studied and has positive enlightening significance for research on some other complex economic and financial systems.

Example 3.3. Consider a nonlinear finance system, whose dynamics are described by the following equation,

$$\begin{cases} \dot{R}(t) = D(t) + (C(t) - a)R(t) \\ \dot{C}(t) = 1 - bC(t) - R^2(t) \\ \dot{D}(t) = -R(t) - cD(t) \end{cases} \tag{3.69}$$

where the parameters are the same to the ones given as in Example 2.2. Following a same coordinate change and modelling method as in Example 2.2, the system (3.69) can be formulated into the following form

$$\begin{cases} \dot{x}_1(t) = x_3(t) + (x_2(t) + \frac{1}{b} - a)x_1(t) \\ \dot{x}_2(t) = -bx_2(t) - x_1^2(t) + u(t) \\ \dot{x}_3(t) = -x_1(t) - cx_3(t) + u(t) \end{cases} \tag{3.70}$$

where $u(t)$ is the control input.

Linearize the nonlinear finance system (3.70) at three operating points $(0,0,0)$ and $(\pm 0.3,0,0)$, then the system (3.70) can be described by the following T-S fuzzy affine model with three fuzzy rules as,

Plant Rule $\mathscr{R}^l$: IF $x_1(t)$ is $\mathscr{F}_1^l$, THEN

$$\begin{cases} \dot{x}(t) = (A_l + \Delta A_l)x(t) + a_l + \Delta a_l + B(u(t) + f(t,x(t))) \\ y(t) = Cx(t),\ l \in \{1,2,3\}, \end{cases} \tag{3.71}$$

where the system matrices are given as in (2.127), and $y(t)$ is the measurement output with $C = \begin{bmatrix} 1 & 0 & 1 \end{bmatrix}$, and the parameter uncertainties ΔA_l and Δa_l are given in the form of (1.3) with matrices H_{l1}, H_{l2}, and H_{l3} given as in (2.128). The fuzzy membership functions are also given as in Figure 3.9 with $d_1 = 0.3$ and $d_2 = 1$.

Based on the method given in this chapter, one can select the transformation matrix T as $T = \begin{bmatrix} 1 & 0 & 0 \\ 0 & 1 & -1 \\ 0 & 1 & 1 \end{bmatrix}$. These three regions can be precisely described by the degenerate ellipsoids as in (2.16) with Q_i and q_i computed by (3.67).

Utilizing Theorem 3.2 with $\alpha = 0.5$, $\tilde{J} = \begin{bmatrix} -1 & 0 \end{bmatrix}$, $\hat{J} = \begin{bmatrix} 0 & -1 & 0 \end{bmatrix}$, feasible solutions can be obtained and the sliding surface gain is $K = -20.5538$ with $P_2 = \mathbf{I}$. Note that when utilizing Theorem 3.2 with PQLFs in (3.38), the characterizing matrices $\hat{G}_i$ and $\hat{F}_i$ can be obtained via (3.68).

With initial condition $x_0 = \begin{bmatrix} 0.5 & 0 & 0.1 \end{bmatrix}^{\mathrm{T}}$ and nonlinear disturbance function $f(t,x(t)) = \sin(4t)y(t) + \begin{bmatrix} 0.1 & 0.3 & 0.2 \end{bmatrix} x(t)$, by adopting the sliding mode controller (3.47) with $\gamma = 0.4$, $\eta = 0.2$, and $\sigma = 0.6$, Figure 3.16 and Figure 3.17, respectively, show the system states and control input. The sliding surface is demonstrated in Figure 3.18.

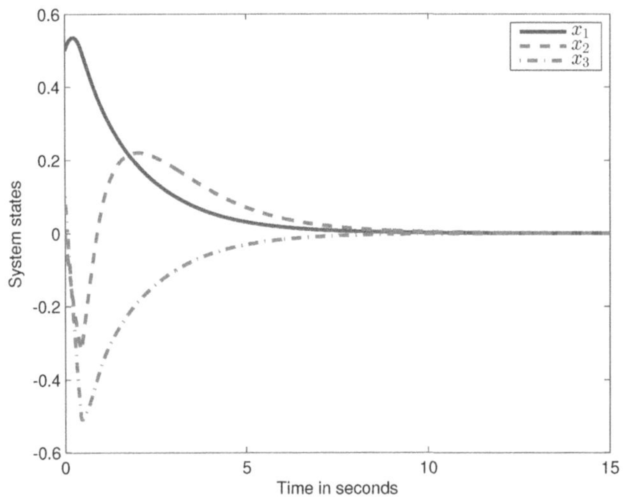

Figure 3.16 The closed-loop system states with sliding mode controller (3.47) in Example 3.3.

When the parameter σ is not accessible, the fuzzy sliding mode controller (3.47) is not applicable. With the same disturbance $f(t,x(t))$ and initial condition, exploiting the adaptive sliding mode controller (3.54) with $\gamma = 0.4$, $\eta = 0.2$, and $\varpi = 0.01$, Figure 3.19 and Figure 3.20 show the system states and the control input, respectively. Figure 3.21 presents the sliding surface.

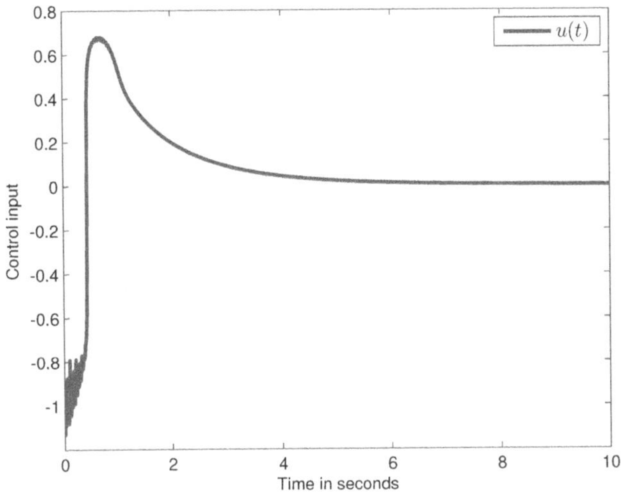

Figure 3.17 The control input with sliding mode controller (3.47) in Example 3.3.

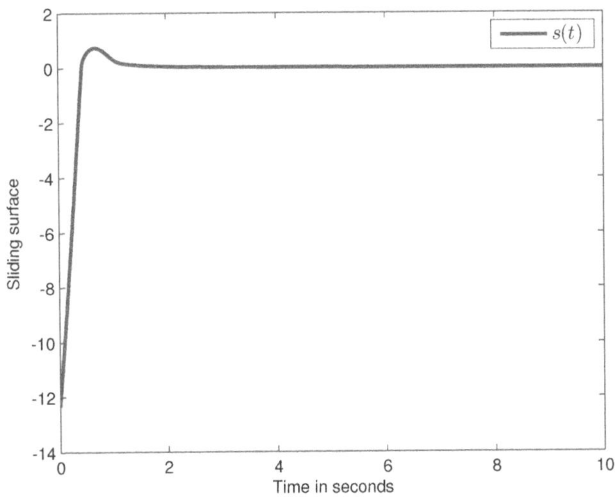

Figure 3.18 The sliding surface with sliding mode controller (3.47) in Example 3.3.

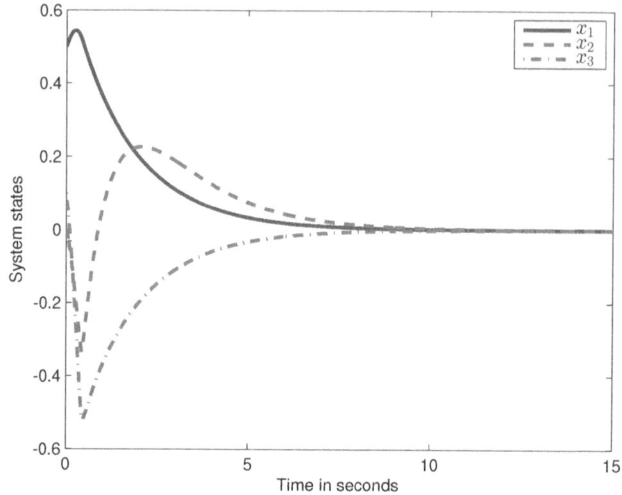

Figure 3.19 The closed-loop system states with adaptive sliding mode controller (3.54) in Example 3.3.

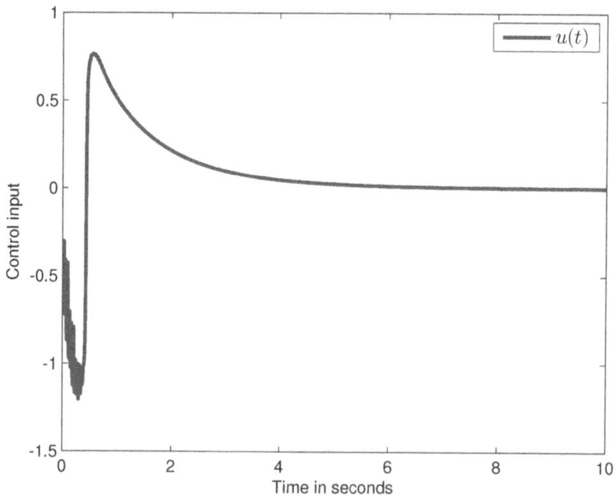

Figure 3.20 The control input with adaptive sliding mode controller (3.54) in Example 3.3.

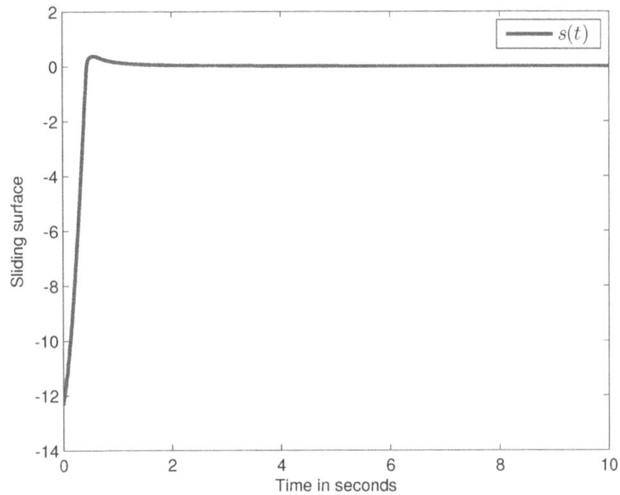

Figure 3.21 The sliding surface with adaptive sliding mode controller (3.54) in Example 3.3.

These simulation results illustrate that the presented sliding mode controller design methods are effective clearly.

3.4 OUTPUT FEEDBACK DYNAMIC SMC DESIGN VIA T-S FUZZY AFFINE MODELS

Note that the previous output feedback sliding mode controller synthesis results are also obtained based on an assumption that each local model of T-S fuzzy systems shares one common control input matrix. To relax this assumption, an output feedback dynamic sliding mode controller design method will be proposed in this section.

3.4.1 MODEL DESCRIPTION: CASE II

Consider a T-S fuzzy affine system characterized by r fuzzy **IF-THEN** rules as follows.

Plant Rule $\mathscr{R}^l$: IF $\theta_1(t)$ is $\mathscr{F}_1^l$ and $\cdots$ and $\theta_\varphi(t)$ is $\mathscr{F}_\varphi^l$, THEN

$$\begin{cases} \dot{x}(t) = (A_l + \Delta A_l)x(t) + a_l + \Delta a_l + (B_l + \Delta B_l)u(t) + D_l w(t) \\ y(t) = Cx(t) \\ z(t) = L_l x(t) + N_l u(t),\ l \in \mathscr{L} := \{1, \ldots, r\}, \end{cases} \tag{3.72}$$

where $\mathscr{R}^l$ represents the l-th fuzzy inference rule; $\mathscr{F}_\phi^l\,(\phi = 1, \ldots, \varphi)$ represents fuzzy sets; r represents the number of inference rules; $\theta(t) := \left[\theta_1(t), \ldots, \theta_\varphi(t)\right]$ represent the measurable premise variables; $x(t) \in \Re^{n_x}$ is the system state; $u(t) \in \Re^{n_u}$ is the control input; $y(t) \in \Re^{n_y}$ is the system measurement output; $w(t) \in \Re^{n_w}$ is the external disturbance belonging to $L_2[0,\infty)$; $z(t) \in \Re^{n_z}$ is the regulated output; $C = C_l$, $l \in \mathscr{L}$ is the common measurement output matrix; ΔA_l, Δa_l and ΔB_l are the parameter uncertainties in the l-th local model satisfying conditions (1.3)–(1.4).

Similar to Chapter 2, the system (3.72) can be reformulated as

$$\begin{cases} \dot{x}(t) = (A(\mu) + \Delta A(\mu))x(t) + a(\mu) + \Delta a(\mu) \\ \qquad\qquad + (B(\mu) + \Delta B(\mu))u(t) + D(\mu)w(t) \\ y(t) = Cx(t) \\ z(t) = L(\mu)x(t) + N(\mu)u(t) \end{cases} \tag{3.73}$$

where

$$\begin{cases} A(\mu) = \sum_{l=1}^{r} \mu_l A_l, \Delta A(\mu) = \sum_{l=1}^{r} \mu_l \Delta A_l, a(\mu) = \sum_{l=1}^{r} \mu_l a_l, \Delta a(\mu) = \sum_{l=1}^{r} \mu_l \Delta a_l, \\ B(\mu) = \sum_{l=1}^{r} \mu_l B_l, \Delta B(\mu) = \sum_{l=1}^{r} \mu_l \Delta B_l, D(\mu) = \sum_{l=1}^{r} \mu_l D_l, L(\mu) = \sum_{l=1}^{r} \mu_l L_l, \\ N(\mu) = \sum_{l=1}^{r} \mu_l N_l. \end{cases} \tag{3.74}$$

The premise variable space can be decomposed by a similar line in Chapter 2, and to characterize the indices adopted for each region S_i, the following set is introduced $\mathscr{N}(i) := \{m | \mu_m[\theta(t)] > 0, \ m \in \mathscr{L}, \ \theta(t) \in S_i, \ i \in \mathscr{I}\}$. Based on the set $\mathscr{N}(i)$, the system (3.73) can be formulated into the piecewise fuzzy affine form as,

$$\begin{cases} \dot{x}(t) = (\mathscr{A}_i + \Delta\mathscr{A}_i)x(t) + \mathrm{a}_i + \Delta\mathrm{a}_i + (\mathscr{B}_i + \Delta\mathscr{B}_i)u(t) + \mathscr{D}_i w(t) \\ y(t) = Cx(t) \\ z(t) = \mathscr{L}_i x(t) + \mathscr{N}_i u(t), \theta(t) \in S_i, i \in \mathscr{I}, \end{cases} \tag{3.75}$$

where

$$\begin{cases} \mathscr{A}_i := \sum_{m \in \mathscr{N}(i)} \mu_m A_m, \Delta\mathscr{A}_i := \sum_{m \in \mathscr{N}(i)} \mu_m \Delta A_m, \\ \mathrm{a}_i := \sum_{m \in \mathscr{N}(i)} \mu_m a_m, \Delta\mathrm{a}_i := \sum_{m \in \mathscr{N}(i)} \mu_m \Delta a_m, \\ \mathscr{B}_i := \sum_{m \in \mathscr{N}(i)} \mu_m B_m, \Delta\mathscr{B}_i := \sum_{m \in \mathscr{N}(i)} \mu_m \Delta B_m, \\ \mathscr{D}_i := \sum_{m \in \mathscr{N}(i)} \mu_m D_m, \mathscr{L}_i := \sum_{m \in \mathscr{N}(i)} \mu_m L_m, \mathscr{N}_i := \sum_{m \in \mathscr{N}(i)} \mu_m N_m \end{cases} \tag{3.76}$$

with $0 < \mu_m[\theta(t)] \leq 1$, $\sum_{m \in \mathscr{N}(i)} \mu_m[\theta(t)] = 1$.

This section will propose a new output feedback dynamic SMC design approach for T-S fuzzy affine system (3.72). It will be shown that the local finite-time reachability of the sliding surface can be ensured. Some new asymptotic stability analysis criteria for the sliding motion will be formulated through CQLF and PQLFs, respectively. For a prescribed level of disturbance attenuation $\gamma > 0$, the sliding motion is ensured to be asymptotically stable with $w(t) \equiv 0$, and for nonzero $w(t) \in L_2[0,\infty)$, it is satisfied that $\int_{t_r}^{\infty} \|z(t)\|^2 dt < \gamma^2 \int_{t_r}^{\infty} \|w(t)\|^2 dt$ under zero initial condition, where t_r represents the time when the system states reach the sliding surface.

3.4.2 SLIDING SURFACE DESIGN

For the fuzzy affine system (3.72) or equivalent (3.75), design the sliding surface based on the measurement output and control input as

$$s(t) = Ky(t) + u(t) = 0, \tag{3.77}$$

where $K \in \Re^{n_u \times n_y}$ is the sliding surface gain to be determined.

With (3.77), one has

$$u(t) = -KCx(t). \tag{3.78}$$

Based on (3.78) and system (3.75), the holonomic sliding mode dynamic system is obtained as

$$\begin{cases} \dot{x}(t) = (\mathscr{A}_i + \Delta\mathscr{A}_i)x(t) + \mathrm{a}_i + \Delta\mathrm{a}_i + (\mathscr{B}_i + \Delta\mathscr{B}_i)u(t) + \mathscr{D}_i w(t) \\ 0 \times \dot{u}(t) = -KCx(t) - u(t) \\ z(t) = \mathscr{L}_i x(t) + \mathscr{N}_i u(t), i \in \mathscr{I}. \end{cases} \tag{3.79}$$

Reformulate system (3.79) as

$$\begin{cases} E\dot{\bar{x}}(t) = (\bar{\mathscr{A}}_i + \Delta\bar{\mathscr{A}}_i)\bar{x}(t) + \bar{\mathrm{a}}_i + \Delta\bar{\mathrm{a}}_i + \bar{\mathscr{D}}_i w(t) \\ z(t) = \bar{\mathscr{L}}_i \bar{x}(t), i \in \mathscr{I}, \end{cases} \tag{3.80}$$

where

$$\begin{cases} E = \begin{bmatrix} \mathbf{I}_{n_x} & \mathbf{0} \\ \mathbf{0} & \mathbf{0} \end{bmatrix}, \bar{x}(t) = \begin{bmatrix} x(t) \\ u(t) \end{bmatrix}, \bar{\mathscr{A}}_i = \begin{bmatrix} \mathscr{A}_i & \mathscr{B}_i \\ -KC & -\mathbf{I} \end{bmatrix}, \Delta\bar{\mathscr{A}}_i = \begin{bmatrix} \Delta\mathscr{A}_i & \Delta\mathscr{B}_i \\ \mathbf{0} & \mathbf{0} \end{bmatrix}, \\ \bar{\mathrm{a}}_i = \begin{bmatrix} \mathrm{a}_i \\ \mathbf{0} \end{bmatrix}, \Delta\bar{\mathrm{a}}_i = \begin{bmatrix} \Delta\mathrm{a}_i \\ \mathbf{0} \end{bmatrix}, \bar{\mathscr{D}}_i = \begin{bmatrix} \mathscr{D}_i \\ \mathbf{0} \end{bmatrix}, \bar{\mathscr{L}}_i = \begin{bmatrix} \mathscr{L}_i & \mathscr{N}_i \end{bmatrix}. \end{cases} \tag{3.81}$$

Via combining the first equation of system (3.75) with the sliding surface (3.77), the sliding mode dynamics can be characterized by the descriptor system (3.80). Therefore, the admissibility of the system (3.80) ensures the existence of the designed sliding surface (3.77). In the following, we will first formulate sufficient criteria for the existence of the designed sliding surface (3.77), and then an output feedback dynamic SMC design approach will be proposed.

3.4.3 STABILITY AND ROBUST PERFORMANCE ANALYSIS OF SLIDING MOTION

The results on the asymptotic stability and robust $\mathscr{H}_\infty$ performance analysis of the sliding motion will be presented through CQLF and PQLFs, respectively. Then a novel fuzzy output feedback dynamic SMC method will be proposed to guarantee the local finite-time convergence of the sliding surface.

3.4.3.1 Stability and Robust Performance Analysis of Sliding Motion through CQLF

The asymptotic stability analysis of sliding motion will be conducted via convex optimization technique. Similar to Chapter 2, it is assumed that one can employ an ellipsoid $\mathscr{E}_i$ to outer approximate each polyhedral region S_i in (3.75), and the relationship given as in (2.83) holds for each region.

Theorem 3.5. The sliding mode dynamic system (3.80) is asymptotically stable with robust $\mathscr{H}_\infty$ performance γ, if there exist matrices $0 < P_1 = P_1^{\mathrm{T}} \in \Re^{n_x \times n_x}$, $P_2 \in \Re^{n_u \times n_u}$, $\bar{K} \in \Re^{n_u \times n_y}$, and scalars $\{\varepsilon_i > 0, i \in \mathscr{I}\}$, $\{\kappa_i < 0, i \in \mathscr{I}_1\}$, guaranteeing the LMIs in (3.82)–(3.83) hold,

$$\begin{bmatrix} \Gamma_1 & \star & \star & \star \\ \Gamma_2 & -P_2 - P_2^{\mathrm{T}} + N_m^{\mathrm{T}} N_m + \varepsilon_i H_{m3}^{\mathrm{T}} H_{m3} & \star & \star \\ D_m^{\mathrm{T}} P_1 & \mathbf{0} & -\gamma^2 \mathbf{I} & \star \\ H_{m1}^{\mathrm{T}} P_1 & \mathbf{0} & \mathbf{0} & -\varepsilon_i \mathbf{I} \end{bmatrix} < 0, \quad i \in \mathscr{I}_0, m \in \mathscr{N}(i), \tag{3.82}$$

$$\begin{bmatrix} \kappa_i(q_i^{\mathrm{T}} q_i - 1) + \varepsilon_i H_{m4}^{\mathrm{T}} H_{m4} & \star & \star & \star & \star \\ \kappa_i Q_i^{\mathrm{T}} q_i + P_1 a_m + \varepsilon_i H_{m2}^{\mathrm{T}} H_{m4} & \Gamma_1 + \kappa_i Q_i^{\mathrm{T}} Q_i & \star & \star & \star \\ \varepsilon_i H_{m3}^{\mathrm{T}} H_{m4} & \Gamma_2 & \Gamma_3 & \star & \star \\ \mathbf{0} & D_m^{\mathrm{T}} P_1 & \mathbf{0} & -\gamma^2 \mathbf{I} & \star \\ \mathbf{0} & H_{m1}^{\mathrm{T}} P_1 & \mathbf{0} & \mathbf{0} & -\varepsilon_i \mathbf{I} \end{bmatrix} < 0, \quad i \in \mathscr{I}_1, m \in \mathscr{N}(i), \tag{3.83}$$

where

$$\begin{cases} \Gamma_1 = \mathtt{Sym}\{P_1 A_m + \alpha J^{\mathrm{T}} \bar{K} C\} + L_m^{\mathrm{T}} L_m + \varepsilon_i H_{m2}^{\mathrm{T}} H_{m2}, \\ \Gamma_2 = -\bar{K} C + B_m^{\mathrm{T}} P_1 + \alpha P_2 J + N_m^{\mathrm{T}} L_m + \varepsilon_i H_{m3}^{\mathrm{T}} H_{m2}, \\ \Gamma_3 = -P_2 - P_2^{\mathrm{T}} + N_m^{\mathrm{T}} N_m + \varepsilon_i H_{m3}^{\mathrm{T}} H_{m3}. \end{cases} \tag{3.84}$$

and $J \in \Re^{n_u \times n_x}$ is an arbitrary matrix, and α is a constant scalar.

Moreover, the gain of the sliding surface (3.77) can be obtained via

$$K = P_2^{-\mathrm{T}} \bar{K}. \tag{3.85}$$

Proof. The more complicated case for $i \in \mathscr{I}_1$ is merely proved without loss of generality. Consider a CQLF as

$$V(t) = \bar{x}^{\mathrm{T}}(t) E P \bar{x}(t), \tag{3.86}$$

where $P = \begin{bmatrix} P_1 & \mathbf{0} \\ P_3 & P_2 \end{bmatrix}$, $0 < P_1 = P_1^{\mathrm{T}} \in \Re^{n_x \times n_x}$, $P_3 \in \Re^{n_u \times n_x}$, and $P_2 \in \Re^{n_u \times n_u}$ is an arbitrary nonsingular matrix.

With the particular structure of E, one has $EP = P^{\mathrm{T}} E$. The sliding motion is asymptotically stable with robust $\mathscr{H}_\infty$ performance γ if

$$\dot{V}(t) + z^{\mathrm{T}}(t) z(t) - \gamma^2 w^{\mathrm{T}}(t) w(t) < 0. \tag{3.87}$$

Define $\upsilon(t) = \begin{bmatrix} 1 & \bar{x}^{\mathrm{T}}(t) & w^{\mathrm{T}}(t) \end{bmatrix}^{\mathrm{T}}$. Along the system (3.80), one can reformulate inequality (3.87) as

$$\upsilon^{\mathrm{T}}(t) \begin{bmatrix} \mathbf{0} & \star & \star \\ P^{\mathrm{T}}(\bar{\mathrm{a}}_i + \Delta \bar{\mathrm{a}}_i) & \mathtt{Sym}\{(\bar{\mathscr{A}}_i + \Delta \bar{\mathscr{A}}_i)^{\mathrm{T}} P\} + \bar{\mathscr{L}}_i^{\mathrm{T}} \bar{\mathscr{L}}_i & \star \\ \mathbf{0} & \bar{\mathscr{D}}_i^{\mathrm{T}} P & -\gamma^2 \mathbf{I} \end{bmatrix} \upsilon(t) < 0. \tag{3.88}$$

With the space partition in (2.83) and utilizing the S-procedure, for $\kappa_i < 0, i \in \mathscr{I}_1$, the following inequality implies (3.88),

$$\begin{bmatrix} \kappa_i(q_i^{\mathrm{T}} q_i - 1) & \star & \star \\ P^{\mathrm{T}}(\bar{\mathrm{a}}_i + \Delta\bar{\mathrm{a}}_i) + \kappa_i \bar{Q}_i^{\mathrm{T}} q_i & \Xi_1 & \star \\ \mathbf{0} & \bar{\mathscr{D}}_i^{\mathrm{T}} P & -\gamma^2 \mathbf{I} \end{bmatrix} < 0, \tag{3.89}$$

where

$$\begin{cases} \Xi_1 = \mathtt{Sym}\{(\bar{\mathscr{A}}_i + \Delta\bar{\mathscr{A}}_i)^{\mathrm{T}} P\} + \bar{\mathscr{L}}_i^{\mathrm{T}} \bar{\mathscr{L}}_i + \kappa_i \bar{Q}_i^{\mathrm{T}} \bar{Q}_i, \\ \bar{Q}_i = \begin{bmatrix} Q_i & \mathbf{0} \end{bmatrix}. \end{cases} \tag{3.90}$$

Define $\bar{K} = P_2^{\mathrm{T}} K$. With the structure of P and Q_i, the following inequality implies (3.89),

$$\begin{bmatrix} \kappa_i(q_i^{\mathrm{T}} q_i - 1) & \star & \star & \star \\ \kappa_i Q_i^{\mathrm{T}} q_i + P_1(\mathrm{a}_i + \Delta\mathrm{a}_i) & \Pi_1 & \star & \star \\ \mathbf{0} & \Pi_2 & \Pi_3 & \star \\ \mathbf{0} & \mathscr{D}_i^{\mathrm{T}} P_1 & \mathbf{0} & -\gamma^2 \mathbf{I} \end{bmatrix} < 0, \tag{3.91}$$

where

$$\begin{cases} \Pi_1 = \mathtt{Sym}\{P_1(\mathscr{A}_i + \Delta\mathscr{A}_i) - P_3^{\mathrm{T}} K C\} + \mathscr{L}_i^{\mathrm{T}} \mathscr{L}_i + \kappa_i Q_i^{\mathrm{T}} Q_i, \\ \Pi_2 = -\bar{K} C + (\mathscr{B}_i + \Delta\mathscr{B}_i)^{\mathrm{T}} P_1 - P_3 + \mathscr{N}_i^{\mathrm{T}} \mathscr{L}_i, \\ \Pi_3 = -P_2 - P_2^{\mathrm{T}} + \mathscr{N}_i^{\mathrm{T}} \mathscr{N}_i. \end{cases} \tag{3.92}$$

Extracting the MFs and Lemma A1 to handle the parameter uncertainties given in (1.3), the following inequality implies (3.91) for $\varepsilon_i > 0, i \in \mathscr{I}$,

$$\begin{bmatrix} \kappa_i(q_i^{\mathrm{T}} q_i - 1) + \varepsilon_i H_{m4}^{\mathrm{T}} H_{m4} & \star & \star & \star & \star \\ \kappa_i Q_i^{\mathrm{T}} q_i + P_1 \mathrm{a}_m + \varepsilon_i H_{m2}^{\mathrm{T}} H_{m4} & \Phi_1 & \star & \star & \star \\ \varepsilon_i H_{m3}^{\mathrm{T}} H_{m4} & \Phi_2 & \Phi_3 & \star & \star \\ \mathbf{0} & D_m^{\mathrm{T}} P_1 & \mathbf{0} & -\gamma^2 \mathbf{I} & \star \\ \mathbf{0} & H_{m1}^{\mathrm{T}} P_1 & \mathbf{0} & \mathbf{0} & -\varepsilon_i \mathbf{I} \end{bmatrix} < 0, \quad i \in \mathscr{I}_1, m \in \mathscr{N}(i), \tag{3.93}$$

where

$$\begin{cases} \Phi_1 = \mathtt{Sym}\{P_1 A_m - P_3^{\mathrm{T}} K C\} + L_m^{\mathrm{T}} L_m + \kappa_i Q_i^{\mathrm{T}} Q_i + \varepsilon_i H_{m2}^{\mathrm{T}} H_{m2}, \\ \Phi_2 = -\bar{K} C + B_m^{\mathrm{T}} P_1 - P_3 + N_m^{\mathrm{T}} L_m + \varepsilon_i H_{m3}^{\mathrm{T}} H_{m2}, \\ \Phi_3 = -P_2 - P_2^{\mathrm{T}} + N_m^{\mathrm{T}} N_m + \varepsilon_i H_{m3}^{\mathrm{T}} H_{m3}. \end{cases} \tag{3.94}$$

For numerical tractability of the stability analysis conditions, P_3 can be specified as $P_3 = -\alpha P_2 J$, where $J \in \Re^{n_u \times n_x}$ is an arbitrary matrix and α is a scalar.

Thus, it can be seen that inequality (3.83) implies (3.93). The proof is then accomplished. ■

To reduce the conservatism, in the following, a new stability and robust performance analysis approach will be given for the sliding motion based on PQLFs.

3.4.3.2 Stability and Robust Performance Analysis of Sliding Motion through PQLFs

A new sufficient condition for the asymptotic stability and robust $\mathscr{H}_\infty$ performance analysis of the sliding motion will be given based on PQLFs. To ensure the PQLFs continuous across boundaries of the region, as in subchapter 2.4, matrices $\tilde{F}_i = \begin{bmatrix} f_i & F_i \end{bmatrix}, i \in \mathscr{I}$, with $f_i = 0$ for $i \in \mathscr{I}_0$ are established to describe the boundary among the regions as in (2.98).

To further reduce the conservatism, the S-procedure will be also utilized by constructing matrices $\tilde{G}_i = \begin{bmatrix} g_i & G_i \end{bmatrix}, i \in \mathscr{I}$, with $g_i = 0$ for $i \in \mathscr{I}_0$ and satisfy (2.99).

Rewrite the system (3.80) into the following compact form,

$$\begin{cases} \tilde{E}\dot{\tilde{x}}(t) = \left(\tilde{\mathscr{A}}_i + \Delta\tilde{\mathscr{A}}_i\right)\tilde{x}(t) + \tilde{\mathscr{D}}_i w(t) \\ z(t) = \tilde{\mathscr{L}}_i \tilde{x}(t),\ i \in \mathscr{I}, \end{cases} \tag{3.95}$$

where

$$\begin{aligned} &\left[\ \tilde{E} \mid \tilde{x}(t) \mid \tilde{\mathscr{A}}_i \mid \Delta\tilde{\mathscr{A}}_i \mid \tilde{\mathscr{D}}_i \mid \tilde{\mathscr{L}}_i\ \right] \\ &= \begin{cases} \left[E \mid \bar{x}(t) \mid \bar{\mathscr{A}}_i \mid \Delta\bar{\mathscr{A}}_i \mid \bar{\mathscr{D}}_i \mid \bar{\mathscr{L}}_i\right], i \in \mathscr{I}_0, \\ \left[\begin{bmatrix} 1 & \mathbf{0} \\ \mathbf{0} & E \end{bmatrix} \middle| \begin{bmatrix} 1 \\ \bar{x}(t) \end{bmatrix} \middle| \begin{bmatrix} 0 & \mathbf{0} \\ \bar{\mathrm{a}}_i & \bar{\mathscr{A}}_i \end{bmatrix} \middle| \begin{bmatrix} 0 & \mathbf{0} \\ \Delta\bar{\mathrm{a}}_i & \Delta\bar{\mathscr{A}}_i \end{bmatrix} \middle| \begin{bmatrix} \mathbf{0} \\ \bar{\mathscr{D}}_i \end{bmatrix} \middle| \begin{bmatrix} \mathbf{0} & \bar{\mathscr{L}}_i \end{bmatrix}\right], i \in \mathscr{I}_1, \end{cases} \end{aligned} \tag{3.96}$$

where $\bar{x}(t)$, E, $\bar{\mathscr{A}}_i$, $\Delta\bar{\mathscr{A}}_i$, $\bar{\mathscr{D}}_i$, and $\bar{\mathscr{L}}_i$ are defined in (3.81).

Theorem 3.6. For a given positive scalar γ, the sliding mode dynamical system (3.95) is asymptotically stable with robust $\mathscr{H}_\infty$ performance γ, if there exist matrices $X = X^{\mathrm{T}}$, $U_i = U_i^{\mathrm{T}} \succeq 0$, $W_i = W_i^{\mathrm{T}} \succeq 0$, $\bar{K} \in \Re^{n_u \times n_y}$, $P_2 \in \Re^{n_u \times n_u}$, and scalars $\varepsilon_i > 0$, $i \in \mathscr{I}$ such that for all $i \in \mathscr{I}$, the LMIs in (3.97)–(3.98) hold,

$$\tilde{F}_i^{\mathrm{T}} X \tilde{F}_i - \tilde{G}_i^{\mathrm{T}} U_i \tilde{G}_i > 0, i \in \mathscr{I}, \tag{3.97}$$

$$\begin{bmatrix} \tilde{\Lambda}_1 & \star & \star & \star \\ \tilde{\Lambda}_2 & \tilde{\Lambda}_3 & \star & \star \\ D_m^{\mathrm{T}} \tilde{R} \tilde{F}_i^{\mathrm{T}} X \tilde{F}_i & \mathbf{0} & -\gamma^2 \mathbf{I} & \star \\ H_{m1}^{\mathrm{T}} \tilde{R} \tilde{F}_i^{\mathrm{T}} X \tilde{F}_i & \mathbf{0} & \mathbf{0} & -\varepsilon_i \mathbf{I} \end{bmatrix} < 0, m \in \mathscr{N}(i), i \in \mathscr{I}, \tag{3.98}$$

where

$$\begin{cases} \tilde{\Lambda}_1 = \mathtt{Sym}\{\tilde{F}_i^{\mathrm{T}} X \tilde{F}_i \tilde{A}_{m1} + \alpha \tilde{J}^{\mathrm{T}} \bar{K} C \tilde{R}\} + \tilde{R}^{\mathrm{T}} L_m^{\mathrm{T}} L_m \tilde{R} + \varepsilon_i \tilde{\Lambda}_{1(1)} \tilde{\Lambda}_{1(1)}^{\mathrm{T}} \\ \qquad + \tilde{G}_i^{\mathrm{T}} W_i \tilde{G}_i, \\ \tilde{\Lambda}_2 = -\bar{K} C \tilde{R} + B_m^{\mathrm{T}} \tilde{R} \tilde{F}_i^{\mathrm{T}} X \tilde{F}_i + \alpha P_2 \tilde{J} + N_m^{\mathrm{T}} L_m \tilde{R} + \varepsilon_i H_{m3}^{\mathrm{T}} \tilde{\Lambda}_{1(1)}^{\mathrm{T}}, \\ \tilde{\Lambda}_3 = -P_2 - P_2^{\mathrm{T}} + N_m^{\mathrm{T}} N_m + \varepsilon_i H_{m3}^{\mathrm{T}} H_{m3}, \\ \left.\tilde{\Lambda}_{1(1)} = H_{m2}^{\mathrm{T}},\ \tilde{A}_{m1} = A_m,\ \tilde{R} = \mathbf{I}_{n_x}, \tilde{J} \in \Re^{n_u \times n_x},\ \right\} \text{if } i \in \mathscr{I}_0, \\ \left.\begin{aligned} &\tilde{\Lambda}_{1(1)} = \begin{bmatrix} H_{m4}^{\mathrm{T}} \\ H_{m2}^{\mathrm{T}} \end{bmatrix}, \tilde{A}_{m1} = \begin{bmatrix} 0 & \mathbf{0} \\ a_m & A_m \end{bmatrix}, \tilde{R} = \begin{bmatrix} \mathbf{0} & \mathbf{I}_{n_x} \end{bmatrix}, \\ &\tilde{J} \in \Re^{n_u \times (n_x + 1)} \end{aligned}\right\} \text{if } i \in \mathscr{I}_1 \end{cases} \tag{3.99}$$

and α is a constant scalar, $\tilde{J} \in \Re^{n_u \times n_x}$, for $i \in \mathscr{I}_0$, and $\tilde{J} \in \Re^{n_u \times (n_x+1)}$, for $i \in \mathscr{I}_1$, are arbitrary matrices.

Moreover, the gain of the sliding surface (3.77) can be obtained via

$$K = P_2^{-\mathrm{T}} \bar{K}. \tag{3.100}$$

Proof. Consider the following PQLF which is continuous among the subspace boundaries

$$V(t) = \tilde{x}^{\mathrm{T}}(t)\tilde{E}\tilde{\mathscr{P}}_i\tilde{x}(t), i \in \mathscr{I}, \tag{3.101}$$

where

$$\tilde{\mathscr{P}}_i = \begin{bmatrix} \tilde{F}_i^{\mathrm{T}} X \tilde{F}_i & \mathbf{0} \\ \tilde{P}_3 & P_2 \end{bmatrix} \tag{3.102}$$

and $X = X^{\mathrm{T}}$, $P_2 \in \Re^{n_u \times n_u}$ is an arbitrary nonsingular matrix, and $\tilde{P}_3 \in \Re^{n_u \times n_x}$, for $i \in \mathscr{I}_0$, $\tilde{P}_3 \in \Re^{n_u \times (n_x+1)}$, for $i \in \mathscr{I}_1$.

The inequality in (3.97) ensures the positive definiteness of PQLF in each region. Based on the PQLF shown in (3.101), if

$$\dot{V}(t) + z^{\mathrm{T}}(t)z(t) - \gamma^2 w^{\mathrm{T}}(t)w(t) < 0, \tag{3.103}$$

the sliding mode dynamical system (3.95) is asymptotically stable with robust $\mathscr{H}_\infty$ performance γ.

With the structure of $\tilde{\mathscr{P}}_i$ shown in (3.102), we have $\tilde{E}\tilde{\mathscr{P}}_i = \tilde{\mathscr{P}}_i^{\mathrm{T}}\tilde{E}$. Based on system (3.95), the inequality (3.103) is derived as

$$\begin{bmatrix} \tilde{x}(t) \\ w(t) \end{bmatrix}^{\mathrm{T}} \begin{bmatrix} \mathtt{Sym}\{(\tilde{\mathscr{A}}_i + \Delta\tilde{\mathscr{A}}_i)^{\mathrm{T}}\tilde{\mathscr{P}}_i\} + \tilde{\mathscr{L}}_i^{\mathrm{T}}\tilde{\mathscr{L}}_i & \star \\ \tilde{\mathscr{D}}_i^{\mathrm{T}}\tilde{\mathscr{P}}_i & -\gamma^2\mathbf{I} \end{bmatrix} \begin{bmatrix} \tilde{x}(t) \\ w(t) \end{bmatrix} < 0, \tag{3.104}$$

where $\tilde{\mathscr{A}}_i$, $\Delta\tilde{\mathscr{A}}_i$, $\tilde{\mathscr{L}}_i$, $\tilde{\mathscr{D}}_i$ are given in (3.96).

Define $\bar{K} = P_2^{\mathrm{T}}K$. Using S-procedure, the following inequality implies (3.104),

$$\begin{bmatrix} \tilde{\Upsilon}_1 & \star & \star \\ \tilde{\Upsilon}_2 & -P_2 - P_2^{\mathrm{T}} + \mathscr{N}_i^{\mathrm{T}}\mathscr{N}_i & \star \\ \mathscr{D}_i^{\mathrm{T}}\tilde{R}\tilde{F}_i^{\mathrm{T}}X\tilde{F}_i & \mathbf{0} & -\gamma^2\mathbf{I} \end{bmatrix} < 0, \tag{3.105}$$

where $W_i \succeq 0$ and

$$\begin{cases} \tilde{\Upsilon}_1 = \mathtt{Sym}\{\tilde{F}_i^{\mathrm{T}}X\tilde{F}_i(\tilde{\mathscr{A}}_{i1} + \Delta\tilde{\mathscr{A}}_{i1}) - \tilde{P}_3^{\mathrm{T}}KC\tilde{R}\} + \tilde{R}^{\mathrm{T}}\mathscr{L}_i^{\mathrm{T}}\mathscr{L}_i\tilde{R} + \tilde{G}_i^{\mathrm{T}}W_i\tilde{G}_i, \\ \tilde{\Upsilon}_2 = -\bar{K}C\tilde{R} + (\mathscr{B}_i^{\mathrm{T}} + \Delta\mathscr{B}_i^{\mathrm{T}})\tilde{R}\tilde{F}_i^{\mathrm{T}}X\tilde{F}_i - \tilde{P}_3 + \mathscr{N}_i^{\mathrm{T}}\mathscr{L}_i\tilde{R}, \\ \left.\tilde{\mathscr{A}}_{i1} = \mathscr{A}_i,\ \Delta\tilde{\mathscr{A}}_{i1} = \Delta\mathscr{A}_i,\ \tilde{R} = \mathbf{I}_{n_x},\ \right\} \text{if } i \in \mathscr{I}_0, \\ \left.\tilde{\mathscr{A}}_{i1} = \begin{bmatrix} 0 & \mathbf{0} \\ \mathrm{a}_i & \mathscr{A}_i \end{bmatrix}, \Delta\tilde{\mathscr{A}}_{i1} = \begin{bmatrix} 0 & \mathbf{0} \\ \Delta\mathrm{a}_i & \Delta\mathscr{A}_i \end{bmatrix}, \tilde{R} = \begin{bmatrix} \mathbf{0} & \mathbf{I}_{n_x} \end{bmatrix}, \right\} i \in \mathscr{I}_1. \end{cases} \tag{3.106}$$

Expand the MFs and utilize Lemma A1 to handle the parameter uncertainties given as in (1.3). Then the following inequality implies (3.105) for scalars $\varepsilon_i > 0$, $i \in \mathscr{I}$,

$$\begin{bmatrix} \tilde{\Sigma}_1 & \star & \star & \star \\ \tilde{\Sigma}_2 & \tilde{\Sigma}_3 & \star & \star \\ D_m^{\mathrm{T}}\tilde{R}\tilde{F}_i^{\mathrm{T}}X\tilde{F}_i & \mathbf{0} & -\gamma^2\mathbf{I} & \star \\ H_{m1}^{\mathrm{T}}\tilde{R}\tilde{F}_i^{\mathrm{T}}X\tilde{F}_i & \mathbf{0} & \mathbf{0} & -\varepsilon_i\mathbf{I} \end{bmatrix} < 0, m \in \mathscr{N}(i), i \in \mathscr{I}, \tag{3.107}$$

where

$$\begin{cases} \tilde{\Sigma}_1 = \mathrm{Sym}\{\tilde{F}_i^{\mathrm{T}} X \tilde{F}_i \tilde{A}_{m1} - \tilde{P}_3^{\mathrm{T}} K C \tilde{R}\} + \tilde{R}^{\mathrm{T}} L_m^{\mathrm{T}} L_m \tilde{R} + \tilde{G}_i^{\mathrm{T}} W_i \tilde{G}_i \\ \qquad + \varepsilon_i \tilde{\Lambda}_{1(1)} \tilde{\Lambda}_{1(1)}^{\mathrm{T}}, \\ \tilde{\Sigma}_2 = -\tilde{K} C \tilde{R} + B_m^{\mathrm{T}} \tilde{R} \tilde{F}_i^{\mathrm{T}} X \tilde{F}_i - \tilde{P}_3 + N_m^{\mathrm{T}} L_m \tilde{R} + \varepsilon_i H_{m3}^{\mathrm{T}} \tilde{\Lambda}_{1(1)}^{\mathrm{T}}, \\ \tilde{\Sigma}_3 = -P_2 - P_2^{\mathrm{T}} + N_m^{\mathrm{T}} N_m + \varepsilon_i H_{m3}^{\mathrm{T}} H_{m3}, \\ \tilde{\Lambda}_{1(1)} = H_{m2}^{\mathrm{T}},\ \tilde{A}_{m1} = A_m, \ \} \text{ if } i \in \mathscr{I}_0, \\ \tilde{\Lambda}_{1(1)} = \begin{bmatrix} H_{m4}^{\mathrm{T}} \\ H_{m2}^{\mathrm{T}} \end{bmatrix}, \tilde{A}_{m1} = \begin{bmatrix} 0 & \mathbf{0} \\ a_m & A_m \end{bmatrix}, \ \Big\} \text{ if } i \in \mathscr{I}_1. \end{cases} \tag{3.108}$$

For numerical tractability, matrix $\tilde{P}_3$ can be specified as

$$\tilde{P}_3 = -\alpha P_2 \tilde{J}, \tag{3.109}$$

where α is a constant scalar, $\tilde{J} \in \Re^{n_u \times n_x}$, for $i \in \mathscr{I}_0$, and $\tilde{J} \in \Re^{n_u \times (n_x+1)}$, for $i \in \mathscr{I}_1$, are arbitrary matrices.

Based on Schur complement, inequality (3.98) implies (3.107). The proof is thus accomplished. ■

Remark 3.4.1. Compared with conventional stability analysis methods for the sliding motion, a descriptor system scheme is utilized to handle the couplings in the sliding surface gain with the control input matrices. Through a state-input augmentation scheme, the asymptotic stability and robust performance analysis of sliding motion can be analyzed in a unified convex optimization framework. It is also worthy of pointing out that the PQLFs-based method used in Theorem 3.6 leads to the conservatism reduction of the solutions.

In the following, a fuzzy output feedback dynamic SMC method will be proposed to drive the resulting closed-loop system states onto the designed sliding surface (3.77) locally in finite time.

3.5 OUTPUT FEEDBACK DYNAMIC SLIDING MODE CONTROLLER DESIGN

In this section, an output feedback dynamic sliding mode controller will be designed to drive the system states onto the designed sliding surface (3.77) in finite time. Then some illustrative simulation examples will be given to show the effectiveness of the proposed methods.

3.5.1 DYNAMIC SLIDING MODE CONTROLLER DESIGN

Similar to [72], a sliding patch Ω is first introduced as

$$\Omega := \{x(t) \in \Re^{n_x} : \|x(t)\| \leq \sigma\}. \tag{3.110}$$

Then a fuzzy output feedback dynamic sliding mode controller in parallel distributed compensation form is synthesized as

$$\dot{u}(t) = -(\eta + \zeta(t))\mathrm{sgn}(s(t)) - \sum_{l=1}^{r} \mu_l K C B_l u(t), \tag{3.111}$$

where $s(t) = Ky(t) + u(t)$ is given in (3.77), and $\eta > 0$ is a given scalar, and

$$\begin{aligned}\zeta(t) = \sum_{l=1}^{r} \mu_l \{ & (\|KCA_l\| + \|KCH_{l1}\| \cdot \|H_{l2}\|)\,\sigma \\ & + \|KCH_{l1}\| \cdot \|H_{l3}\| \cdot \|u(t)\| + \|KCD_l\| \varpi(t) \\ & + \|KCa_l\| + \|KCH_{l1}\| \cdot \|H_{l4}\| \}\end{aligned} \tag{3.112}$$

with $\varpi(t)$ being the uniform upper-bound of $w(t)$.

Note that as the proposed output feedback sliding mode controller (3.111) is in an output feedback dynamic form, there is no need to make a singular valuable decomposition (SVD) to the control input matrices B_l as in [94, 95] etc. Thus, the parameter uncertainties are permitted to exist in the control input matrices. The designed dynamic SMC scheme (3.111) tends to be more applicable.

Theorem 3.7. Given the sliding surface (3.77) and consider the fuzzy affine system (3.72) or equivalently (3.75). The trajectories of the closed-loop system states can be forced onto the sliding surface (3.77) in the region Ω in finite time by the fuzzy output feedback dynamic sliding mode controller in (3.111).

Proof. Construct the Lyapunov candidate as

$$V(t) = \frac{1}{2} s^{\mathrm{T}}(t) s(t). \tag{3.113}$$

Based on the sliding surface in (3.77), it can be seen that

$$\dot{s}(t) = KC\dot{x}(t) + \dot{u}(t). \tag{3.114}$$

Based on the controller (3.111), we have

$$\begin{aligned}\dot{V}(t) &= s^{\mathrm{T}}(t)\dot{s}(t) \\ &= s^{\mathrm{T}}(t) \sum_{l=1}^{r} \mu_l KC \{ (A_l + \Delta A_l)x(t) + a_l + \Delta a_l + (B_l + \Delta B_l)u(t) \\ &\quad + D_l w(t) \} - (\eta + \zeta(t))\|s(t)\| - s^{\mathrm{T}}(t) \sum_{l=1}^{r} \mu_l KCB_l u(t) \\ &= s^{\mathrm{T}}(t) \sum_{l=1}^{r} \mu_l KC \{ (A_l + \Delta A_l)x(t) + a_l + \Delta a_l \\ &\quad + \Delta B_l u(t) + D_l w(t) \} - \zeta(t)\|s(t)\| - \eta\|s(t)\|.\end{aligned} \tag{3.115}$$

Considering the parameter uncertainties in (1.3), we have the following inequality based on the sliding patch Ω given in (3.110),

$$\begin{aligned}\dot{V}(t) \leq \sum_{l=1}^{r} \mu_l \|s(t)\| \{ & (\|KCA_l\| + \|KCH_{l1}\| \cdot \|H_{l2}\|)\,\sigma + \|KCa_l\| \\ & + \|KCH_{l1}\| \cdot \|H_{l4}\| + \|KCH_{l1}\| \cdot \|H_{l3}\| \cdot \|u(t)\| + \|KCD_l\| \varpi(t) \} \\ & - \eta\|s(t)\| - \zeta(t)\|s(t)\|.\end{aligned} \tag{3.116}$$

Noticing that

$$\begin{aligned}\zeta(t) = \sum_{l=1}^{r} \mu_l \{ & (\|KCA_l\| + \|KCH_{l1}\| \cdot \|H_{l2}\|)\,\sigma \\ & + \|KCH_{l1}\| \cdot \|H_{l3}\| \cdot \|u(t)\| + \|KCD_l\| \varpi(t) \\ & + \|KCa_l\| + \|KCH_{l1}\| \cdot \|H_{l4}\| \},\end{aligned} \tag{3.117}$$

we have

$$\dot{V}(t) \leq -\eta \|s(t)\| < 0, \tag{3.118}$$

which indicates that system states can converge onto the designed sliding surface (3.77) in the sliding patch Ω in finite time. Thus, the proof is accomplished. ■

Most existing fuzzy output feedback SMC design approaches, such as [94, 95], require that each local model of system (3.72) shares one common input matrix, while the output feedback dynamic sliding mode controller in (3.111) does not. Note that the designed controller (3.111) can also deal with mismatched disturbances.

3.5.2 ILLUSTRATIVE EXAMPLES

In this subsection, two simulation examples will be given to show the effectiveness of the proposed output feedback dynamic SMC design approaches.

Example 3.4. Consider a fuzzy affine system in the form of (3.72) with three rules, and the system matrices are

$$\begin{cases} \left[\begin{array}{c|c|c} A_1 & A_2 & A_3 \end{array}\right] = \left[\begin{array}{cc|cc|cc} -0.2 & 0.2 & 1.5 & 0.1 & 1.0 & 0.1 \\ 0.2 & 0.1 & 0.2 & -1.0 & 0.2 & -1 \end{array}\right], \\ \left[\begin{array}{c|c|c} a_1 & a_2 & a_3 \end{array}\right] = \left[\begin{array}{c|c|c} 0 & 0 & 0 \\ 0 & 0.5 & -0.5 \end{array}\right], \\ \left[\begin{array}{c|c|c} B_1 & B_2 & B_3 \end{array}\right] = \left[\begin{array}{c|c|c} 0.8 & 1.0 & 1.0 \\ 1.0 & 1.0 & 1.5 \end{array}\right], D_l = \begin{bmatrix} 1.0 \\ 1.0 \end{bmatrix}, C = \begin{bmatrix} 1.0 & 0.5 \end{bmatrix}, \\ L_1 = \begin{bmatrix} 0.05 & 0.01 \end{bmatrix}, L_2 = \begin{bmatrix} 0.08 & 0.01 \end{bmatrix}, L_3 = \begin{bmatrix} 0.03 & 0.01 \end{bmatrix}, \\ N_l = 0.1,\ l = \{1,2,3\} \end{cases} \tag{3.119}$$

and the parameter uncertainties ΔA_l, Δa_l, and ΔB_l are given in the form of (1.3) with

$$\begin{cases} H_{l1} = \begin{bmatrix} 0.1 \\ 0.1 \end{bmatrix}, H_{l2} = \begin{bmatrix} 0.12 & 0.1 \end{bmatrix}, \\ H_{l3} = 0.08, H_{l4} = 0.1, l = \{1,2,3\}. \end{cases} \tag{3.120}$$

The MFs are presented in Figure 3.9 with $d_1 = 50$ and $d_2 = 300$. It can be observed from Figure 3.9 that the premise variable space is divided into three regions as in (3.66) with $n_x = 2$, where S_2 is a crisp region with $\mathcal{N}(2) = \{1\}$, and S_1 and S_3 are fuzzy regions with $\mathcal{N}(1) = \{1,2\}$ and $\mathcal{N}(3) = \{1,3\}$. These three regions can be accurately described by degenerate ellipsoids as in (2.82) and Q_i and q_i can be easily computed correspondingly.

We aim to design a sliding surface (3.77) and a fuzzy output feedback dynamic sliding mode controller (3.111) to ensure that the closed-loop system states can be forced onto the designed sliding surface (3.77) locally in finite time and the sliding motion is asymptotically stable with a satisfactory robust $\mathcal{H}_\infty$ performance γ. *As the control input matrices B_l in each local model given in (3.119) are not the same and possess uncertainty terms, the sliding mode controller design approaches given in [94, 95, 72] are not applicable.* It is also shown that no feasible solution is obtained through the CQLF-based method given in Theorem 3.5. By applying the PQLFs-based method presented in Theorem 3.6 with $\alpha = -3$, $\tilde{J} = \begin{bmatrix} 0 & 1 & 0 \end{bmatrix}$ for $i \in \mathscr{I}_1$, and $\tilde{J} = \begin{bmatrix} 1 & 0 \end{bmatrix}$ for $i \in \mathscr{I}_0$, one indeed obtains feasible solutions, and the sliding surface gain is calculated as $K = 22.4630$ with robust $\mathscr{H}_\infty$ performance $\gamma_{\min} = 1.7891$. It is noted that when applying Theorem 3.6 with PQLFs in (3.101), the characterizing matrices $\tilde{G}_i$ and $\tilde{F}_i$ can be obtained via (3.68) with $\rho = \begin{bmatrix} 1 & 0 \end{bmatrix}$ and $n_x = 2$.

To verify the validity of the proposed control method, simulations have been conducted. With the sliding surface $s(t) = 22.4630y(t) + u(t)$, initial condition $x_0 = \begin{bmatrix} 0.6 & -0.3 \end{bmatrix}$, $u(0) = 0$, and external disturbance $w(t) = 55e^{-0.1t}\sin(t)$, by applying the output feedback dynamic SMC method shown in (3.111) with $\eta = 1.2$, $\sigma = 0.7$, and $\varpi(t) = 55e^{-0.1t}$, the trajectories of the closed-loop system states and the control input are presented in Figure 3.22 and Figure 3.23, respectively. With zero initial condition, the $\mathscr{H}_\infty$ performance is presented in Figure 3.24. The ratio $\frac{\sqrt{\int_0^{t_f} z^T(t)z(t)dt}}{\sqrt{\int_0^{t_f} w^T(t)w(t)dt}}$ is about 0.23, which is lower than the minimum disturbance attenuation level $\gamma_{\min} = 1.7891$.

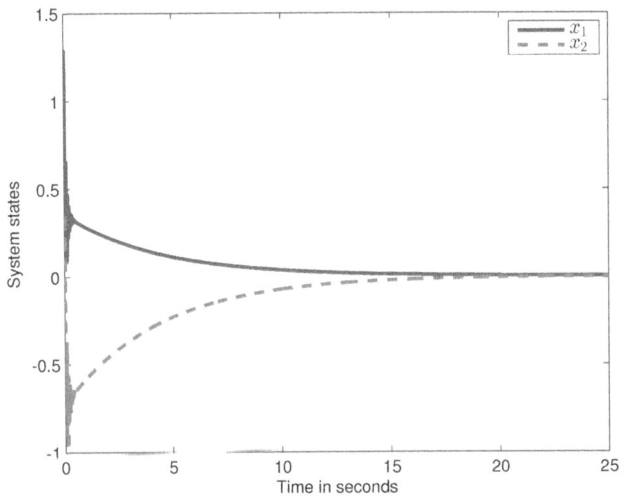

Figure 3.22 The closed-loop system states with dynamic sliding mode controller (3.111) in Example 3.4.

Example 3.5. Consider an inverted pendulum system same to the one in Example 2.3. The following motion equations characterize the pendulum system as,

$$\begin{cases} \dot{x}_1(t) = x_2(t) \\ \dot{x}_2(t) = \dfrac{g\sin(x_1) - amlx_2^2\sin(2x_1)/2 - a\cos(x_1)u}{4l/3 - aml\cos^2(x_1)} + w(t) \end{cases} \tag{3.121}$$

where the parameters are given as in Example 2.3.

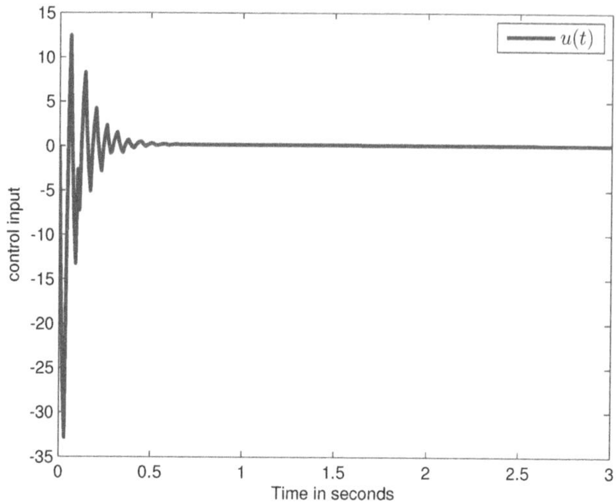

Figure 3.23 The control input with dynamic sliding mode controller (3.111) in Example 3.4.

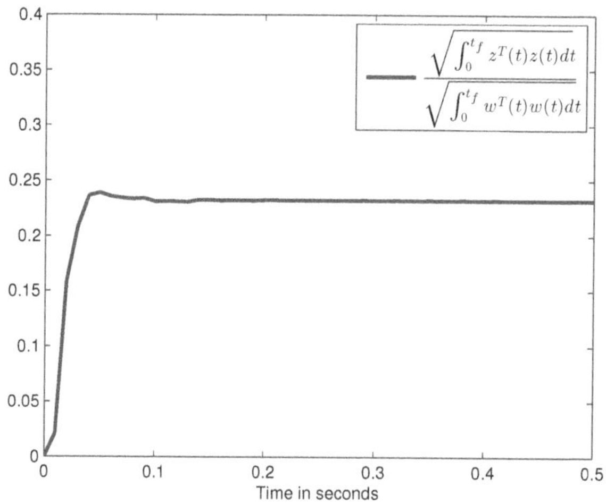

Figure 3.24 The sliding surface with dynamic sliding mode controller (3.111) in Example 3.4.

Linearize the nonlinear inverted pendulum (3.121) at three operating points $\left(-\frac{\pi}{3},0\right)^{\mathrm{T}}$, $(0,0)^{\mathrm{T}}$, and $\left(\frac{\pi}{3},0\right)^{\mathrm{T}}$. The measurement output $y(t)$ and the regulated output $z(t)$ are given as $y(t) = 0.1x_1(t) + x_2(t)$ and $z(t) = x_1(t) + x_2(t) + 0.05u(t)$, respectively, and then system (3.121) can be described by a T-S fuzzy affine model with three rules as,

Plant Rule $\mathscr{R}^l$: IF $x_1(t)$ is $\mathscr{F}_1^l$, THEN

$$\begin{cases} \dot{x}(t) = (A_l + \Delta A_l)x(t) + a_l + \Delta a_l + (B_l + \Delta B_l)u(t) + D_l w(t) \\ y(t) = Cx(t) \\ z(t) = L_l x(t) + N_l u(t),\ l \in \{1,2,3\}, \end{cases} \tag{3.122}$$

where the system matrices are

$$\begin{cases} A_1 = A_3 = \begin{bmatrix} 0 & 1 \\ 5.8512 & 0 \end{bmatrix}, A_2 = \begin{bmatrix} 0 & 1 \\ 17.2941 & 0 \end{bmatrix}, a_1 = \begin{bmatrix} 0 \\ -7.0992 \end{bmatrix}, \\ a_2 = \begin{bmatrix} 0 \\ 0 \end{bmatrix}, a_3 = \begin{bmatrix} 0 \\ 7.0992 \end{bmatrix}, B_1 = B_3 = \begin{bmatrix} 0 \\ -0.0779 \end{bmatrix}, B_2 = \begin{bmatrix} 0 \\ -0.1765 \end{bmatrix}, \\ D_l = \begin{bmatrix} 0 \\ 1 \end{bmatrix}, C = \begin{bmatrix} 1 & 0.1 \end{bmatrix}, L_l = \begin{bmatrix} 1 & 1 \end{bmatrix}, N_l = 0.01, l \in \{1,2,3\} \end{cases} \tag{3.123}$$

and the parameter uncertainties ΔA_l, Δa_l, and ΔB_l are given in the form of (1.3) with matrices H_{l1}, H_{l2}, and H_{l3} given as in (2.124).

The MFs of this example are also presented in Figure 3.9 with $d_1 = \frac{\pi}{12}$ and $d_2 = \frac{\pi}{2}$. By applying the PQLFs-based method presented in Theorem 3.6 with $\alpha = 440$, $\tilde{J} = \begin{bmatrix} 0 & 1 & 1 \end{bmatrix}$ for $i \in \mathscr{I}_1$, and $\tilde{J} = \begin{bmatrix} 1 & 1 \end{bmatrix}$ for $i \in \mathscr{I}_0$, one obtains feasible solutions, and the sliding surface gain is computed as $K = -911.7137$ with robust $\mathscr{H}_\infty$ performance $\gamma_{\min} = 4.2761$. It is noted that when applying Theorem 3.6 with PQLFs in (3.101), matrices $\tilde{G}_i$ and $\tilde{F}_i$ can be calculated via (3.68) with $\rho = \begin{bmatrix} 1 & 0 \end{bmatrix}$ and $n_x = 2$.

To verify the validity of the proposed approach, simulations are conducted under different initial conditions. With initial condition $x_0 = \begin{bmatrix} \frac{\pi}{4} & 0 \end{bmatrix}^{\mathrm{T}}$ and exogenous disturbance $w(t) = 110\mathrm{e}^{-0.08t}\sin(0.03t)$, by applying the fuzzy output feedback dynamic sliding mode controller in (3.111) with $\eta = 5$, $\sigma = 0.1$, and $\varpi(t) = 110\mathrm{e}^{-0.08t}$, the closed-loop system states and the control input are presented in Figure 3.25 and Figure 3.26, respectively. With zero initial condition, the $\mathscr{H}_\infty$ performance is presented in Figure 3.27. The ratio $\frac{\sqrt{\int_0^{t_f} z^{\mathrm{T}}(t)z(t)dt}}{\sqrt{\int_0^{t_f} w^{\mathrm{T}}(t)w(t)dt}}$ is about 0.34, which is lower than the minimum disturbance attenuation level $\gamma_{\min} = 4.2761$.

The simulation results illustrate that the proposed output feedback dynamic sliding mode controller design method is effective clearly.

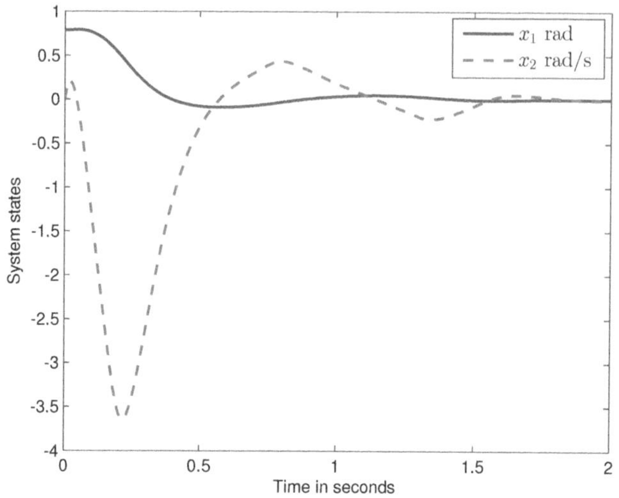

Figure 3.25 The closed-loop system states with dynamic sliding mode controller (3.111) in Example 3.5.

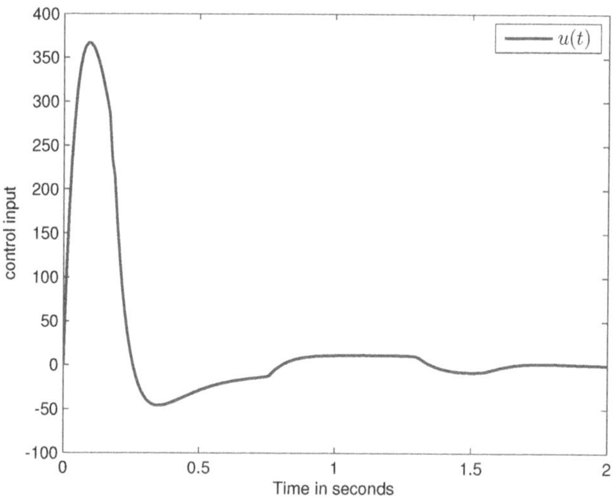

Figure 3.26 The control input with dynamic sliding mode controller (3.111) in Example 3.5.

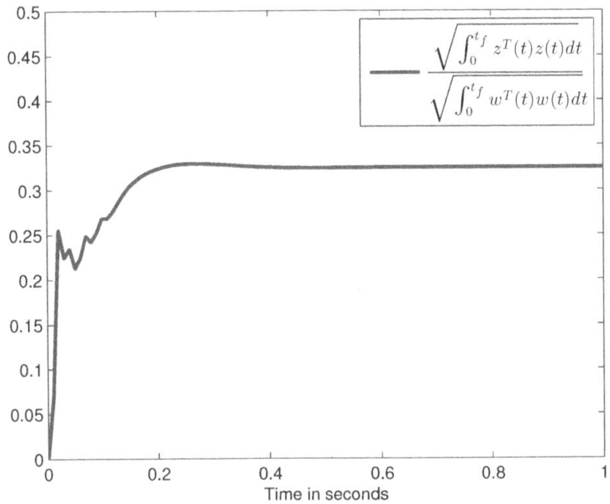

Figure 3.27 Response of the ratio $\frac{\sqrt{\int_0^{t_f} z^T(t)z(t)dt}}{\sqrt{\int_0^{t_f} w^T(t)w(t)dt}}$ with dynamic sliding mode controller (3.111) in Example 3.5.

3.6 CONCLUSION

This chapter investigates the output feedback SMC problems for continuous-time nonlinear 1-D systems subject to exogenous disturbance and mismatched uncertainties based on T-S fuzzy affine models. Associated with the sliding surface, the sliding mode dynamical features can be described by a singular system. Based on CQLF and PQLFs, respectively, stability and robust performance analysis of the sliding motion are conducted, and the sliding surface gain can be calculated under a convex optimization framework. Then some new output feedback fuzzy sliding mode controller and output feedback dynamic sliding mode controller design methods are proposed to guarantee that closed-loop system states can be driven onto the sliding surface in finite time locally. Finally, illustrative simulation examples are shown to verify the effectiveness of the proposed methods.

4 State Feedback Fuzzy SMC of Discrete-Time Nonlinear Systems

4.1 INTRODUCTION

This chapter focuses on the state feedback SMC problem for discrete-time uncertain nonlinear 1-D systems subject to mismatched parameter uncertainties and external nonlinear disturbance based on T-S fuzzy affine models. Via sufficiently exploring the dynamical properties of the original system and combining with the sliding surface function, a descriptor system is established to characterize the holonomic sliding mode dynamics. Then based on piecewise quadratic Lyapunov functions (PQLFs) and some convexification techniques, the asymptotical stability conditions for the sliding motion are derived under a convex optimization framework, and the stability analysis results on the sliding motion tend to be less conservative. *It is worth mentioning that different from the continuous-time situation, the discrete-time PQLFs will jump from one region at the time k to another region at the next time $k+1$, which is known as region transition feature* and will be elaborated in this chapter in detail. Considering the local models sharing one common control input matrix or not, two-state feedback SMC design methods will be proposed, which can both drive the system states into a neighbourhood of the sliding surface. Simulation examples are presented to validate the effectiveness of the proposed methods.

4.2 STATE FEEDBACK SMC DESIGN VIA T-S FUZZY AFFINE MODELS

4.2.1 MODEL DESCRIPTION: CASE I

Consider the first equation of the discrete-time T-S fuzzy affine system described as in (1.2) with a common control input matrix B, fuzzy **IF-THEN** rules as,

Plant Rule $\mathscr{R}^l$: IF $\theta_1(k)$ is $\mathscr{F}_1^l$ and $\cdots$ and $\theta_\varphi(k)$ is $\mathscr{F}_\varphi^l$, THEN

$$x(k+1) = (A_l + \Delta A_l)x(k) + a_l + \Delta a_l + B(u(k) + f(x(k),k)) \tag{4.1}$$

where B is the common control input matrix with full-column rank; $f(x(k),k) \in \Re^{n_u}$ is the nonlinear disturbance. ΔA_l and Δa_l are the parameter uncertainties in the l-th local model satisfying the condition in (1.3) as $\begin{bmatrix} \Delta A_l & \Delta a_l \end{bmatrix} = H_{l1}\Delta_l(k)\begin{bmatrix} H_{l2} & H_{l3} \end{bmatrix}, l \in \{1,\ldots,r\}$, where H_{l1}, H_{l2} and H_{l3} are known real-valued

DOI: 10.1201/9781032695716-4

matrices and $\Delta_l^{\mathrm{T}}(k)\Delta_l(k) \leq \mathbf{I}$. It is assumed that the external disturbance function $f(x(k),k)$ satisfies

$$\|f(x(k),k)\| \leq \gamma \|x(k)\|, \tag{4.2}$$

where $\gamma > 0$ is a positive constant.

Similar to Chapter 2, utilizing a singleton fuzzifier, product inference and center-average defuzzifier, the system (4.1) can be inferred as

$$x(k+1) = (A(\mu)+\Delta A(\mu))x(k) + a(\mu) + \Delta a(\mu) + B(u(k) + f(x(k),k)) \tag{4.3}$$

where

$$\begin{cases} A(\mu) = \sum\limits_{l=1}^{r} \mu_l A_l, \Delta A(\mu) = \sum\limits_{l=1}^{r} \mu_l \Delta A_l, \\ a(\mu) = \sum\limits_{l=1}^{r} \mu_l a_l, \Delta a(\mu) = \sum\limits_{l=1}^{r} \mu_l \Delta a_l. \end{cases} \tag{4.4}$$

It is worthy of mentioning that the fuzzy affine models possess more powerful capability to approximate nonlinear systems with higher accuracy compared with the fuzzy systems with linear local models [19]. Correspondingly, the synthesis and analysis for T-S fuzzy affine systems also tend to be more complicated generally. In this chapter, some new state feedback SMC synthesis results for discrete-time uncertain T-S fuzzy affine system (4.1) or equivalently (4.3) will be proposed via a descriptor-system-based method, and it will be demonstrated that the resulting closed-loop system states can be driven into a neighbourhood of the sliding surface.

4.2.2 SLIDING SURFACE DESIGN

For the system (4.1) or equivalently (4.3), the sliding surface is designed as

$$s(k) = Kx(k), \tag{4.5}$$

where $K \in \Re^{n_u \times n_x}$ is the sliding surface gain to be determined.

As the control input matrix B in the T-S fuzzy affine system (4.1) is with full column rank, then there always exists a full-row rank matrix $M \in \Re^{(n_x - n_u)\times n_x}$ satisfying $MB = 0$. Thus, one can construct a nonsingular transformation matrix T as

$$T = \begin{bmatrix} M \\ B^{\mathrm{T}} \end{bmatrix}. \tag{4.6}$$

Based on (4.6), it is satisfied that $MT^{-1} = \begin{bmatrix} \mathbf{I}_{n_u} & \mathbf{0}_{(n_x-n_u)\times n_u} \end{bmatrix}$. Through a state transformation with matrix T, the system dynamics in (4.3) can be formulated as,

$$\begin{aligned} Tx(k+1) = {} & T(A(\mu)+\Delta A(\mu))x(k) + T(a(\mu)+\Delta a(\mu)) \\ & + TB(u(k) + f(x(k),k)). \end{aligned} \tag{4.7}$$

The premise variable space can be decomposed by a similar line in Chapter 2. The following set $\mathscr{N}(i) := \{s|\mu_s[\theta(k)] > 0,\ \theta(k) \in S_i,\ s \in \mathscr{L},\ i \in \mathscr{I}\}$ is introduced to denote the indices for local models used in the interpolation in each region S_i.

Based on the set $\mathscr{N}(i)$, the system (4.7) is formulated as,

$$\begin{aligned} Tx(k+1) = {} & T(\mathbf{A}_i + \Delta\mathbf{A}_i)x(k) + T(\mathbf{a}_i + \Delta\mathbf{a}_i) \\ & + TB(u(k) + f(x(k),k)), \theta(k) \in S_i, i \in \mathscr{I}, \end{aligned} \tag{4.8}$$

where

$$\begin{cases} \mathbf{A}_i := \sum\limits_{s\in\mathscr{N}(i)} \mu_s A_s, \Delta\mathbf{A}_i := \sum\limits_{s\in\mathscr{N}(i)} \mu_s \Delta A_s, \\ \mathbf{a}_i := \sum\limits_{s\in\mathscr{N}(i)} \mu_s a_s, \Delta\mathbf{a}_i := \sum\limits_{s\in\mathscr{N}(i)} \mu_s \Delta a_s \end{cases} \tag{4.9}$$

with $\sum_{s\in\mathscr{N}(i)} \mu_s[\theta(k)] = 1$, $0 < \mu_s[\theta(k)] \leq 1$. Notice that $\mathscr{N}(i)$ contains only one element for a crisp region, that is to say, for all $i \in \mathscr{I}_0$, $\mathrm{a}_i + \Delta\mathrm{a}_i = 0$.

With $MB = 0$, it follows from (4.8) that,

$$\begin{cases} Mx(k+1) = M(\mathbf{A}_i + \Delta\mathbf{A}_i)x(k) + M(\mathbf{a}_i + \Delta\mathbf{a}_i) \\ B^{\mathrm{T}}x(k+1) = B^{\mathrm{T}}(\mathbf{A}_i + \Delta\mathbf{A}_i)x(k) \\ \qquad + B^{\mathrm{T}}(\mathbf{a}_i + \Delta\mathbf{a}_i) + B^{\mathrm{T}}B(u(k) + f(x(k),k))\ \theta(k) \in S_i, i \in \mathscr{I}, \end{cases} \tag{4.10}$$

where $B^{\mathrm{T}}B$ is nonsingular. It is easy to see that the first equation in (4.10) describes the sliding mode dynamics of the system (4.3) or (4.7). Noticing that the sliding surface function satisfies $s(k) = 0$, the holonomic sliding mode dynamical equation can be obtained as

$$Ex(k+1) = (\mathscr{A}_i + \Delta\mathscr{A}_i)x(k) + \mathrm{a}_i + \Delta\mathrm{a}_i,\ \theta(k) \in S_i, i \in \mathscr{I}, \tag{4.11}$$

where

$$\begin{cases} E = \begin{bmatrix} M \\ 0 \end{bmatrix}, \mathscr{A}_i + \Delta\mathscr{A}_i = \begin{bmatrix} M(\mathbf{A}_i + \Delta\mathbf{A}_i) \\ K \end{bmatrix}, \\ \mathrm{a}_i + \Delta\mathrm{a}_i = \begin{bmatrix} M(\mathbf{a}_i + \Delta\mathbf{a}_i) \\ \mathbf{0} \end{bmatrix}. \end{cases} \tag{4.12}$$

Via a state transformation $z(k) = Tx(k)$ and together with the fact that $MT^{-1} = \begin{bmatrix} \mathbf{I}_{n_u} & \mathbf{0}_{(n_x-n_u)\times n_u} \end{bmatrix}$, the system in (4.11) can be derived as,

$$\tilde{E}z(k+1) = (\tilde{\mathscr{A}_i} + \Delta\tilde{\mathscr{A}_i})z(k) + \tilde{\mathrm{a}}_i + \Delta\tilde{\mathrm{a}}_i,\ \tilde{\theta}(k) \in \tilde{S}_i, i \in \mathscr{I}, \tag{4.13}$$

where $\tilde{\theta}(k)$ and $\tilde{S}_i$, respectively, refer to the new premise variables and partitioned space with the transformation matrix T, and

$$\begin{cases} \tilde{E} = \begin{bmatrix} \mathbf{I}_{n_x-n_u} & \mathbf{0} \\ \mathbf{0} & \mathbf{0}_{n_u} \end{bmatrix}, \\ \tilde{\mathscr{A}_i} + \Delta\tilde{\mathscr{A}_i} = \begin{bmatrix} M(\mathbf{A}_i + \Delta\mathbf{A}_i)T_1 & M(\mathbf{A}_i + \Delta\mathbf{A}_i)T_2 \\ KT_1 & KT_2 \end{bmatrix}, \\ \tilde{\mathrm{a}}_i + \Delta\tilde{\mathrm{a}}_i = \begin{bmatrix} M(\mathbf{a}_i + \Delta\mathbf{a}_i) \\ \mathbf{0} \end{bmatrix}, T^{-1} = \begin{bmatrix} T_1 & T_2 \end{bmatrix} \end{cases} \tag{4.14}$$

and $T_1 \in \Re^{n_x\times(n_x-n_u)}$, $T_2 \in \Re^{n_x\times n_u}$ are known constant matrices.

Combining the designed sliding surface (4.5) with the first equation in system (4.10), the sliding motion is fully described by the descriptor system (4.13). The existence of the proposed sliding surface is ensured by the admissibility of the system (4.13). In the following section, we will firstly give sufficient conditions for the existence of sliding surface (4.5), and then a state feedback fuzzy SMC design method will be proposed.

4.2.3 STABILITY ANALYSIS OF SLIDING MOTION

Based on PQLFs, new results on the asymptotic stability analysis of the sliding motion will be proposed in this subsection. Then a state feedback sliding mode controller will be designed to drive the system states into a neighbourhood of the sliding surface (4.5).

Firstly, the asymptotic stability of the sliding motion will be analyzed under the framework of PQLFs. For future use, a new set Ω is introduced in order to characterize region transitions as follows,

$$\Omega := \{(i,j)|\tilde{\theta}(k)\in\tilde{S}_i, \tilde{\theta}(k+1)\in\tilde{S}_j, i,j\in\mathscr{I}\}. \tag{4.15}$$

Similar to [58] and Chapter 2, it is assumed in this subsection that an ellipsoid $\mathscr{E}_i := \{z|\|Q_i z + q_i\| \leq 1\}$ is used to outer approximate each polyhedral region $\tilde{S}_i$ in (4.13), and the relationship described in (2.19) holds for each region.

Based on PQLFs, the sliding motion stability analysis result is presented as follows.

Theorem 4.1. The sliding mode dynamical system (4.13) is asymptotically stable if there exist matrices $0 < P_{i1} = P_{i1}^T \in \Re^{(n_x-n_u)\times(n_x-n_u)}$, $P_{i2} \in \Re^{n_u\times(n_x-n_u)}$, $P_{i3} \in \Re^{n_u\times n_u}$, $i \in \mathscr{I}$, and $\hat{K} \in \Re^{n_u\times n_x}$, $\{G_{i1(1)}, G_{i2(1)}\} \in \Re^{(n_x-n_u)\times(n_x-n_u)}$, $\{G_{i1(2)}, G_{i2(2)}\} \in \Re^{n_u\times(n_x-n_u)}$, $G_{(3)} \in \Re^{n_u\times n_u}$, and two sets of scalars $\{\varepsilon_i > 0, i \in \mathscr{I}\}$, $\{\rho_i < 0, i \in \mathscr{I}_1\}$, ensuring the LMIs in (4.16)–(4.17) hold,

$$\begin{bmatrix} \Sigma_{11} + \varepsilon_i T^{-\mathrm{T}} H_{s2}^{\mathrm{T}} H_{s2} T^{-1} & \star & \star \\ \Sigma_{21} - G_{i2}^{\mathrm{T}} & \Sigma_{22} & \star \\ \Sigma_{31} & \Sigma_{32} & -\varepsilon_i \mathbf{I} \end{bmatrix} < 0, i \in \mathscr{I}_0, s \in \mathscr{N}(i), \tag{4.16}$$

$$\begin{bmatrix} \rho_i(q_i^{\mathrm{T}} q_i - 1) + \varepsilon_i H_{s3}^{\mathrm{T}} H_{s3} & \star & \star & \star \\ \rho_i Q_i^{\mathrm{T}} q_i + \Sigma_{10} + \varepsilon_i T^{-\mathrm{T}} H_{s2}^{\mathrm{T}} H_{s3} & \rho_i Q_i^{\mathrm{T}} Q_i + \Sigma_{11} + \varepsilon_i T^{-\mathrm{T}} H_{s2}^{\mathrm{T}} H_{s2} T^{-1} & \star & \star \\ \Sigma_{20} & \Sigma_{21} - G_{i2}^{\mathrm{T}} & \Sigma_{22} & \star \\ \mathbf{0} & \Sigma_{31} & \Sigma_{32} & -\varepsilon_i \mathbf{I} \end{bmatrix} < 0,$$

$$i \in \mathscr{I}_1, s \in \mathscr{N}(i), \tag{4.17}$$

where

$$\begin{cases} \Sigma_{11} = \mathrm{Sym}\left\{ \begin{bmatrix} G_{i2(1)} \\ G_{i2(2)} \end{bmatrix} MA_s T^{-1} + \begin{bmatrix} \mathbf{0} \\ \rho_2 \mathbf{I} \end{bmatrix} \hat{K} T^{-1} \right\} - \tilde{E}^{\mathrm{T}} P_i \tilde{E}, \\ \Sigma_{21} = \begin{bmatrix} G_{i1(1)} \\ G_{i1(2)} \end{bmatrix} MA_s T^{-1} + \begin{bmatrix} J \\ \rho_1 \mathbf{I} \end{bmatrix} \hat{K} T^{-1}, \Sigma_{22} = P_j - G_{i1} - G_{i1}^{\mathrm{T}}, \\ \Sigma_{31} = H_{s1}^{\mathrm{T}} M^{\mathrm{T}} \begin{bmatrix} G_{i2(1)}^{\mathrm{T}} & G_{i2(2)}^{\mathrm{T}} \end{bmatrix}, \Sigma_{32} = H_{s1}^{\mathrm{T}} M^{\mathrm{T}} \begin{bmatrix} G_{i1(1)}^{\mathrm{T}} & G_{i1(2)}^{\mathrm{T}} \end{bmatrix}, \\ \Sigma_{10} = \begin{bmatrix} G_{i2(1)} \\ G_{i2(2)} \end{bmatrix} Ma_s, \Sigma_{20} = \begin{bmatrix} G_{i1(1)} \\ G_{i1(2)} \end{bmatrix} Ma_s, \\ P_i = \begin{bmatrix} P_{i1} & \star \\ P_{i2} & P_{i3} \end{bmatrix}, T^{-1} = \begin{bmatrix} T_1 & T_2 \end{bmatrix}, \\ G_{i1} = \begin{bmatrix} G_{i1(1)} & JG_{(3)} \\ G_{i1(2)} & \rho_1 G_{(3)} \end{bmatrix}, G_{i2} = \begin{bmatrix} G_{i2(1)} & \mathbf{0} \\ G_{i2(2)} & \rho_2 G_{(3)} \end{bmatrix}, \end{cases} \tag{4.18}$$

in which $J \in \Re^{(n_x - n_u)\times n_u}$ is an arbitrary matrix, ρ_1, ρ_2 are constant scalars, and $T_1 \in \Re^{n_x \times (n_x - n_u)}$, $T_2 \in \Re^{n_x \times n_u}$ are known constant matrices given in (4.18).

Moreover, the gain of the sliding surface (4.5) can be calculated by

$$K = G_{(3)}^{-1} \hat{K}. \tag{4.19}$$

Proof. Note that the condition (4.16) for $i \in \mathscr{I}_0$ is a special case of the condition (4.17) for $i \in \mathscr{I}_1$. Without loss of generality, merely the more complex case for $i \in \mathscr{I}_1$ will be proved. For system (4.13), consider the following PQLF as

$$V(k) = z^{\mathrm{T}}(k) \tilde{E}^{\mathrm{T}} P_i \tilde{E} z(k), \tag{4.20}$$

where

$$P_i = \begin{bmatrix} P_{i1} & \star \\ P_{i2} & P_{i3} \end{bmatrix}, \; i \in \mathscr{I} \tag{4.21}$$

and $P_{i3} \in \Re^{n_u \times n_u}$, $P_{i2} \in \Re^{n_u \times (n_x - n_u)}$, $0 < P_{i1} = P_{i1}^T \in \Re^{(n_x - n_u)\times(n_x - n_u)}$.

Along the trajectories of system (4.13) and based on (4.20), the following inequality ensures the asymptotic stability of the sliding mode dynamical system (4.13),

$$\Delta V(k) = \left(\tilde{E} z(k+1)\right)^{\mathrm{T}} P_j \left(\tilde{E} z(k+1)\right) - z^{\mathrm{T}}(k) \tilde{E}^{\mathrm{T}} P_i \tilde{E} z(k) < 0, (i,j) \in \Omega. \tag{4.22}$$

Define the following augmented vector

$$\vartheta(k) = \begin{bmatrix} 1 & z^{\mathrm{T}}(k) & z^{\mathrm{T}}(k+1)\tilde{E}^{\mathrm{T}} \end{bmatrix}^{\mathrm{T}}, \tag{4.23}$$

and then the inequality (4.22) can be rewritten as,

$$\vartheta^{\mathrm{T}}(k) \begin{bmatrix} 0 & \star & \star \\ \mathbf{0} & -\tilde{E}^{\mathrm{T}} P_i \tilde{E} & \star \\ \mathbf{0} & \mathbf{0} & P_j \end{bmatrix} \vartheta(k) < 0, \; (i,j) \in \Omega. \tag{4.24}$$

Utilizing the S-procedure with the partition information described as in (2.19), for $\rho_i < 0$, $i \in \mathscr{I}_1$, the following inequality implies (4.24)

$$\vartheta^{\mathrm{T}}(k)\begin{bmatrix} \rho_i(q_i^{\mathrm{T}}q_i-1) & \star & \star \\ \rho_i Q_i^{\mathrm{T}}q_i & -\tilde{E}^{\mathrm{T}}P_i\tilde{E}+\rho_i Q_i^{\mathrm{T}}Q_i & \star \\ \mathbf{0} & \mathbf{0} & P_j \end{bmatrix}\vartheta(k)<0, i\in\mathscr{I}_1,(i,j)\in\Omega. \tag{4.25}$$

Based on system (4.13), the following equality holds for any matrix $\mathscr{G}_i$, $i \in \mathscr{I}$,

$$\begin{aligned} 0 &= 2\mathscr{G}_i\left\{\tilde{\mathrm{a}}_i+\Delta\tilde{\mathrm{a}}_i+(\tilde{\mathscr{A}_i}+\Delta\tilde{\mathscr{A}_i})z(k)-\tilde{E}z(k+1)\right\} \\ &= 2\mathscr{G}_i\begin{bmatrix} \tilde{\mathrm{a}}_i+\Delta\tilde{\mathrm{a}}_i & \tilde{\mathscr{A}_i}+\Delta\tilde{\mathscr{A}_i} & -\mathbf{I} \end{bmatrix}\vartheta(k). \end{aligned} \tag{4.26}$$

Combining (4.25) with (4.26), one has,

$$\vartheta^{\mathrm{T}}(k)\left\{\begin{bmatrix} \rho_i(q_i^{\mathrm{T}}q_i-1) & \star & \star \\ \rho_i Q_i^{\mathrm{T}}q_i & \rho_i Q_i^{\mathrm{T}}Q_i-\tilde{E}^{\mathrm{T}}P_i\tilde{E} & \star \\ \mathbf{0} & \mathbf{0} & P_j \end{bmatrix}+\mathtt{Sym}\{\mathscr{G}_i\mathscr{A}_i\}\right\}\vartheta(k)<0, \quad i\in\mathscr{I}_1,(i,j)\in\Omega, \tag{4.27}$$

where $\mathscr{A}_i = \begin{bmatrix} \tilde{\mathrm{a}}_i+\Delta\tilde{\mathrm{a}}_i & \tilde{\mathscr{A}_i}+\Delta\tilde{\mathscr{A}_i} & -\mathbf{I} \end{bmatrix}$.

If $\Delta V(k) < 0$, it follows from (4.27) that

$$\begin{bmatrix} \rho_i(q_i^{\mathrm{T}}q_i-1) & \star & \star \\ \rho_i Q_i^{\mathrm{T}}q_i & \rho_i Q_i^{\mathrm{T}}Q_i-\tilde{E}^{\mathrm{T}}P_i\tilde{E} & \star \\ \mathbf{0} & \mathbf{0} & P_j \end{bmatrix}+\mathtt{Sym}\{\mathscr{G}_i\mathscr{A}_i\}<0, \quad i\in\mathscr{I}_1,(i,j)\in\Omega. \tag{4.28}$$

Extracting the MFs, the following inequality implies (4.28),

$$\begin{bmatrix} \rho_i(q_i^{\mathrm{T}}q_i-1) & \star & \star \\ \rho_i Q_i^{\mathrm{T}}q_i & \rho_i Q_i^{\mathrm{T}}Q_i-\tilde{E}^{\mathrm{T}}P_i\tilde{E} & \star \\ \mathbf{0} & \mathbf{0} & P_j \end{bmatrix}+\mathtt{Sym}\{\mathscr{G}_i\mathbb{A}_s\}<0, \quad s\in\mathscr{N}(i),i\in\mathscr{I}_1,(i,j)\in\Omega, \tag{4.29}$$

where

$$\begin{cases} \mathbb{A}_s = \begin{bmatrix} \tilde{a}_s+\Delta\tilde{a}_s & \tilde{A}_s+\Delta\tilde{A}_s & -\mathbf{I} \end{bmatrix}, \\ \tilde{a}_s+\Delta\tilde{a}_i = \begin{bmatrix} M(a_s+\Delta a_s) \\ \mathbf{0} \end{bmatrix}, \tilde{A}_s+\Delta\tilde{A}_s = \begin{bmatrix} M(A_s+\Delta A_s) \\ K \end{bmatrix}T^{-1}. \end{cases} \tag{4.30}$$

It can be seen from (4.30) that the sliding surface gain does not exist in the first row of $\tilde{a}_s+\Delta\tilde{a}_s$ and $\tilde{A}_s+\Delta\tilde{A}_s$. For numerical tractability of the stability analysis conditions, $\mathscr{G}_i$ can be specified as $\mathscr{G}_i = \begin{bmatrix} \mathbf{0} & G_{i2}^{\mathrm{T}} & G_{i1}^{\mathrm{T}} \end{bmatrix}^{\mathrm{T}}$, where the slack variable matrices G_{i1} and G_{i2} are specified as,

$$G_{i1} = \begin{bmatrix} G_{i1(1)} & JG_{(3)} \\ G_{i1(2)} & \rho_1 G_{(3)} \end{bmatrix}, G_{i2} = \begin{bmatrix} G_{i2(1)} & \mathbf{0} \\ G_{i2(2)} & \rho_2 G_{(3)} \end{bmatrix}, \tag{4.31}$$

where ρ_1 and ρ_2 are scalar parameters, and $J \in \Re^{(n_x-n_u)\times n_u}$ is an arbitrary matrix.

Define

$$\hat{K} = G_{(3)}K. \tag{4.32}$$

Based on (4.31)–(4.32), inequality (4.29) can be derived as,

$$\begin{bmatrix} \rho_i(q_i^{\mathrm{T}}q_i - 1) & \star & \star \\ \rho_i Q_i^{\mathrm{T}} q_i + \Gamma_{10} & \Gamma_{11} & \star \\ \Gamma_{20} & \Gamma_{21} - G_{i2}^{\mathrm{T}} & P_j - G_{i1} - G_{i1}^{\mathrm{T}} \end{bmatrix} < 0, s \in \mathscr{N}(i), i \in \mathscr{I}_1, (i,j) \in \Omega, \tag{4.33}$$

where

$$\begin{cases} \Gamma_{11} = \rho_i Q_i^{\mathrm{T}} Q_i - \tilde{E}^{\mathrm{T}} P_i \tilde{E} \\ \qquad + \mathrm{Sym}\left\{ \begin{bmatrix} G_{i2(1)} \\ G_{i2(2)} \end{bmatrix} M(A_s + \Delta A_s) T^{-1} + \begin{bmatrix} \mathbf{0} \\ \rho_2 \mathbf{I} \end{bmatrix} \hat{K} T^{-1} \right\}, \\ \Gamma_{21} = \begin{bmatrix} G_{i1(1)} \\ G_{i1(2)} \end{bmatrix} M(A_s + \Delta A_s) T^{-1} + \begin{bmatrix} J \\ \rho_1 \mathbf{I} \end{bmatrix} \hat{K} T^{-1}, \\ \Gamma_{10} = \begin{bmatrix} G_{i2(1)} \\ G_{i2(2)} \end{bmatrix} M(a_s + \Delta a_s), \Gamma_{20} = \begin{bmatrix} G_{i1(1)} \\ G_{i1(2)} \end{bmatrix} M(a_s + \Delta a_s). \end{cases} \tag{4.34}$$

Adopting Lemma A1 to tackle the parameter uncertainties in (4.33), then inequality (4.17) implies inequality (4.33) for a set of scalars $\varepsilon_i > 0$, $i \in \mathscr{I}$ based on Schur complement.

In addition, the condition in (4.17) indicates that

$$G_{(3)} + G_{(3)}^{\mathrm{T}} - P_{j3} > 0, \tag{4.35}$$

which implies that $G_{(3)}$ is invertible. The sliding surface gain can thus be attained via (4.19). The proof is then accomplished. ■

Remark 4.2.1. Through the PQLF $V(k) = z^{\mathrm{T}}(k)\tilde{E}^{\mathrm{T}}P_i\tilde{E}z(k)$ in (4.20) with the piecewise Lyapunov matrices P_i defined in (4.21), sufficient conditions for sliding motion stability analysis are presented in Theorem 4.1. It is noted that if $P_{i1} = P_{j1} \equiv P_1 \in \Re^{(n_x - n_u)\times(n_x - n_u)}$, $P_{i2} = P_{j2} \equiv P_2 \in \Re^{n_u \times (n_x - n_u)}$, and $P_{i3} = P_{j3} \equiv P_3 \in \Re^{n_u \times n_u}$, the corresponding stability analysis results can be easily obtained based on the CQLF as $V(k) = z^{\mathrm{T}}(k)\tilde{E}^{\mathrm{T}}P\tilde{E}z(k)$ with $P = \begin{bmatrix} P_1 & \star \\ P_2 & P_3 \end{bmatrix}$. Thus, the CQLF can be viewed as a special case of the PQLFs which tend to be more general.

In the following, a state feedback sliding mode controller design method will be proposed to drive the closed-loop system state trajectories into a neighbourhood the sliding surface (4.5).

4.3 STATE FEEDBACK SLIDING MODE CONTROLLER DESIGN

4.3.1 SLIDING MODE CONTROLLER DESIGN

Theorem 4.2. For the T-S fuzzy affine system (4.1) or equivalently (4.3) with the sliding surface given in (4.5), the resulting closed-loop system state trajectories can be driven into a neighbourhood of the sliding surface (4.5) by the following state feedback sliding mode controller,

$$u(k) = -(KB)^{-1}\left\{\eta s(k) + \sum_{l=1}^{r} \mu_l K(A_l x(k) + a_l) + \zeta(k)\mathrm{sgn}(s(k))\right\}, \tag{4.36}$$

where

$$\zeta(k) \geq \sum_{l=1}^{r} \mu_l \|KH_{l1}\| \{\|H_{l2}\| \cdot \|x(k)\| + \|H_{l3}\|\} + \gamma\|KB\| \cdot \|x(k)\| \tag{4.37}$$

with given small positive scalars η.

Proof. Construct the following Lyapunov function

$$V(k) = \frac{1}{2}s^{\mathrm{T}}(k)s(k). \tag{4.38}$$

The incremental $\Delta V(k)$ is

$$\Delta V(k) = V(k+1) - V(k) = s^{\mathrm{T}}(k)\Delta s(k) + \Pi, \tag{4.39}$$

where

$$\Delta s(k) = s(k+1) - s(k),\ \Pi = \frac{1}{2}\Delta s^{\mathrm{T}}(k)\Delta s(k). \tag{4.40}$$

Based on the state feedback sliding mode controller (4.36) and considering the matching condition in (4.2), then (4.39) can be formulated as

$$\begin{aligned}
\Delta V(k) &= s^{\mathrm{T}}(k)\{Kx(k+1) - s(k)\} + \Pi \\
&= s^{\mathrm{T}}(k)\Big\{\sum_{l=1}^{r} \mu_l K[(A_l + \Delta A_l)x(k) + a_l + \Delta a_l + B(u(k) + f(x(k),k))] \\
&\qquad -s(k)\Big\} + \Pi \\
&= s^{\mathrm{T}}(k)\Big\{\sum_{l=1}^{r} \mu_l K(\Delta A_l x(k) + \Delta a_l) + KBf(x(k),k) - (\eta+1)s(k) \\
&\qquad -\zeta(k)\mathrm{sgn}(s(k))\Big\} + \Pi \\
&\leq \sum_{l=1}^{r} \mu_l \{\|KH_{l1}\|(\|H_{l2}\| \cdot \|x(k)\| + \|H_{l3}\|) + \gamma\|KB\| \cdot \|x(k)\|\}\|s(k)\| \\
&\qquad -(\eta+1)\|s(k)\|^2 - \zeta(k)\|s(k)\| + \Pi.
\end{aligned} \tag{4.41}$$

Noticing that

$$\zeta(k) \geq \sum_{l=1}^{r} \mu_l \|KH_{l1}\| \{\|H_{l2}\| \cdot \|x(k)\| + \|H_{l3}\|\} + \gamma\|KB\| \cdot \|x(k)\|, \tag{4.42}$$

then one has

$$\Delta V(k) \leq -(\eta+1)\|s(k)\|^2 + \Pi. \tag{4.43}$$

Then $\Delta V(k) < 0$ can be ensured by tuning the parameter η appropriately. It can be shown that the state trajectories of the discrete-time T-S fuzzy affine system (4.1) can be driven into a neighbourhood of the sliding surface (4.5) by the sliding mode controller (4.36). Thus, the proof is completed. ■

4.3.2 ILLUSTRATIVE EXAMPLES

Example 4.1. Consider a simplified pitch model of the helicopter [101]. In this example, the helicopter model is characterized by the following equation

$$\begin{cases} \dot{\mathbf{x}}_1 = \mathbf{x}_2 \\ \dot{\mathbf{x}}_2 = \dfrac{1}{I_{yy}}\left[-mI_{cgx}g\cos(\mathbf{x}_1) - mI_{cgz}g\sin(\mathbf{x}_1) - F_{vM}\mathbf{x}_2 + u\right] \end{cases} \tag{4.44}$$

where $\mathbf{x}_1$ and $\mathbf{x}_2$ are the pitch angle and pitch rate, respectively. The nominal plant parameters are given as in Table 4.1.

Table 4.1
Nominal parameters of the simplified pitch model of helicopter

Parameter	Nominal Value	Parameter	Nominal Value
m	0.9941kg	I_{yy}	0.0283kg·m^2
F_{vM}	0.0041Nm/(rad/s)	I_{cgx}	0.0134m
g	9.81m/s^2	I_{cgz}	0.0289m

Change the coordinates with $x(t) = \mathbf{x}(t) - \mathbf{x}_e$, where $\mathbf{x}_e(t) = \begin{bmatrix} -1.3768 & 0 \end{bmatrix}^{\mathrm{T}}$. Linearizing the nonlinear helicopter system at three operating points $x_{o1} = \begin{pmatrix} -\frac{\pi}{5} & 0 \end{pmatrix}^{\mathrm{T}}, x_{o2} = \begin{pmatrix} 0 & 0 \end{pmatrix}^{\mathrm{T}}, x_{o3} = \begin{pmatrix} \frac{\pi}{5} & 0 \end{pmatrix}^{\mathrm{T}}$, and discretizing the helicopter system with sampling period $T = 0.05$s, and then nonlinear helicopter system (4.44) can be described by the following T-S fuzzy affine model with three rules,

Plant Rule $\mathscr{R}^l$: IF $x_1(k)$ is $\mathscr{F}_1^l$, THEN

$$x(k+1) = (A_l + \Delta A_l)x(k) + a_l + \Delta a_l + B\left(u(k) + f(x(k),k)\right),\ l \in \{1,2,3\}, \tag{4.45}$$

where the system matrices are

$$\begin{cases} A_1 = \begin{bmatrix} 1 & 0.05 \\ 0.0001 & 0.9928 \end{bmatrix}, A_2 = \begin{bmatrix} 1 & 0.05 \\ -0.3225 & 0.9928 \end{bmatrix}, A_3 = \begin{bmatrix} 1 & 0.05 \\ -0.5743 & 0.9928 \end{bmatrix}, \\ a_1 = \begin{bmatrix} 0 \\ 0.5490 \end{bmatrix}, a_2 = \begin{bmatrix} 0 \\ 0 \end{bmatrix}, a_3 = \begin{bmatrix} 0 \\ -0.2602 \end{bmatrix}, B = \begin{bmatrix} 0 \\ 1.7651 \end{bmatrix}, \end{cases} \tag{4.46}$$

and the parameter uncertainties ΔA_l and Δa_l are given in the form of (1.3) with

$$\begin{cases} H_{l1} = \begin{bmatrix} 0 \\ 0.02 \end{bmatrix}, H_{l2} = \begin{bmatrix} 0.01 & 0.01 \end{bmatrix}, \\ H_{l3} = 0.01,\ l \in \{1,2,3\}. \end{cases} \tag{4.47}$$

Figure 4.1 shows the MFs in Example 4.1 with $d_1 = \dfrac{\pi}{5}$ and $d_2 = \dfrac{3\pi}{5}$. It follows from Figure 4.1 that the premise variable space can be decomposed into three regions

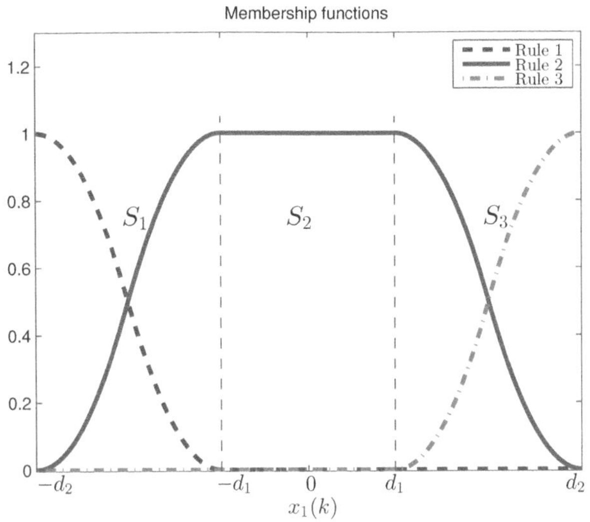

Figure 4.1 Membership functions of Example 4.1-Example 4.4.

as in (3.66) with $n_x = 2$, where S_2 is a crisp region with $\mathcal{N}(2) = \{2\}$, and S_1 and S_3 are fuzzy regions with $\mathcal{N}(1) = \{1,2\}$ and $\mathcal{N}(3) = \{2,3\}$.

With the approach proposed in this chapter, the transformation matrix T can be chosen as $T = \begin{bmatrix} 1 & 0 \\ 0 & 1.7651 \end{bmatrix}$, and for this case, $z_1(k) = x_1(k)$. These three regions can be accurately characterized by degenerate ellipsoids as in (2.16) with

$$\begin{cases} Q_1 = Q_3 = \begin{bmatrix} \dfrac{2}{d_2 - d_1} & 0 \end{bmatrix}, Q_2 = \begin{bmatrix} \dfrac{1}{d_1} & 0 \end{bmatrix}, \\ q_1 = \dfrac{d_1 + d_2}{d_2 - d_1}, q_2 = 0, q_3 = \dfrac{d_1 + d_2}{d_1 - d_2}. \end{cases} \tag{4.48}$$

We aim to design a sliding surface (4.5) and a state feedback sliding mode controller (4.36) such that the closed-loop system states can be driven into a neighbourhood of the sliding surface (4.5) and the sliding motion is asymptotically stable. Applying Theorem 4.1 with $\rho_1 = 1$, $\rho_2 = 0$ and $J = -10$, the gain of the sliding surface (4.5) can be obtained as $K = \begin{bmatrix} 3.6119 & 1.3779 \end{bmatrix}$.

To show the effectiveness of the proposed SMC design method, simulations are conducted with different initial conditions. Under initial condition $x_0 = \begin{bmatrix} \pi/3 & 0 \end{bmatrix}^{\mathrm{T}}$ and $f(x(k),k) = \begin{bmatrix} 0.1\cos(k) & 0.2 \end{bmatrix} x(k)$, using the state feedback sliding mode controller in (4.36) with $\gamma = 0.2$ and $\eta = 0.05$, the resultant closed-loop system states and control input are shown in Figure 4.2 and Figure 4.3, respectively. The sliding surface is shown in Figure 4.4.

To further show the effectiveness of the proposed state feedback sliding mode controller design method, simulation studies on an autonomous land vehicle system will be presented in the following example.

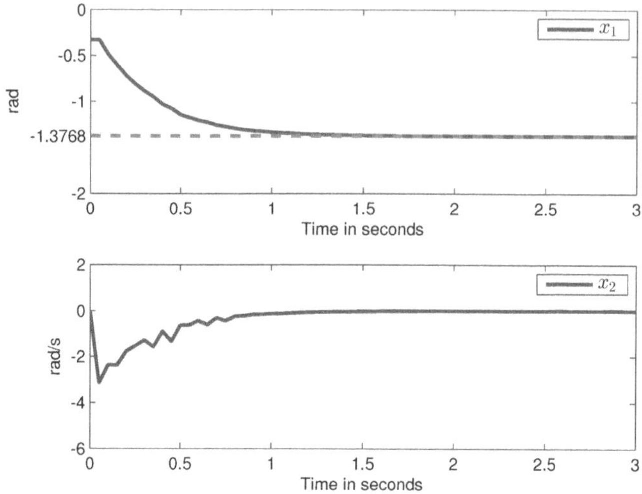

Figure 4.2 The closed-loop system states with sliding mode controller (4.36) in Example 4.1.

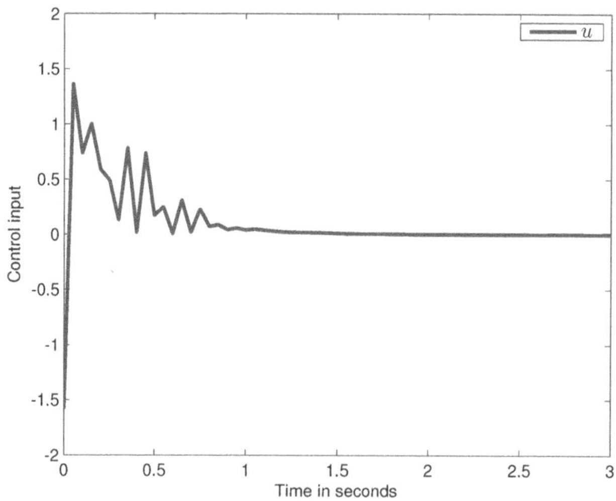

Figure 4.3 The control input with sliding mode controller (4.36) in Example 4.1.

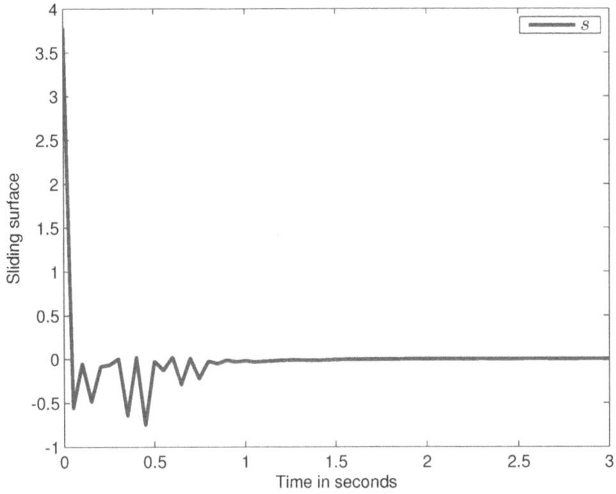

Figure 4.4 The sliding surface with sliding mode controller (4.36) in Example 4.1.

Example 4.2. Consider an autonomous land vehicle system same to the one in Example 3.2, which is also described by the following motion equation

$$\begin{bmatrix} \dot{\theta}(t) \\ \dot{\psi}(t) \\ \dot{\mathbf{y}}(t) \end{bmatrix} = \begin{bmatrix} 0 & 1 & 0 \\ 0 & -\frac{k}{l} & 0 \\ 0 & 0 & 0 \end{bmatrix} \begin{bmatrix} \theta(t) \\ \psi(t) \\ \mathbf{y}(t) \end{bmatrix} + \begin{bmatrix} 0 \\ 0 \\ v_0 \sin(\theta(t)) \end{bmatrix} + \begin{bmatrix} 0 \\ \frac{1}{l} \\ 0 \end{bmatrix} (u(t) + f(x(t),t)), \tag{4.49}$$

where the parameters are the same to the ones in Example 3.2.

Denote the state variables as $x_1(t) = \theta(t)$, $x_2(t) = \psi(t)$ and $x_3(t) = \mathbf{y}(t)$. Linearize the system (4.49) around three operating points $\begin{pmatrix} 0 & 0 & 0 \end{pmatrix}^{\mathrm{T}}$, $\begin{pmatrix} -\frac{2\pi}{5} & 0 & 0 \end{pmatrix}^{\mathrm{T}}$, and $\begin{pmatrix} \frac{2\pi}{5} & 0 & 0 \end{pmatrix}^{\mathrm{T}}$, and discretize the autonomous land vehicle system (4.49) with sampling period $T = 0.1$s, and then the system (4.49) can be characterized by the following T-S fuzzy affine model as,

Plant Rule $\mathscr{R}^l$: IF $x_1(k)$ is $\mathscr{F}_1^l$, THEN

$$x(k+1) = (A_l + \Delta A_l)x(k) + a_l + \Delta a_l + B(u(k) + f(x(k),k)),\ l \in \{1,2,3\}, \tag{4.50}$$

where the system matrices are

$$\begin{cases} A_1 = A_3 = \begin{bmatrix} 1 & 0.1 & 0 \\ 0 & 0.999 & 0 \\ 0.0309 & 0 & 1 \end{bmatrix}, A_2 = \begin{bmatrix} 1 & 0.1 & 0 \\ 0 & 0.999 & 0 \\ 0.1 & 0 & 0.1 \end{bmatrix}, \\ a_1 = \begin{bmatrix} 0 \\ 0 \\ -0.0563 \end{bmatrix}, a_2 = \begin{bmatrix} 0 \\ 0 \\ 0 \end{bmatrix}, a_3 = \begin{bmatrix} 0 \\ 0 \\ 0.0563 \end{bmatrix}, B = \begin{bmatrix} 0 \\ 0.1 \\ 0 \end{bmatrix}, \end{cases} \tag{4.51}$$

and the parameter uncertainties ΔA_l and Δa_l are given in the form of (1.3) with matrices H_{l1}, H_{l2}, and H_{l3} given as in (3.65).

The MFs of this example is also shown in Figure 4.1 with $d_1 = \frac{\pi}{15}$ and $d_2 = \frac{3\pi}{5}$. Based on the method given in Theorem 4.1, the transformation matrix T can be chosen as $T = \begin{bmatrix} 1 & 0 & 0 \\ -2 & 0 & -1 \\ 0 & 0.1 & 0 \end{bmatrix}$, and for this case, $z_1(k) = x_1(k)$. One can describe the corresponding three regions precisely by degenerate ellipsoids as in (2.16) with Q_i and q_i computed by (3.67).

Via using Theorem 4.1 with $\rho_1 = 1$, $\rho_2 = 0$, and $J = \begin{bmatrix} -1 & 2 \end{bmatrix}$, the gain of the sliding surface (4.5) can be obtained as $K = \begin{bmatrix} 2.9379 & 0.3556 & 1.1243 \end{bmatrix}$.

To show the effectiveness of the proposed SMC design method, simulations are conducted with different initial conditions. With initial condition $x_0 = \begin{bmatrix} \frac{\pi}{4} & 0 & 0 \end{bmatrix}^{\mathrm{T}}$ and disturbance $f(x(k),k) = \begin{bmatrix} 0.1\cos(k) & 0.2 & 0.3\sin(k) \end{bmatrix} x(k)$, adopting the sliding mode controller in (4.36) with $\gamma = 0.3$ and $\eta = 1.2$, Figure 4.5 and Figure 4.6 show the time responses of the system states and control input, respectively. The sliding surface is presented in Figure 4.7.

The simulation results clearly illustrate that the proposed sliding mode controller design method is effective.

4.4 STATE FEEDBACK DYNAMIC SMC DESIGN VIA T-S FUZZY AFFINE MODELS

In view of the system model in (4.1), it is worth mentioning that the aforementioned state feedback sliding mode controller design method and stability analysis results are proposed with a constraint that each local model of the T-S fuzzy system shares a common control input matrix, which is restrictive for some practical systems. In the following, a state feedback dynamic sliding mode controller design method will be proposed to remove this restriction.

4.4.1 MODEL DESCRIPTION: CASE II

Consider the first and third equation of the discrete-time T-S fuzzy affine system described as in (1.2), which are characterized by r fuzzy **IF-THEN** rules as,

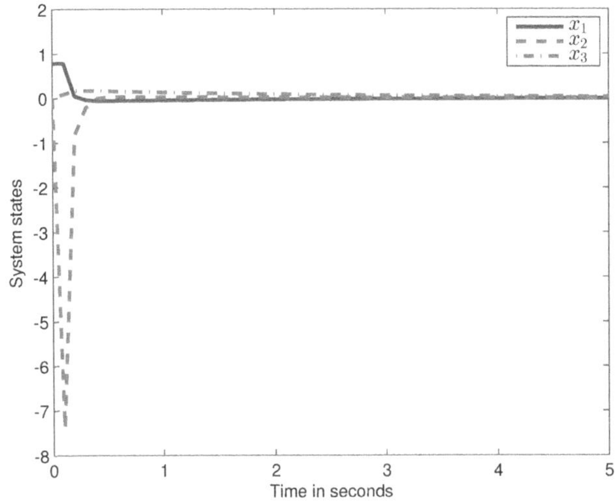

Figure 4.5 The closed-loop system states with sliding mode controller (4.36) in Example 4.2.

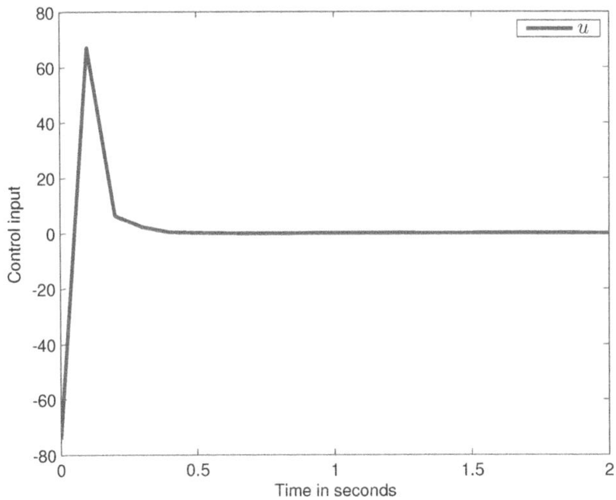

Figure 4.6 The control input with sliding mode controller (4.36) in Example 4.2.

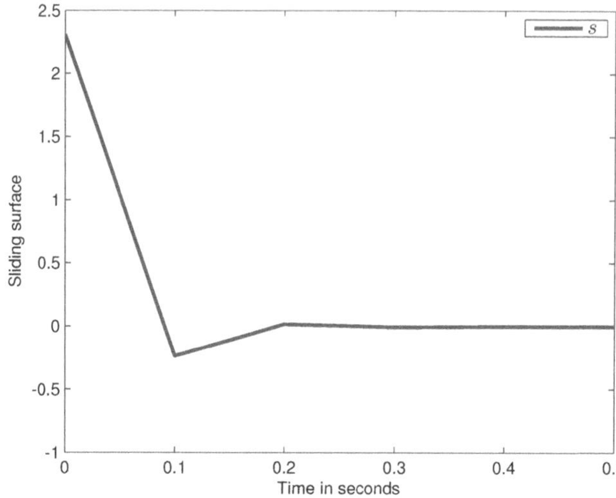

Figure 4.7 The sliding surface with sliding mode controller (4.36) in Example 4.2.

Plant Rule $\mathscr{R}^l$: IF $\theta_1(k)$ is $\mathscr{F}_1^l$ and $\cdots$ and $\theta_\varphi(k)$ is $\mathscr{F}_\varphi^l$, THEN

$$\begin{cases} x(k+1) = (A_l + \Delta A_l)x(k) + a_l + \Delta a_l + (B_l + \Delta B_l)u(k) + D_l w(k) \\ z(k) = L_l x(k) + N_l u(k),\ l \in \mathscr{L} := \{1, \ldots, r\}, \end{cases} \tag{4.52}$$

where $\mathscr{R}^l$ is the l-th fuzzy inference rule; $\mathscr{F}_\phi^l(\phi = 1, \ldots, \varphi)$ is fuzzy sets; r is the number of inference rules; $\theta(k) := [\theta_1(k), \ldots, \theta_\varphi(k)]$ are the measurable premise variables; $x(k) \in \Re^{n_x}$ is the system state; $u(k) \in \Re^{n_u}$ is the control input; $w(k) \in \Re^{n_w}$ is the external disturbance belonging to $l_2[0, \infty)$; $z(k) \in \Re^{n_z}$ is the regulated output; ΔA_l, ΔB_l and Δa_l are the parameter uncertainties satisfying conditions in (1.3)–(1.4).

Compared with the T-S fuzzy affine system model described as in (4.1), the control input matrices B_l of the T-S fuzzy affine system (4.52) are not common and contain parameter uncertainties. The external disturbance $w(k)$ also does not exist in a matched form.

Similar to Chapter 2, the global T-S fuzzy affine model can be obtained as,

$$\begin{cases} x(k+1) = (A(\mu) + \Delta A(\mu))x(k) + a(\mu) + \Delta a(\mu) + (B(\mu) + \Delta B(\mu))u(k) \\ \qquad + D(\mu)w(k) \\ z(k) = L(\mu)x(k) + N(\mu)u(k) \end{cases} \tag{4.53}$$

where

$$\left\{\begin{array}{l} A(\mu)=\sum\limits_{l=1}^{r}\mu_lA_l,\Delta A(\mu)=\sum\limits_{l=1}^{r}\mu_l\Delta A_l,a(\mu)=\sum\limits_{l=1}^{r}\mu_la_l,\Delta a(\mu)=\sum\limits_{l=1}^{r}\mu_l\Delta a_l,\\ B(\mu)=\sum\limits_{l=1}^{r}\mu_lB_l,\Delta B(\mu)=\sum\limits_{l=1}^{r}\mu_l\Delta B_l,D(\mu)=\sum\limits_{l=1}^{r}\mu_lD_l,L(\mu)=\sum\limits_{l=1}^{r}\mu_lL_l,\\ N(\mu)=\sum\limits_{l=1}^{r}\mu_lN_l. \end{array}\right. \tag{4.54}$$

The premise variable space can be decomposed by a similar line in Chapter 2. Introduce the following set $\mathcal{N}(i) := \{m|\mu_m[\theta(k)]>0,\ m\in\mathcal{L},\ \theta(k)\in S_i,\ i\in\mathcal{I}\}$ to describe the indices for local models utilized within the interpolation in each region S_i Based on the set $\mathcal{N}(i)$, the system (4.53) can be formulated as

$$\left\{\begin{array}{l} x(k+1)=(\mathcal{A}_i+\Delta\mathcal{A}_i)x(k)+\mathrm{a}_i+\Delta\mathrm{a}_i+(\mathcal{B}_i+\Delta\mathcal{B}_i)u(k)+\mathcal{D}_iw(k)\\ z(k)=\mathcal{L}_ix(k)+\mathcal{N}_iu(k),\theta(k)\in S_i,i\in\mathcal{I}, \end{array}\right. \tag{4.55}$$

where

$$\left\{\begin{array}{l} \mathcal{A}_i:=\sum\limits_{m\in\mathcal{N}(i)}\mu_mA_m,\Delta\mathcal{A}_i:=\sum\limits_{m\in\mathcal{N}(i)}\mu_m\Delta A_m,\mathrm{a}_i:=\sum\limits_{m\in\mathcal{N}(i)}\mu_ma_m,\\ \mathcal{B}_i:=\sum\limits_{m\in\mathcal{N}(i)}\mu_mB_m,\Delta\mathcal{B}_i:=\sum\limits_{m\in\mathcal{N}(i)}\mu_m\Delta B_m,\Delta\mathrm{a}_i:=\sum\limits_{m\in\mathcal{N}(i)}\mu_m\Delta a_m,\\ \mathcal{D}_i:=\sum\limits_{m\in\mathcal{N}(i)}\mu_mD_m,\mathcal{L}_i:=\sum\limits_{m\in\mathcal{N}(i)}\mu_mL_m,\mathcal{N}_i:=\sum\limits_{m\in\mathcal{N}(i)}\mu_mN_m, \end{array}\right. \tag{4.56}$$

in which $0<\mu_m[\theta(k)]\le 1$, $\sum_{m\in\mathcal{N}(i)}\mu_m[\theta(k)]=1$.

In the following, a state feedback dynamic sliding mode controller design approach will be proposed for T-S fuzzy affine system (4.52) to guarantee that the closed-loop system states can be driven into a neighbourhood of the sliding surface. Some asymptotic stability analysis conditions for the sliding motion will be proposed by using the CQLF-based and PQLFs-based methods, respectively. For a prescribed level of disturbance attenuation $\gamma>0$, asymptotical stability of the sliding motion can be ensured when $w(l)\equiv 0$, and for nonzero $w(k)\in l_2[0,\infty)$, it is satisfied that $\sum_{k=0}^{\infty}z^{\mathrm{T}}(k)z(k)<\gamma^2\sum_{k=0}^{\infty}w^{\mathrm{T}}(k)w(k)$ under zero initial condition $x(k)=0$.

4.4.2 SLIDING SURFACE DESIGN

For the T-S fuzzy affine system (4.52), design the sliding surface based on the system state and control input as

$$s(k)=Kx(k)+u(k)=0, \tag{4.57}$$

where $K\in\Re^{n_u\times n_x}$ is the sliding surface gain to be determined.

In view of (4.57), one has

$$u(k)=-Kx(k). \tag{4.58}$$

4.4.3 STABILITY AND ROBUST PERFORMANCE ANALYSIS OF SLIDING MOTION

In this section, the asymptotic stability and robust $\mathscr{H}_\infty$ performance analysis of the sliding motion will be conducted based on a common quadratic Lyapunov function (CQLF) and piecewise quadratic Lyapunov functions (PQLFs), respectively.

4.4.3.1 Stability and Robust Performance Analysis of Sliding Motion through CQLF

Based on (4.58) and the system (4.55), the sliding motion can be derived into the following form as

$$\begin{cases} x(k+1) = (\mathbb{A}_i + \Delta\mathbb{A}_i)x(k) + \mathrm{a}_i + \Delta\mathrm{a}_i + \mathscr{D}_i w(k) \\ z(k) = \mathbb{L}_i x(k), i \in \mathscr{I}, \end{cases} \tag{4.59}$$

where

$$\begin{cases} \mathbb{A}_i = \mathscr{A}_i - \mathscr{B}_i K, \Delta\mathbb{A}_i = \Delta\mathscr{A}_i - \Delta\mathscr{B}_i K, \\ \mathbb{L}_i = \mathscr{L}_i - \mathscr{N}_i K. \end{cases} \tag{4.60}$$

Similar to Chapter 2, it is assumed that an ellipsoid $\mathscr{E}_i$ can be used to outer approximate each polyhedral region S_i in (4.55) as $\mathscr{E}_i := \{x \| \|Q_i x + q_i\| \leq 1\}$, and the relationship given as in (2.83) holds for each region.

By using a CQLF-based method, sufficient conditions for the asymptotic stability and robust $\mathscr{H}_\infty$ performance analysis of the sliding mode dynamical system (4.59) will be proposed in Theorem 4.3.

Theorem 4.3. The sliding mode dynamic system (4.59) is asymptotically stable with robust $\mathscr{H}_\infty$ performance γ, if there exist matrices $0 < \bar{P} = \bar{P}^{\mathrm{T}} \in \Re^{n_x \times n_x}$, $\bar{K} \in \Re^{n_u \times n_x}$, and two sets of scalars $\{\varepsilon_i > 0, i \in \mathscr{I}\}$, $\{\kappa_i < 0, i \in \mathscr{I}_1\}$, such that the following conditions hold,

$$\begin{bmatrix} -\bar{P} & \star & \star & \star & \star \\ \mathbf{0} & -\gamma^2\mathbf{I} & \star & \star & \star \\ L_n\bar{P} - N_n\bar{K} & \mathbf{0} & -\mathbf{I} & \star & \star \\ A_n\bar{P} - B_n\bar{K} & D_n & \mathbf{0} & \Xi_{44} & \star \\ H_{n2}\bar{P} - H_{n4}\bar{K} & \mathbf{0} & \mathbf{0} & \mathbf{0} & -\varepsilon_i\mathbf{I} \end{bmatrix} < 0, i \in \mathscr{I}_0, n \in \mathscr{N}(i), \tag{4.61}$$

$$\begin{bmatrix} \kappa_i(q_i q_i^{\mathrm{T}} - \mathbf{I}) & \star & \star & \star & \star & \star \\ \bar{P}Q_i^{\mathrm{T}} & -\bar{P} & \star & \star & \star & \star \\ \mathbf{0} & \mathbf{0} & -\gamma^2\mathbf{I} & \star & \star & \star \\ \mathbf{0} & L_n\bar{P} - N_n\bar{K} & \mathbf{0} & -\mathbf{I} & \star & \star \\ \kappa_i a_n q_i^{\mathrm{T}} & A_n\bar{P} - B_n\bar{K} & D_n & \mathbf{0} & \Xi_{44} + \kappa_i a_n a_n^{\mathrm{T}} & \star \\ \kappa_i H_{n3} q_i^{\mathrm{T}} & H_{n2}\bar{P} - H_{n4}\bar{K} & \mathbf{0} & \mathbf{0} & \kappa_i H_{n3} a_n^{\mathrm{T}} & \kappa_i H_{n3} H_{n3}^{\mathrm{T}} - \varepsilon_i \mathbf{I} \end{bmatrix} < 0,$$
$$i \in \mathscr{I}_1, n \in \mathscr{N}(i), \tag{4.62}$$

where $\Xi_{44} = -\bar{P} + \varepsilon_i H_{n1} H_{n1}^{\mathrm{T}}$.

The gain of the sliding surface (4.57) can be attained by

$$K = \bar{K}\bar{P}^{-1}. \tag{4.63}$$

Proof. Without loss of generality, only the more complicated case for $i \in \mathscr{I}_1$ will be proved . For system (4.59), consider a CQLF as

$$V(k) = x^{\mathrm{T}}(k)Px(k), \tag{4.64}$$

where $0 < P = P^{\mathrm{T}} \in \Re^{n_x \times n_x}$.

The system (4.59) is asymptotically stable with robust $\mathscr{H}_\infty$ performance γ if the following inequality holds,

$$V(k+1) - V(k) + z^{\mathrm{T}}(k)z(k) - \gamma^2 w^{\mathrm{T}}(k)w(k) < 0. \tag{4.65}$$

Define $\upsilon(k) = \begin{bmatrix} 1 & \bar{x}^{\mathrm{T}}(k) & w^{\mathrm{T}}(k) \end{bmatrix}^{\mathrm{T}}$. Along the trajectories of system (4.59), rewrite the inequality (4.65) as

$$\begin{bmatrix} 1 \\ x(k) \\ w(k) \end{bmatrix}^{\mathrm{T}} \left\{ \begin{bmatrix} \mathbf{0} \\ \mathrm{a}_i^{\mathrm{T}} + \Delta\mathrm{a}_i^{\mathrm{T}} \\ \mathbb{A}_i^{\mathrm{T}} + \Delta\mathbb{A}_i^{\mathrm{T}} \\ \mathscr{D}_i^{\mathrm{T}} \end{bmatrix} P(\star) + \begin{bmatrix} \mathbf{0} & \star & \star \\ \mathbf{0} & P - \mathbb{L}_i^{\mathrm{T}}\mathbb{L}_i & \star \\ \mathbf{0} & \mathbf{0} & -\gamma^2\mathbf{I} \end{bmatrix} \right\} \begin{bmatrix} 1 \\ x(k) \\ w(k) \end{bmatrix} < 0. \tag{4.66}$$

Noticing the partition information described as in (2.83) and using the S-procedure, then the following inequality implies (4.66) for $\rho_i < 0, i \in \mathscr{I}_1$,

$$\begin{bmatrix} \rho_i(q_i^{\mathrm{T}}q_i - 1) & \star & \star & \star \\ \rho_i Q_i^{\mathrm{T}} q_i & P + \mathbb{L}_i^{\mathrm{T}}\mathbb{L}_i + \rho_i Q_i^{\mathrm{T}} Q_i & \star & \star \\ \mathbf{0} & \mathbf{0} & -\gamma^2\mathbf{I} & \star \\ \mathrm{a}_i + \Delta\mathrm{a}_i & \mathbb{A}_i + \Delta\mathbb{A}_i & \mathscr{D}_i & -P^{-1} \end{bmatrix} < 0. \tag{4.67}$$

Extract the MFs in (4.67) and use Lemma A1 to handle the parameter uncertainties. Multiply $\mathrm{diag}\{1, \bar{P}, \mathbf{I}_{(n_w+n_z+n_x+p_2)}\}$ on both sides of (4.67) with $\bar{P} = P^{-1}$, and then one has,

$$\begin{bmatrix} \rho_i(q_i^{\mathrm{T}}q_i - 1) & \star & \star & \star & \star & \star \\ \rho_i \bar{P} Q_i^{\mathrm{T}} q_i & -P + \rho_i \bar{P} Q_i^{\mathrm{T}} Q_i \bar{P} & \star & \star & \star & \star \\ \mathbf{0} & \mathbf{0} & -\gamma^2\mathbf{I} & \star & \star & \star \\ \mathbf{0} & L_n\bar{P} - N_n\bar{K} & \mathbf{0} & -\mathbf{I} & \star & \star \\ a_n & A_n\bar{P} - B_n\bar{K} & D_n & \mathbf{0} & -\bar{P} + \varepsilon_i H_{n1}H_{n1}^{\mathrm{T}} & \star \\ H_{n3} & H_{n2}\bar{P} - H_{n4}\bar{K} & \mathbf{0} & \mathbf{0} & \mathbf{0} & -\varepsilon_i\mathbf{I} \end{bmatrix} < 0, \quad i \in \mathscr{I}_1, \tag{4.68}$$

where $\bar{K} = K\bar{P}$.

In view of the terms $\rho_i \bar{P} Q_i^{\mathrm{T}} Q_i \bar{P}$ and $\rho_i \bar{P} Q_i^{\mathrm{T}} q_i$ in (4.68), scalars ρ_i are coupled with the Lyapunov matrix $\bar{P}$. In order to solve the inequality (4.68) by a linearization procedure, the key is to decouple the Lyapunov matrix $\bar{P}$ from the scalars ρ_i firstly. Based on Schur complement, inequality (4.68) can be rewritten as,

$$\begin{bmatrix} -P+\rho_i\bar{P}Q_i^{\mathrm{T}}Q_i\bar{P} & \star & \star & \star & \star \\ \mathbf{0} & -\gamma^2\mathbf{I} & \star & \star & \star \\ L_n\bar{P}-N_n\bar{K} & \mathbf{0} & -\mathbf{I} & \star & \star \\ A_n\bar{P}-B_n\bar{K} & D_n & \mathbf{0} & -\bar{P}+\varepsilon_iH_{n1}H_{n1}^{\mathrm{T}} & \star \\ H_{n2}\bar{P}-H_{n4}\bar{K} & \mathbf{0} & \mathbf{0} & \mathbf{0} & -\varepsilon_i\mathbf{I} \end{bmatrix} + \begin{bmatrix} \rho_i\bar{P}Q_i^{\mathrm{T}}q_i \\ \mathbf{0} \\ \mathbf{0} \\ a_n \\ H_{n3} \end{bmatrix} (\rho_i(1-q_i^{\mathrm{T}}q_i))^{-1}(\star) < 0. \tag{4.69}$$

Consider the Matrix Inversion Lemma described given as in (1.10), and one has,

$$\begin{cases} \rho_i\bar{P}Q_i^{\mathrm{T}}q_i(1-q_i^{\mathrm{T}}q_i)^{-1}q_i^{\mathrm{T}}Q_i\bar{P} = \rho_i\bar{P}Q_i^{\mathrm{T}}(\mathbf{I}-q_iq_i^{\mathrm{T}})^{-1}Q_i\bar{P} - \rho_i\bar{P}Q_i^{\mathrm{T}}Q_i\bar{P}, \\ a_n(1-q_i^{\mathrm{T}}q_i)^{-1}q_i^{\mathrm{T}}Q_i\bar{P} = a_nq_i^{\mathrm{T}}(\mathbf{I}-q_iq_i^{\mathrm{T}})^{-1}Q_i\bar{P}, \\ H_{n3}(1-q_i^{\mathrm{T}}q_i)^{-1}q_i^{\mathrm{T}}Q_i\bar{P} = H_{n3}q_i^{\mathrm{T}}(\mathbf{I}-q_iq_i^{\mathrm{T}})^{-1}Q_i\bar{P}, \\ \rho_i^{-1}a_n(1-q_i^{\mathrm{T}}q_i)^{-1}a_n^{\mathrm{T}} = \rho_i^{-1}a_na_n^{\mathrm{T}} + \rho_i^{-1}a_nq_i^{\mathrm{T}}(\mathbf{I}-q_iq_i^{\mathrm{T}})^{-1}q_ia_n^{\mathrm{T}}, \\ \rho_i^{-1}H_{n3}(1-q_i^{\mathrm{T}}q_i)^{-1}H_{n3}^{\mathrm{T}} = \rho_i^{-1}H_{n3}H_{n3}^{\mathrm{T}} + \rho_i^{-1}H_{n3}q_i^{\mathrm{T}}(\mathbf{I}-q_iq_i^{\mathrm{T}})^{-1}q_iH_{n3}^{\mathrm{T}}, \\ \rho_i^{-1}a_n(1-q_i^{\mathrm{T}}q_i)^{-1}H_{n3}^{\mathrm{T}} = \rho_i^{-1}a_nH_{n3}^{\mathrm{T}} + \rho_i^{-1}a_nq_i^{\mathrm{T}}(\mathbf{I}-q_iq_i^{\mathrm{T}})^{-1}q_iH_{n3}^{\mathrm{T}}. \end{cases} \tag{4.70}$$

Introduce $\kappa_i = \rho_i^{-1}$ and use Schur complement, condition (4.62) can be obtained based on (4.69)–(4.70). This ends the proof. ■

Via a CQLF-based method, sufficient conditions for asymptotic stability and robust $\mathscr{H}_\infty$ performance analysis of the sliding motion are proposed in Theorem 4.3. Alternatively, by using a PQLFs-based approach and a descriptor system, new results on the asymptotic stability and robust performance analysis of the sliding motion will be obtained nextly, and the sliding surface gain can also be calculated under a convex optimization framework.

4.4.3.2 Stability and Robust Performance Analysis of Sliding Motion through PQLF

To analyze the sliding motion through a PQLFs-based approach, the sliding motion can be described by the following descriptor system based on system (4.55) and (4.58),

$$\begin{cases} x(k+1) = (\mathscr{A}_i+\Delta\mathscr{A}_i)x(k) + \mathrm{a}_i + \Delta\mathrm{a}_i + (\mathscr{B}_i+\Delta\mathscr{B}_i)u(k) + \mathscr{D}_iw(k) \\ 0\times u(k+1) = -Kx(k) - u(k) \\ z(k) = \mathscr{L}_ix(k) + \mathscr{N}_iu(k), i\in\mathscr{I}. \end{cases} \tag{4.71}$$

Reformulate the system (4.71) as

$$\begin{cases} E\bar{x}(k+1) = (\bar{\mathscr{A}}_i+\Delta\bar{\mathscr{A}}_i)\bar{x}(k) + \bar{\mathrm{a}}_i + \Delta\bar{\mathrm{a}}_i + \bar{\mathscr{D}}_iw(k) \\ z(k) = \bar{\mathscr{L}}_i\bar{x}(k), i\in\mathscr{I}, \end{cases} \tag{4.72}$$

where

$$\begin{cases} E = \begin{bmatrix} \mathbf{I}_{n_x} & \mathbf{0} \\ \mathbf{0} & \mathbf{0} \end{bmatrix}, \bar{x}(k) = \begin{bmatrix} x(k) \\ u(k) \end{bmatrix}, \bar{\mathscr{A}}_i = \begin{bmatrix} \mathscr{A}_i & \mathscr{B}_i \\ -K & -\mathbf{I} \end{bmatrix}, \Delta\bar{\mathscr{A}}_i = \begin{bmatrix} \Delta\mathscr{A}_i & \Delta\mathscr{B}_i \\ \mathbf{0} & \mathbf{0} \end{bmatrix}, \\ \bar{\mathrm{a}}_i = \begin{bmatrix} \mathrm{a}_i \\ \mathbf{0} \end{bmatrix}, \Delta\bar{\mathrm{a}}_i = \begin{bmatrix} \Delta\mathrm{a}_i \\ \mathbf{0} \end{bmatrix}, \bar{\mathscr{D}}_i = \begin{bmatrix} \mathscr{D}_i \\ \mathbf{0} \end{bmatrix}, \bar{\mathscr{L}}_i = \begin{bmatrix} \mathscr{L}_i & \mathscr{N}_i \end{bmatrix}. \end{cases} \tag{4.73}$$

By combining the first equation of system (4.55) with the sliding surface (4.58), the descriptor system (4.72) can characterize the dynamics of the sliding motion. For future use, a new set $\Omega := \{(i,j)|x(k) \in S_i, x(k+1) \in S_j, i,j \in \mathscr{I}\}$ is introduced to describe region transitions.

Theorem 4.4. The sliding mode dynamical system (4.72) is asymptotically stable with robust $\mathscr{H}_\infty$ performance γ, if there exist matrices $0 < P_i = P_i^{\mathrm{T}} \in \Re^{(n_x+n_u)\times(n_x+n_u)}$, $\{Y_{i1}, W_{i1}\} \in \Re^{n_x \times n_x}$, $\{Y_{i2}, W_{i2}\} \in \Re^{n_u \times n_x}$, $i \in \mathscr{I}$, $Y_0 \in \Re^{n_u \times n_u}$, $\bar{K} \in \Re^{n_u \times n_x}$, and scalars $\{\varepsilon_i > 0, i \in \mathscr{I}\}$, $\{\kappa_i < 0, i \in \mathscr{I}_1\}$ exist, such that the following conditions hold,

$$\begin{bmatrix} \Psi_{11} & \star & \star & \star \\ \Psi_{21} - W_i^{\mathrm{T}} & P_j - Y_i - Y_i^{\mathrm{T}} & \star & \star \\ \Psi_{31} & \Psi_{32} & -\gamma^2\mathbf{I} & \star \\ \Psi_{41} & \Psi_{42} & \mathbf{0} & -\varepsilon_i\mathbf{I} \end{bmatrix} < 0, i \in \mathscr{I}_0, m \in \mathscr{N}(i), \tag{4.74}$$

$$\begin{bmatrix} \kappa_i(q_i^{\mathrm{T}}q_i - 1) + \varepsilon_i H_{m3}^{\mathrm{T}} H_{m3} & \star & \star & \star & \star \\ \kappa_i \bar{Q}_i^{\mathrm{T}} q_i + \Psi_{10} & \rho_i \bar{Q}_i^{\mathrm{T}} \bar{Q}_i + \Psi_{11} & \star & \star & \star \\ \begin{bmatrix} Y_{i1} \\ Y_{i2} \end{bmatrix} a_m & \Psi_{21} - W_i^{\mathrm{T}} & P_j - Y_i - Y_i^{\mathrm{T}} & \star & \star \\ \mathbf{0} & \Psi_{31} & \Psi_{32} & -\gamma^2\mathbf{I} & \star \\ \mathbf{0} & \Psi_{41} & \Psi_{42} & \mathbf{0} & -\varepsilon_i\mathbf{I} \end{bmatrix} < 0,$$
$$i \in \mathscr{I}_1, m \in \mathscr{N}(i), \tag{4.75}$$

where

$$\begin{cases} \Psi_{11} = \mathtt{Sym}\left\{ \begin{bmatrix} W_{i1} \\ W_{i2} \end{bmatrix} \begin{bmatrix} A_m & B_m \end{bmatrix} - \begin{bmatrix} J_2 \\ \rho_2\mathbf{I} \end{bmatrix} \begin{bmatrix} \bar{K} & Y_0 \end{bmatrix} \right\} + \varepsilon_i \begin{bmatrix} H_{m2}^{\mathrm{T}} \\ H_{m4}^{\mathrm{T}} \end{bmatrix} (\star) \\ \qquad -E^{\mathrm{T}} P_i E + \bar{L}_m^{\mathrm{T}} \bar{L}_m, \\ \Psi_{21} = \begin{bmatrix} Y_{i1} \\ Y_{i2} \end{bmatrix} \begin{bmatrix} A_m & B_m \end{bmatrix} - \begin{bmatrix} J_1 \\ \rho_1\mathbf{I} \end{bmatrix} \begin{bmatrix} \bar{K} & Y_0 \end{bmatrix}, \\ \Psi_{31} = D_m^{\mathrm{T}} \begin{bmatrix} W_{i1}^{\mathrm{T}} & W_{i2}^{\mathrm{T}} \end{bmatrix}, \Psi_{32} = D_m^{\mathrm{T}} \begin{bmatrix} Y_{i1}^{\mathrm{T}} & Y_{i2}^{\mathrm{T}} \end{bmatrix}, \\ \Psi_{41} = H_{m1}^{\mathrm{T}} \begin{bmatrix} W_{i1}^{\mathrm{T}} & W_{i2}^{\mathrm{T}} \end{bmatrix}, \Psi_{42} = H_{m1}^{\mathrm{T}} \begin{bmatrix} Y_{i1}^{\mathrm{T}} & Y_{i2}^{\mathrm{T}} \end{bmatrix}, \\ \Psi_{10} = \begin{bmatrix} W_{i1} \\ W_{i2} \end{bmatrix} a_m + \varepsilon_i \begin{bmatrix} H_{m2}^{\mathrm{T}} \\ H_{m4}^{\mathrm{T}} \end{bmatrix} H_{m3}, \\ Y_i = \begin{bmatrix} Y_{i1} & J_1 Y_0 \\ Y_{i2} & \rho_1 Y_0 \end{bmatrix}, W_i = \begin{bmatrix} W_{i1} & J_2 Y_0 \\ W_{i2} & \rho_2 Y_0 \end{bmatrix}, \\ \bar{L}_m = \begin{bmatrix} L_m & N_m \end{bmatrix}, \bar{Q}_i = \begin{bmatrix} Q_i & \mathbf{0}_{1\times n_u} \end{bmatrix}, \end{cases} \tag{4.76}$$

and $\{J_1, J_2\} \in \Re^{n_x \times n_u}$ are arbitrary matrices, and $\rho_1 \neq 0$, ρ_2 are constant scalars.

The gain of the sliding surface (4.57) can be calculated by

$$K = Y_0^{-1}\bar{K}. \tag{4.77}$$

Proof. Notice that the condition (4.74) for $i \in \mathscr{I}_0$ is actually a special case of the condition (4.75) for $i \in \mathscr{I}_1$. Thus, without loss of generality, only the more complicated case for $i \in \mathscr{I}_1$ will be proved. Consider the following PQLF as

$$V(k) = \bar{x}^{\mathrm{T}}(k)E^{\mathrm{T}}P_iE\bar{x}(k), i \in \mathscr{I}, \tag{4.78}$$

where $0 < P_i = P_i^{\mathrm{T}} \in \Re^{(n_x+n_u)\times(n_x+n_u)}$.

Based on the PQLF in (4.78), if it is satisfied that

$$V(k+1) - V(k) + z^{\mathrm{T}}(k)z(k) - \gamma^2 w^{\mathrm{T}}(k)w(k) < 0, \tag{4.79}$$

the sliding mode dynamical system (4.72) is asymptotically stable with robust $\mathscr{H}_\infty$ performance γ.

Then rewrite the inequality (4.79) as,

$$\begin{bmatrix} 1 \\ \bar{x}(k) \\ E\bar{x}(k+1) \\ w(k) \end{bmatrix}^{\mathrm{T}} \begin{bmatrix} 0 & \star & \star & \star \\ \mathbf{0} & \bar{\mathscr{L}}_i^{\mathrm{T}}\bar{\mathscr{L}}_i - E^{\mathrm{T}}P_iE & \star & \star \\ \mathbf{0} & \mathbf{0} & P_j & \star \\ \mathbf{0} & \mathbf{0} & \mathbf{0} & -\gamma^2\mathbf{I} \end{bmatrix} \begin{bmatrix} 1 \\ \bar{x}(k) \\ E\bar{x}(k+1) \\ w(k) \end{bmatrix} < 0, \quad (i,j) \in \Omega. \tag{4.80}$$

Using the S-procedure based on the partition information described as in (2.83), the following inequality implies (4.80) for $\rho_i < 0$, $i \in \mathscr{I}_1$,

$$\begin{bmatrix} 1 \\ \bar{x}(k) \\ E\bar{x}(k+1) \\ w(k) \end{bmatrix}^{\mathrm{T}} \Pi \begin{bmatrix} 1 \\ \bar{x}(k) \\ E\bar{x}(k+1) \\ w(k) \end{bmatrix} < 0, i \in \mathscr{I}_1, (i,j) \in \Omega, \tag{4.81}$$

where

$$\begin{cases} \Pi = \begin{bmatrix} \kappa_i(q_i^{\mathrm{T}}q_i - 1) & \star & \star & \star \\ \kappa_i\bar{Q}_i^{\mathrm{T}}q_i & \kappa_i\bar{Q}_i^{\mathrm{T}}\bar{Q}_i + \bar{\mathscr{L}}_i^{\mathrm{T}}\bar{\mathscr{L}}_i - E^{\mathrm{T}}P_iE & \star & \star \\ \mathbf{0} & \mathbf{0} & P_j & \star \\ \mathbf{0} & \mathbf{0} & \mathbf{0} & -\gamma^2\mathbf{I} \end{bmatrix}, \\ \bar{Q}_i = \begin{bmatrix} Q_i & \mathbf{0}_{1\times n_u} \end{bmatrix}. \end{cases} \tag{4.82}$$

Along the state trajectories of system (4.72), the following condition holds for arbitrary matrices $\mathscr{G}_i$, $i \in \mathscr{I}$,

$$\begin{aligned} 0 &= 2\mathscr{G}_i\left\{\bar{\mathrm{a}}_i + \Delta\bar{\mathrm{a}}_i + (\bar{\mathscr{A}}_i + \Delta\bar{\mathscr{A}}_i)\bar{x}(k) + \bar{\mathscr{D}}_iw(k) - E\bar{x}(k+1)\right\} \\ &= 2\mathscr{G}_i\begin{bmatrix} \bar{\mathrm{a}}_i + \Delta\bar{\mathrm{a}}_i & \bar{\mathscr{A}}_i + \Delta\bar{\mathscr{A}}_i & -\mathbf{I} & \bar{\mathscr{D}}_i \end{bmatrix} \begin{bmatrix} 1 \\ \bar{x}(k) \\ E\bar{x}(k+1) \\ w(k) \end{bmatrix}. \end{aligned} \tag{4.83}$$

Combining (4.81) with (4.83), one has,

$$\begin{bmatrix} 1 \\ \bar{x}(k) \\ E\bar{x}(k+1) \\ w(k) \end{bmatrix}^{\mathrm{T}} (\Pi + \mathrm{Sym}\{\mathscr{G}_i\mathscr{A}_i\}) \begin{bmatrix} 1 \\ \bar{x}(k) \\ E\bar{x}(k+1) \\ w(k) \end{bmatrix} < 0, i \in \mathscr{I}_1, (i,j) \in \Omega, \tag{4.84}$$

where

$$\mathscr{A}_i = \begin{bmatrix} \bar{\mathrm{a}}_i + \Delta\bar{\mathrm{a}}_i & \bar{\mathscr{A}}_i + \Delta\bar{\mathscr{A}}_i & -\mathbf{I} & \bar{\mathscr{D}}_i \end{bmatrix}. \tag{4.85}$$

It follows from (4.85) that the following inequality implies (4.79),

$$\Pi + \mathrm{Sym}\{\mathscr{G}_i\mathscr{A}_i\} < 0, i \in \mathscr{I}_1, (i,j) \in \Omega. \tag{4.86}$$

Expand the MFs in (4.86), and then one has,

$$\begin{bmatrix} \kappa_i(q_i^{\mathrm{T}}q_i - 1) & \star & \star & \star \\ \kappa_i\bar{Q}_i^{\mathrm{T}}q_i & \kappa_i\bar{Q}_i^{\mathrm{T}}\bar{Q}_i + \bar{L}_m^{\mathrm{T}}\bar{L}_m - E^{\mathrm{T}}P_iE & \star & \star \\ \mathbf{0} & \mathbf{0} & P_j & \star \\ \mathbf{0} & \mathbf{0} & \mathbf{0} & -\gamma^2\mathbf{I} \end{bmatrix} + \mathrm{Sym}\{\mathscr{G}_n\mathscr{A}_m\} < 0,$$
$$m \in \mathscr{N}(i), i \in \mathscr{I}_1, (i,j) \in \Omega, \tag{4.87}$$

where

$$\begin{cases} \mathscr{A}_m = \begin{bmatrix} \bar{a}_m + \Delta\bar{a}_m & \bar{A}_m + \Delta\bar{A}_m & -\mathbf{I} & \bar{D}_m \end{bmatrix}, \\ \bar{A}_m + \Delta\bar{A}_m = \begin{bmatrix} A_m + \Delta A_m & B_m + \Delta B_m \\ -K & -\mathbf{I} \end{bmatrix}, \\ \bar{a}_m + \Delta\bar{a}_m = \begin{bmatrix} a_m + \Delta a_m \\ \mathbf{0} \end{bmatrix}, \bar{L}_m = \begin{bmatrix} L_m & N_m \end{bmatrix}. \end{cases} \tag{4.88}$$

It can be seen from (4.88) that the sliding surface gain K does not exist in the first row of $\bar{a}_m + \Delta\bar{a}_m$ and $\bar{A}_m + \Delta\bar{A}_m$. For numerical convexification, specify $\mathscr{G}_i = \begin{bmatrix} \mathbf{0} & W_i^{\mathrm{T}} & Y_i^{\mathrm{T}} & \mathbf{0} \end{bmatrix}^{\mathrm{T}}$, where the slack variable matrices Y_i and W_i are given as,

$$Y_i = \begin{bmatrix} Y_{i1} & J_1Y_0 \\ Y_{i2} & \rho_1Y_0 \end{bmatrix}, W_i = \begin{bmatrix} W_{i1} & J_2Y_0 \\ W_{i2} & \rho_2Y_0 \end{bmatrix} \tag{4.89}$$

with $\rho_1 \neq 0$ and $\rho_2 \neq 0$ being scalar parameters, and $\{J_1, J_2\} \in \Re^{n_x \times n_u}$ are arbitrary matrices.

Define

$$\bar{K} = Y_0K. \tag{4.90}$$

Based on (4.89) and (4.90), the inequality (4.87) can be derived as,

$$\begin{bmatrix} \kappa_i(q_i^{\mathrm{T}}q_i - 1) & \star & \star & \star \\ \kappa_i\bar{Q}_i^{\mathrm{T}}q_i + \Lambda_{10} & \Lambda_{11} & \star & \star \\ \Lambda_{20} & \Lambda_{21} - W_i^{\mathrm{T}} & P_j - Y_i - Y_i^{\mathrm{T}} & \star \\ \mathbf{0} & \Lambda_{31} & \Lambda_{32} & -\gamma^2\mathbf{I} \end{bmatrix} < 0, \tag{4.91}$$
$$m \in \mathscr{N}(i), i \in \mathscr{I}_1, (i,j) \in \Omega,$$

where

$$\begin{cases} \Lambda_{11} = \mathrm{Sym}\left\{\begin{bmatrix} W_{i1} \\ W_{i2} \end{bmatrix}\begin{bmatrix} A_m+\Delta A_m & B_m+\Delta B_m \end{bmatrix} - \begin{bmatrix} J_2 \\ \rho_2\mathbf{I} \end{bmatrix}\begin{bmatrix} \bar{K} & Y_0 \end{bmatrix}\right\} \\ \quad +\kappa_i\bar{Q}_i^{\mathrm{T}}\bar{Q}_i - E^{\mathrm{T}}P_iE + \bar{L}_m^{\mathrm{T}}\bar{L}_m, \\ \Lambda_{21} = \begin{bmatrix} Y_{i1} \\ Y_{i2} \end{bmatrix}\begin{bmatrix} A_m+\Delta A_m & B_m+\Delta B_m \end{bmatrix} - \begin{bmatrix} J_1 \\ \rho_1\mathbf{I} \end{bmatrix}\begin{bmatrix} \bar{K} & Y_0 \end{bmatrix}, \\ \Lambda_{31} = D_m^{\mathrm{T}}\begin{bmatrix} W_{i1}^{\mathrm{T}} & W_{i2}^{\mathrm{T}} \end{bmatrix}, \Lambda_{32} = D_m^{\mathrm{T}}\begin{bmatrix} Y_{i1}^{\mathrm{T}} & Y_{i2}^{\mathrm{T}} \end{bmatrix}, \\ \Lambda_{10} = \begin{bmatrix} W_{i1} \\ W_{i2} \end{bmatrix}(a_m+\Delta a_m), \Lambda_{20} = \begin{bmatrix} V_{n1} \\ V_{n2} \end{bmatrix}(a_m+\Delta a_m). \end{cases} \tag{4.92}$$

Utilizing Lemma A1 to tackle the parameter uncertainties in (4.91), then for scalars $\varepsilon_i > 0$, the inequality (4.75) implies (4.91) based on Schur complement.

Moreover, the conditions in (4.75)–(4.76) indicate $\rho_1 Y_0 + \rho_1 Y_0^{\mathrm{T}} - P_{j3} > 0$, which means that Y_0 is nonsingular. Therefore, the sliding surface gain can be calculated by (4.77). The proof is ended. ■

Via augmenting the control input with system states, a singular piecewise affine system method is proposed to handle the coupling problem between the system matrices and the sliding surface gain. By means of introducing auxiliary matrix multipliers, stability and robust performance analysis for the sliding motion can be carried out by using some convexification techniques.

4.5 STATE FEEDBACK DYNAMIC SLIDING MODE CONTROLLER DESIGN

4.5.1 DYNAMIC SLIDING MODE CONTROLLER DESIGN

Theorem 4.5. For the discrete-time T-S fuzzy affine system (4.52) or equivalently (4.55) with the sliding surface given as in (4.57), the resulting closed-loop system state trajectories can be driven a neighbourhood of the sliding surface (4.57) by the following state feedback dynamic sliding mode controller,

$$\begin{aligned} u(k+1) = u(k) + Kx(k) - \sum_{l=1}^{r}\mu_l K(A_l x(k) + a_l + B_l u(k)) \\ -(\sigma + \vartheta(k))\mathrm{sgn}(s(k)), \end{aligned} \tag{4.93}$$

where $\sigma > 0$ is a given positive constant, and

$$\begin{aligned} \vartheta(k) = \sum_{l=1}^{r}\mu_l \{\|KH_{l1}\| \cdot \|H_{l2}\| \cdot \|x(k)\| + \|KH_{l1}\| \cdot \|H_{l3}\| \\ + \|KH_{l1}\| \cdot \|H_{l4}\| \cdot \|u(k)\| + \|KD_l\| \varpi(k)\}, \end{aligned} \tag{4.94}$$

where $\varpi(k)$ represents the uniform upper-bound of external disturbance $w(k)$.

Proof. Consider the following Lyapunov function,

$$V(k) = \frac{1}{2}s^{\mathrm{T}}(k)s(k). \tag{4.95}$$

The incremental $\Delta V(k)$ is obtained as

$$\begin{aligned}\Delta V(k) &= V(k+1) - V(k) \\ &= s^{\mathrm{T}}(k)\Delta s(k) + \Gamma,\end{aligned} \tag{4.96}$$

where $\Delta s(k) = s(k+1) - s(k)$, $s(k+1) = Kx(k+1) + u(k+1)$, $\Gamma = \frac{1}{2}\Delta s^{\mathrm{T}}(k)\Delta s(k)$.

Substitute the dynamic sliding mode controller (4.93) into (4.96), and denote $\Delta u(k) = u(k+1) - u(k)$. Then one has,

$$\begin{aligned}\Delta V(k) &= s^{\mathrm{T}}(k)\left\{Kx(k+1) + u(k+1) - Kx(k) - u(k)\right\} + \Gamma \\ &= s^{\mathrm{T}}(k)\left\{\sum_{l=1}^{r}\mu_l K\left[(A_l + \Delta A_l)x(k) + a_l + \Delta a_l + (B_l + \Delta B_l)u(k) + D_l w(k)\right]\right. \\ &\qquad \left. - Kx(k) + \Delta u(k)\right\} + \Gamma \\ &= s^{\mathrm{T}}(k)\sum_{l=1}^{r}\mu_l K\left\{\Delta A_l x(k) + \Delta a_l + \Delta B_l u(k) + D_l w(k)\right\} - \vartheta(k)\|s(k)\| \\ &\quad - \sigma\|s(k)\| + \Gamma \\ &\leq \|s(k)\|\sum_{l=1}^{r}\mu_l\left\{\|K\Delta A_l\|\cdot\|x(k)\| + \|K\Delta a_l\| + \|K\Delta B_l\|\cdot\|u(k)\|\right. \\ &\qquad \left. + \|KD_l\|\varpi(k)\right\} - \vartheta(k)\|s(k)\| - \sigma\|s(k)\| + \Gamma \\ &\leq \|s(k)\|\sum_{l=1}^{r}\mu_l\left\{\|KH_{l1}\|\cdot\|H_{l2}\|\cdot\|x(k)\| + \|KH_{l1}\|\cdot\|H_{l3}\|\right. \\ &\quad \left. + \|KH_{l1}\|\cdot\|H_{l4}\|\cdot\|u(k)\| + \|KD_l\|\varpi(k)\right\} - \vartheta(k)\|s(k)\| - \sigma\|s(k)\| + \Gamma.\end{aligned} \tag{4.97}$$

Notice that

$$\begin{aligned}\vartheta(k) = \sum_{l=1}^{r}\mu_l &\left\{\|KH_{l1}\|\cdot\|H_{l2}\|\cdot\|x(k)\| + \|KH_{l1}\|\cdot\|H_{l3}\|\right. \\ &\left. + \|KH_{l1}\|\cdot\|H_{l4}\|\cdot\|u(k)\| + \|KD_l\|\varpi(k)\right\},\end{aligned} \tag{4.98}$$

and then one has

$$\Delta V(k) \leq -\sigma\|s(k)\| + \Gamma. \tag{4.99}$$

By tuning the parameter σ appropriately, $\Delta V(k) < 0$ can be ensured. Thus, the states of the discrete-time T-S fuzzy system (4.52) can be driven into a neighbourhood of the sliding surface (4.57) and maintained hereafter by the sliding mode controller (4.93). The proof is completed. ■

It is noted that the sliding mode controller (4.93) is actually in a dynamic form, thus, there is no need to conduct a singular valuable decomposition to the control input matrices B_l for the SMC design. This feature relaxes the constraint that each local model of T-S fuzzy systems shares a common control input matrix, and the parameter uncertainties ΔB_l are also allowed to exist.

4.5.2 ILLUSTRATIVE EXAMPLES

In this subsection, illustrative simulation examples will be given to show the effectiveness of the developed state feedback dynamic SMC design approaches.

Example 4.3. Consider an inverted pendulum system same to the one in Example 2.3 firstly. The dynamics of the inverted pendulum system are described as follows,

$$\begin{cases} \dot{x}_1(t) = x_2(t) \\ \dot{x}_2(t) = \dfrac{g\sin(x_1) - amlx_2^2\sin(2x_1)/2 - a\cos(x_1)u}{4l/3 - aml\cos^2(x_1)} + w(t) \end{cases} \tag{4.100}$$

where the parameters are given as in Example 2.3.

Linearize the nonlinear pendulum system (4.100) around $\begin{pmatrix} -\frac{\pi}{3} & 0 \end{pmatrix}^{\mathrm{T}}$, $\begin{pmatrix} 0 & 0 \end{pmatrix}^{\mathrm{T}}$, and $\begin{pmatrix} \frac{\pi}{3} & 0 \end{pmatrix}^{\mathrm{T}}$, and discretize the pendulum system with sampling period $T = 0.05$s. The regulated output $z(k)$ is given as $z(k) = x_1(k) + x_2(k) + 0.02u(k)$, and then the system (4.100) can be described by a T-S fuzzy affine model with three rules as,

Plant Rule $\mathscr{R}^l$: IF $x_1(k)$ is $\mathscr{F}_1^l$, THEN

$$\begin{cases} x(k+1) = (A_l + \Delta A_l)x(k) + a_l + \Delta a_l + (B_l + \Delta B_l)u(k) + D_l w(k) \\ z(k) = L_l x(k) + N_l u(k),\ l \in \{1,2,3\}, \end{cases} \tag{4.101}$$

where the system matrices are

$$\begin{cases} A_1 = A_3 = \begin{bmatrix} 1 & 0.05 \\ 0.2926 & 1 \end{bmatrix}, A_2 = \begin{bmatrix} 1 & 0.05 \\ 0.8467 & 1 \end{bmatrix}, a_1 = \begin{bmatrix} 0 \\ -0.3550 \end{bmatrix}, \\ a_2 = \begin{bmatrix} 0 \\ 0 \end{bmatrix}, a_3 = \begin{bmatrix} 0 \\ 0.3550 \end{bmatrix}, B_1 = B_3 = \begin{bmatrix} 0 \\ -0.0039 \end{bmatrix}, B_2 = \begin{bmatrix} 0 \\ -0.0088 \end{bmatrix}, \\ D_l = \begin{bmatrix} 0 \\ 0.05 \end{bmatrix}, L_l = \begin{bmatrix} 1 & 1 \end{bmatrix}, N_l = 0.02, l \in \{1,2,3\} \end{cases} \tag{4.102}$$

and the parameter uncertainties ΔA_l, Δa_l and ΔB_l are given in the form of (1.3) with matrices H_{l1}, H_{l2}, and H_{l3} given as in (2.124).

The MFs of this example are also presented as in Figure 4.1 with $d_1 = \dfrac{\pi}{12}$ and $d_2 = \dfrac{\pi}{2}$. It follows from Figure 4.1 that the premise variable space can be decomposed into three regions, where S_2 is a crisp region with $\mathscr{N}(2) = \{2\}$, and S_1 and S_3 are fuzzy regions with $\mathscr{N}(1) = \{1,2\}$ and $\mathscr{N}(3) = \{2,3\}$. These three regions can be also accurately described by degenerate ellipsoids as in (2.82) with $Q_1 = Q_3 = \begin{bmatrix} \dfrac{2}{d_2 - d_1} & 0 \end{bmatrix}$, $Q_2 = \begin{bmatrix} \dfrac{1}{d_1} & 0 \end{bmatrix}$, $q_1 = \dfrac{d_1 + d_2}{d_2 - d_1}$, $q_2 = 0$, and $q_3 = \dfrac{d_1 + d_2}{d_1 - d_2}$.

The aim is to design a dynamic sliding mode controller (4.93) to drive the system states of the inverted pendulum (4.101) into a neighbourhood of the sliding surface (4.57) and guarantee the asymptotical stability of the sliding motion with robust $\mathscr{H}_\infty$ performance γ. By applying the CQLF-based method presented in Theorem 4.3, feasible solutions can be obtained with the robust $\mathscr{H}_\infty$ performance $\gamma_{\min} = 0.5691$ and the sliding surface gain $K = \begin{bmatrix} -923.9470 & -198.0239 \end{bmatrix}$. Alternatively, the PQLFs-based method presented in Theorem 4.4 can be also used to obtain the feasible solutions. Choosing $\rho_1 = 1, \rho_2 = 0, J_1 = \begin{bmatrix} 1 & 1 \end{bmatrix}^{\mathrm{T}}, J_2 = \begin{bmatrix} 1 & 0 \end{bmatrix}^{\mathrm{T}}$, the sliding surface gain can be computed as $K = \begin{bmatrix} -824.0609 & -192.2279 \end{bmatrix}$ with robust $\mathscr{H}_\infty$ performance $\gamma_{\min} = 0.7747$.

To verify the validity of the proposed approach, simulations are conducted under different initial conditions. For the sliding surface gain $K = \begin{bmatrix} -923.9470 & -198.0239 \end{bmatrix}$ computed by Theorem 4.3, with initial condition $x_0 = \begin{bmatrix} \frac{\pi}{3} & 0 \end{bmatrix}^{\mathrm{T}}$ and exogenous disturbance $w(k) = 50\mathrm{e}^{-0.5k}\sin(0.8k)$, apply the state feedback dynamic sliding mode controller in (4.93) with $\sigma = 0.6$, the closed-loop system states and the control input are presented in Figure 4.8 and Figure 4.9, respectively. Under zero initial condition, the $\mathscr{H}_\infty$ performance is presented in Figure 4.10. The ratio $\frac{\sqrt{\sum_{t=0}^{\lambda} z^{\mathrm{T}}(k)z(k)}}{\sqrt{\sum_{k=0}^{\lambda} w^{\mathrm{T}}(k)w(k)}}$ is about 0.5, which is lower than the minimum disturbance attenuation level $\gamma_{\min} = 0.5691$.

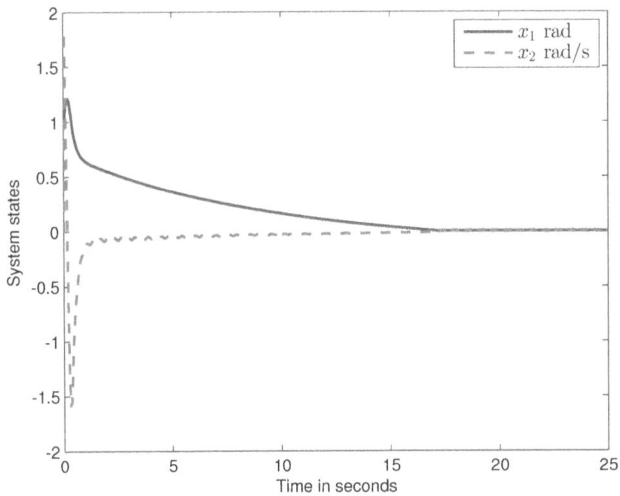

Figure 4.8 The closed-loop system states in Example 4.3 with dynamic sliding mode controller (4.93) and sliding surface gain obtained by Theorem 4.3.

For the sliding surface gain $K = \begin{bmatrix} -824.0609 & -192.2279 \end{bmatrix}$ computed by Theorem 4.4, under the same initial condition and external disturbance, use the dynamic sliding mode controller in (4.93) with $\sigma = 0.6$, the closed-loop system states and the control input are presented in Figure 4.11 and Figure 4.12, respectively. Under zero initial condition, the $\mathscr{H}_\infty$ performance is presented in Figure 4.13. The ratio

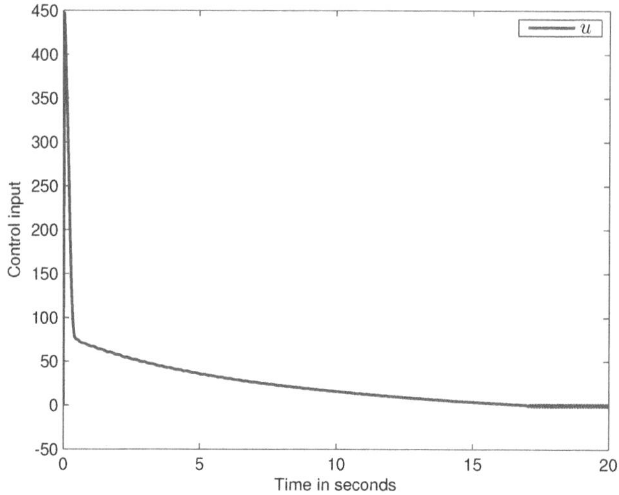

Figure 4.9 The control input in Example 4.3 with dynamic sliding mode controller (4.93) and sliding surface gain obtained by Theorem 4.3.

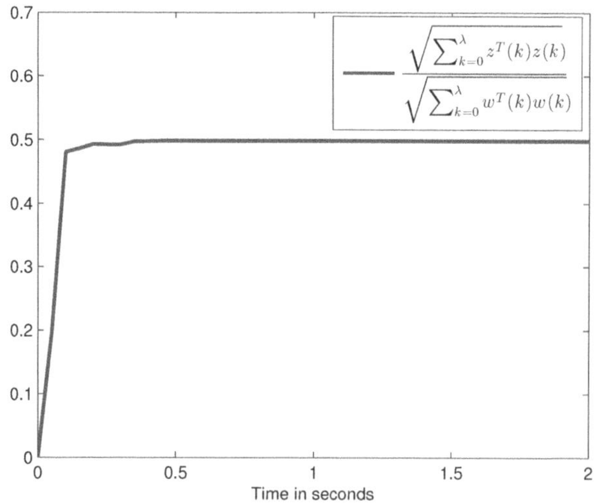

Figure 4.10 Response of the ratio $\frac{\sqrt{\sum_{k=0}^{\lambda} z^{\mathrm{T}}(k)z(k)}}{\sqrt{\sum_{k=0}^{\lambda} w^{\mathrm{T}}(k)w(k)}}$ in Example 4.3 with dynamic sliding mode controller (4.93) and sliding surface gain obtained by Theorem 4.3.

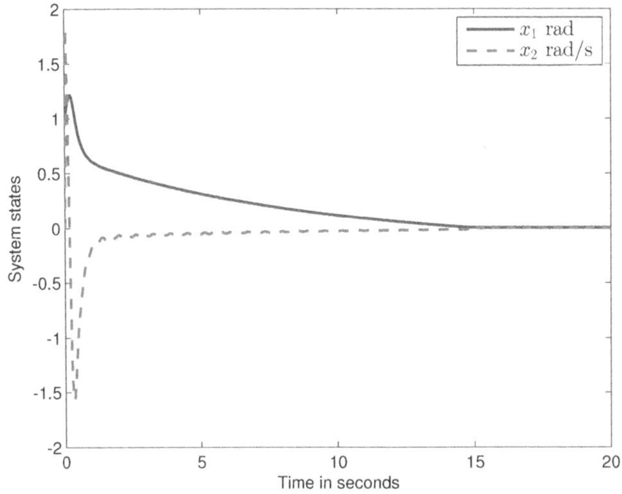

Figure 4.11 The closed-loop system states in Example 4.3 with dynamic sliding mode controller (4.93) and sliding surface gain obtained by Theorem 4.4.

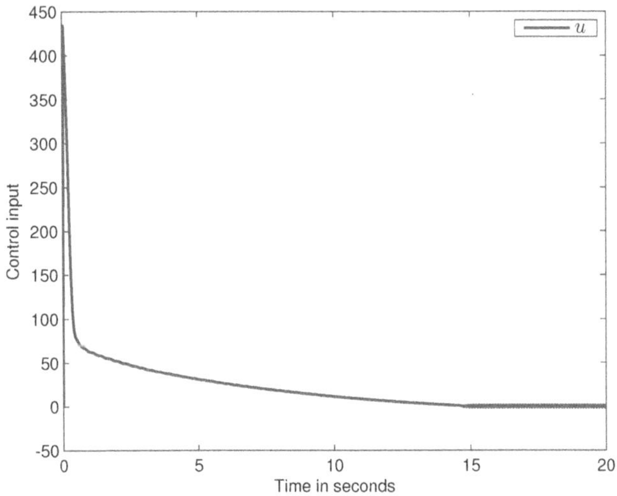

Figure 4.12 The control input in Example 4.3 with dynamic sliding mode controller (4.93) and sliding surface gain obtained by Theorem 4.4.

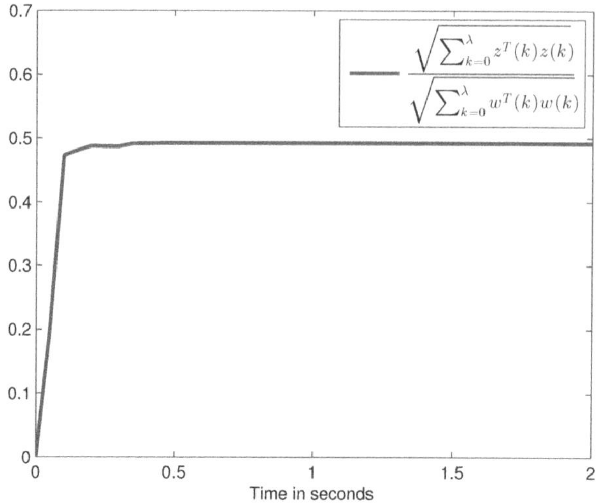

Figure 4.13 Response of the ratio $\frac{\sqrt{\Sigma_{k=0}^{\lambda} z^{\mathrm{T}}(k)z(k)}}{\sqrt{\Sigma_{k=0}^{\lambda} w^{\mathrm{T}}(k)w(k)}}$ in Example 4.3 with dynamic sliding mode controller (4.93) and sliding surface gain obtained by Theorem 4.4.

$\frac{\sqrt{\Sigma_{k=0}^{\lambda} z^{\mathrm{T}}(k)z(k)}}{\sqrt{\Sigma_{k=0}^{\lambda} w^{\mathrm{T}}(k)w(k)}}$ is about 0.49, which is lower than the minimum disturbance attenuation level $\gamma_{\min} = 0.7747$.

In the following, simulation on an autonomous land vehicle system borrowed from [25] will be conducted to show the effectiveness of the proposed dynamic SMC design method.

Example 4.4. Consider an autonomous land vehicle system same to the one given as in Example 4.2, and the dynamics of the cart are described as follows,

$$\begin{bmatrix} \dot{\theta}(t) \\ \dot{\psi}(t) \\ \dot{\mathbf{y}}(t) \end{bmatrix} = \begin{bmatrix} 0 & 1 & 0 \\ 0 & -\frac{d}{l} & 0 \\ 0 & 0 & 0 \end{bmatrix} \begin{bmatrix} \theta(t) \\ \psi(t) \\ \mathbf{y}(t) \end{bmatrix} + \begin{bmatrix} 0 \\ 0 \\ v_0 \sin(\theta(t)) \end{bmatrix} + \begin{bmatrix} 0 \\ \frac{1}{l} \\ 0 \end{bmatrix} u(t) + \begin{bmatrix} 0 \\ 1 \\ 0 \end{bmatrix} w(t), \tag{4.103}$$

where the parameters are the same to the ones in Example 4.2 and $w(t)$ is the external disturbance.

Denote the state variables as $x_1(t) = \theta(t)$, $x_2(t) = \psi(t)$ and $x_3(t) = \mathbf{y}(t)$, and the regulated output is given as $z(t) = x_1(t) + x_3(t) + u(t)$. Linearize the system (4.103) around three operating points $\begin{pmatrix} 0 & 0 & 0 \end{pmatrix}^{\mathrm{T}}$, $\begin{pmatrix} \frac{2\pi}{5} & 0 & 0 \end{pmatrix}^{\mathrm{T}}$, and $\begin{pmatrix} \frac{2\pi}{5} & 0 & 0 \end{pmatrix}^{\mathrm{T}}$, and discretize the autonomous land vehicle system with sampling period $T = 0.1$s, and then the system (4.103) can be characterized by the following T-S fuzzy affine model as

Plant Rule $\mathscr{R}^l$: IF $x_1(k)$ is $\mathscr{F}_1^l$, THEN

$$\left\{\begin{array}{l} x(k+1) = (A_l + \Delta A_l)x(k) + a_l + \Delta a_l + (B_l + \Delta B_l)u(k) + D_l w(k) \\ z(k) = L_l x(k) + N_l u(k),\ l \in \{1,2,3\}, \end{array}\right. \tag{4.104}$$

where the system matrices are

$$\left\{\begin{array}{l} A_1 = A_3 = \begin{bmatrix} 1 & 0.1 & 0 \\ 0 & 0.999 & 0 \\ 0.0309 & 0 & 1 \end{bmatrix}, A_2 = \begin{bmatrix} 1 & 0.1 & 0 \\ 0 & 0.999 & 0 \\ 0.1 & 0 & 0.1 \end{bmatrix}, \\ a_1 = \begin{bmatrix} 0 \\ 0 \\ -0.0563 \end{bmatrix}, a_2 = \begin{bmatrix} 0 \\ 0 \\ 0 \end{bmatrix}, a_3 = \begin{bmatrix} 0 \\ 0 \\ 0.0563 \end{bmatrix}, B_l = \begin{bmatrix} 0 \\ 1 \\ 0 \end{bmatrix}, D_l = \begin{bmatrix} 0 \\ 1 \\ 0 \end{bmatrix}, \end{array}\right. \tag{4.105}$$

and the parameter uncertainties ΔA_l, Δa_l and ΔB_l are given in the form of (1.3) with matrices H_{l1}, H_{l2}, and H_{l3} given as in (3.65).

The MFs of this example is shown in Figure 4.1 with $d_1 = \frac{\pi}{15}$ and $d_2 = \frac{3\pi}{5}$. By applying the CQLF-based method presented in Theorem 4.3, the gain of the sliding surface (4.57) can be obtained as $K = \begin{bmatrix} 9.9177 & 6.8206 & 2.1039 \end{bmatrix}$ with robust $\mathscr{H}_\infty$ performance $\gamma_{\min} = 1.4656$. By using the PQLFs-based results presented in Theorem 4.4 with $\rho_1 = 1$, $\rho_2 = 0$, $J_1 = \begin{bmatrix} 1 & 1 & 1 \end{bmatrix}$, and $J_2 = \begin{bmatrix} 1 & 0 & 0 \end{bmatrix}$, the gain of the sliding surface (4.57) can be obtained as $K = \begin{bmatrix} 13.7638 & 7.4001 & 5.1154 \end{bmatrix}$ with robust $\mathscr{H}_\infty$ performance $\gamma_{\min} = 1.7479$.

To show the effectiveness of the proposed state feedback dynamic SMC design method, simulations are conducted with different initial conditions. With initial condition $x_0 = \begin{bmatrix} \frac{\pi}{3} & 0 & 0 \end{bmatrix}^{\mathrm{T}}$ and external disturbance $w(k) = 10\mathrm{e}^{-0.8k}\sin(0.6k)$, for the sliding surface gain $K = \begin{bmatrix} 13.7638 & 7.4001 & 5.1154 \end{bmatrix}$, using the dynamic sliding mode controller in (4.93) with $\sigma = 0.1$, Figure 4.14 and Figure 4.15 show the time responses of the system states and control input, respectively. Under zero initial condition, the $\mathscr{H}_\infty$ performance is presented in Figure 4.16. The ratio $\frac{\sqrt{\sum_{k=0}^{\lambda} z^{\mathrm{T}}(k)z(k)}}{\sqrt{\sum_{k=0}^{\lambda} w^{\mathrm{T}}(k)w(k)}}$ is about 1.21, which is lower than the minimum disturbance attenuation level $\gamma_{\min} = 1.7479$.

Nextly, we will consider a nonlinear continuous stirred tank reactor (CSTR) system to show the effectiveness of the proposed dynamic SMC design method.

Example 4.5. Consider a nonlinear continuous stirred tank reactor (CSTR) system same to the one in Example 2.1, whose dynamics are described as follows,

$$\left\{\begin{array}{l} \dot{T} = M_0(T_n - T) + M_1 k_0 \mathrm{e}^{-\frac{M_2}{T}} C_A - M_3 T + M_3 \mathbf{u} \\ \dot{C}_A = M_0(C_0 - C_A) - M_0 \mathrm{e}^{-\frac{M_2}{T}} C_A \end{array}\right. \tag{4.106}$$

and the parameters are given as in Example 2.1. At steady input $\mathbf{u}_o = 300\mathrm{K}$, the CSTR system has one equilibrium $\mathbf{x}_o = \begin{bmatrix} 350, & 0.5 \end{bmatrix}^{\mathrm{T}}$. Specially, linearize the nonlinear CSTR system at $\mathbf{x}_{o1} = [324.4 \quad 0.875]^{\mathrm{T}}$, $\mathbf{x}_{o2} = [350 \quad 0.5]^{\mathrm{T}}$, $\mathbf{x}_{o3} =$

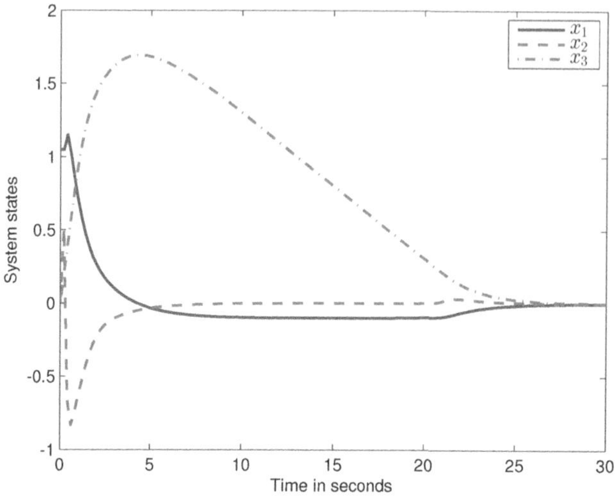

Figure 4.14 The closed-loop system states in Example 4.4 with sliding mode controller (4.93) and sliding surface gain obtained by Theorem 4.4.

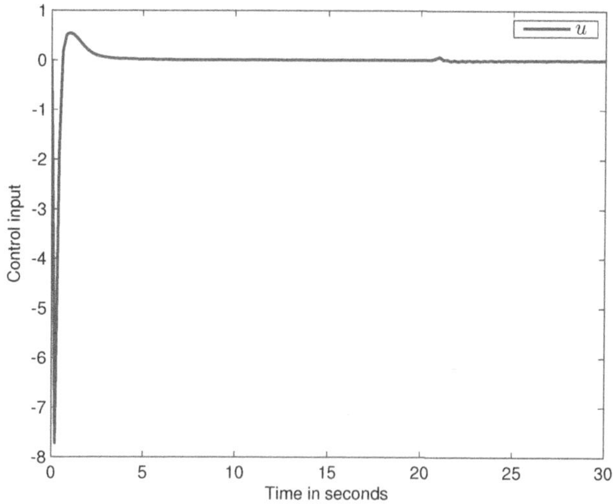

Figure 4.15 The control input in Example 4.4 with dynamic sliding mode controller (4.93) and sliding surface gain obtained by Theorem 4.4.

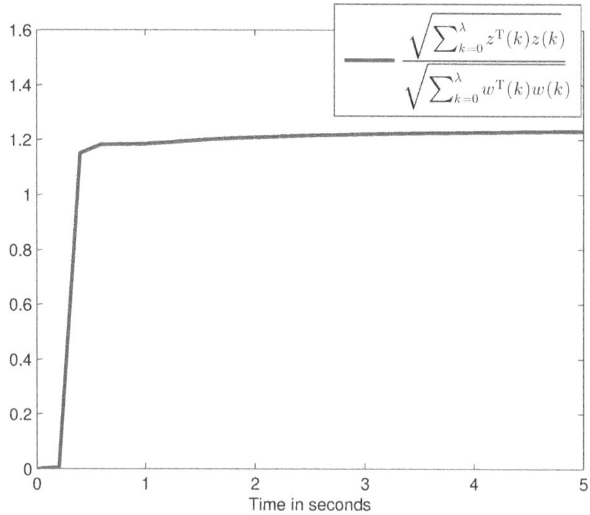

Figure 4.16 Response of the ratio $\frac{\sqrt{\sum_{k=0}^{\lambda} z^{\mathrm{T}}(k)z(k)}}{\sqrt{\sum_{k=0}^{\lambda} w^{\mathrm{T}}(k)w(k)}}$ in Example 4.4 with dynamic sliding mode controller (4.93) and sliding surface gain obtained by Theorem 4.4.

$[370.65 \quad 0.2]^{\mathrm{T}}$, and denote $x = \mathbf{x} - \mathbf{x}_{o2}$, $u = \mathbf{u} - \mathbf{u}_o$. Discretizing the CSTR system with sampling period $T = 0.05$min and considering the external disturbance, and then the CSTR system (4.106) can be described by the following T-S fuzzy affine model as,

Plant Rule $\mathscr{R}^l$: IF $x_1(k)$ is $\mathscr{F}_1^l$, THEN

$$\begin{cases} x(k+1) = (A_l + \Delta A_l)x(k) + a_l + \Delta a_l + B_l u(k) + D_l w(k) \\ z(k) = L_l x(k) + N_l u(k), \ l \in \{1,2,3\}, \end{cases} \tag{4.107}$$

where the system matrices are

$$\begin{cases} A_1 = \begin{bmatrix} 0.9512 & 1.4545 \\ -0.0005 & 0.9430 \end{bmatrix}, A_2 = \begin{bmatrix} 1.2190 & 10.4595 \\ -0.0018 & 0.9001 \end{bmatrix}, \\ A_3 = \begin{bmatrix} 1.3818 & 42.1122 \\ -0.0026 & 0.7487 \end{bmatrix}, a_1 = \begin{bmatrix} -1.7943 \\ 0.0084 \end{bmatrix}, a_2 = \begin{bmatrix} 0 \\ 0 \end{bmatrix}, \\ a_3 = \begin{bmatrix} 4.7483 \\ -0.0224 \end{bmatrix}, B_l = \begin{bmatrix} 0.1046 \\ 0 \end{bmatrix}, D_l = \begin{bmatrix} 0.05 \\ 0 \end{bmatrix}, \\ L_l = \begin{bmatrix} 1 & 0.1 \end{bmatrix}, N_l = 0.1, \ l \in \{1,2,3\}, \end{cases} \tag{4.108}$$

and the parameter uncertainties ΔA_l and Δa_l are given in the form of (1.3) with matrices H_{l1}, H_{l2}, and H_{l3} given as in (2.62).

Figure 4.17 shows the fuzzy MFs of Example 4.5 with $d_1 = -50$, $d_2 = -25.6$, $d_3 = 20.65$ and $d_4 = 50$, and it is found from Figure 4.17 that where S_2 is a fuzzy

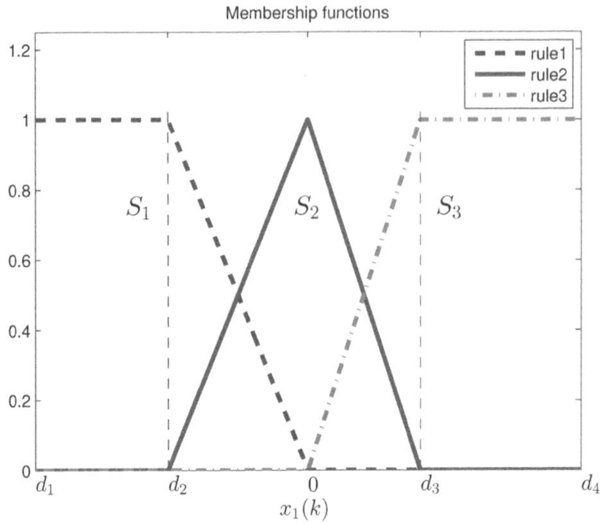

Figure 4.17 Membership functions of Example 4.5.

region with $\mathcal{N}(2) = \{1,2,3\}$, and S_1 and S_3 are crisp regions with $\mathcal{N}(1) = \{1\}$ and $\mathcal{N}(3) = \{3\}$. These three regions can be precisely described by degenerate ellipsoids as in (2.82) with $Q_1 = \begin{bmatrix} \frac{2}{d_2 - d_1} & 0 \end{bmatrix}$, $Q_2 = \begin{bmatrix} \frac{2}{d_3 - d_2} & 0 \end{bmatrix}$, $Q_3 = \begin{bmatrix} \frac{2}{d_4 - d_3} & 0 \end{bmatrix}$ $q_1 = \frac{d_1 + d_2}{d_1 - d_2}$, $q_2 = \frac{d_3 + d_2}{d_2 - d_3}$, $q_3 = \frac{d_4 + d_3}{d_3 - d_4}$.

The purpose is to design a sliding surface (4.57) and a state feedback dynamic sliding mode controller (4.93) to drive the state trajectories of the CSTR system in (4.107) into a neighbourhood of the sliding surface (4.57) and guarantee the asymptotical stability of the sliding motion with robust $\mathcal{H}_\infty$ performance γ. By applying the CQLF-based method presented in Theorem 4.3, feasible solutions can be obtained with the sliding surface gain $K = \begin{bmatrix} 12.3899 & -22.4890 \end{bmatrix}$ and the robust $\mathcal{H}_\infty$ performance $\gamma_{\min} = 0.6049$. Alternatively, one can also use the PQLFs-based method proposed in Theorem 4.4 to obtain the feasible solutions. With $\rho_1 = 1$, $\rho_2 = 0$, $J_1 = \begin{bmatrix} 3 & 1 \end{bmatrix}^{\mathrm{T}}$, $J_2 = \begin{bmatrix} 1 & 0 \end{bmatrix}^{\mathrm{T}}$, the sliding surface gain can be computed as $K = \begin{bmatrix} 12.1694 & 139.6697 \end{bmatrix}$ with robust $\mathcal{H}_\infty$ performance $\gamma_{\min} = 0.7730$.

To verify the validity of the proposed dynamic SMC approach, simulations are conducted. With initial condition $x_0 = \begin{bmatrix} -15 & 1.5 \end{bmatrix}^{\mathrm{T}}$ and external disturbance $w(k) = 170\mathrm{e}^{-0.8k}\sin(0.03k)$, for the sliding surface gain $K = \begin{bmatrix} 13.0858 & 91.4271 \end{bmatrix}$ computed by Theorem 4.4, utilizing the dynamic sliding mode controller (4.93) with $\sigma = 0.3$, the closed-loop system states and the control input are shown in Figure 4.18 and Figure 4.19, respectively. The sliding surface is shown in Figure 4.20.

These simulation results clearly show that effectiveness of the proposed state feedback dynamic sliding mode controller design method.

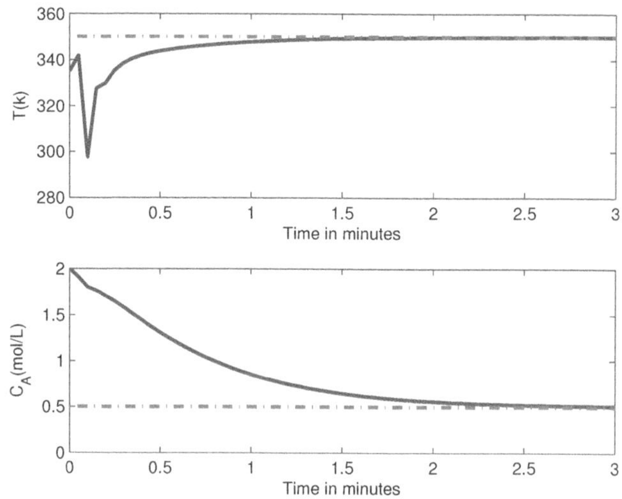

Figure 4.18 The closed-loop system states in Example 4.5 with dynamic sliding mode controller (4.93) and sliding surface gain obtained by Theorem 4.4.

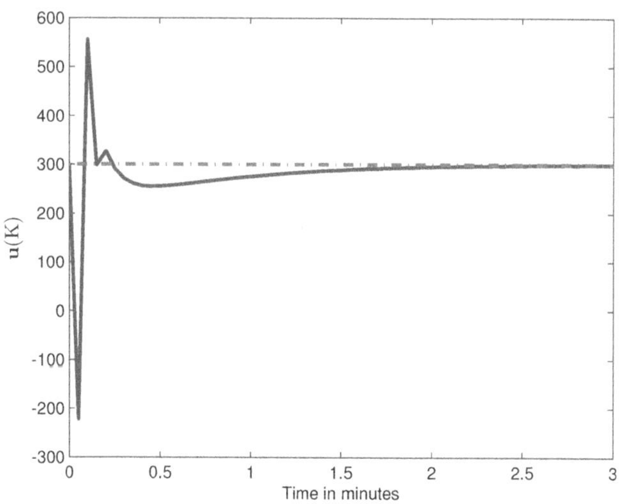

Figure 4.19 The control input in Example 4.5 with dynamic sliding mode controller (4.93) and sliding surface gain obtained by Theorem 4.4.

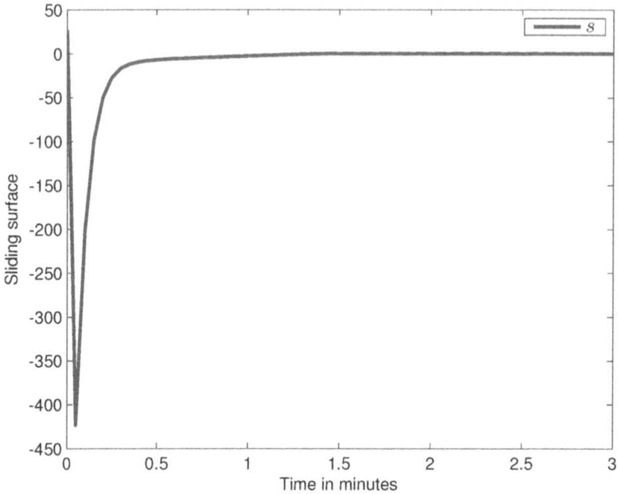

Figure 4.20 The sliding surface in Example 4.5 with dynamic sliding mode controller (4.93) and sliding surface gain obtained by Theorem 4.4.

4.6 CONCLUSION

This chapter addresses the state feedback SMC problem for a class of discrete-time nonlinear systems by T-S fuzzy affine models. Using CQLF-based and PQLFs-based analysis methods, sufficient conditions for the asymptotic stability analysis of the sliding motion are proposed under a convex optimization framework. Based on the full system states, two new sliding mode controller design methods are proposed to drive the resultant closed-loop system states into a neighbourhood of the sliding surface. Simulation studies are finally given to show the effectiveness of the proposed SMC design schemes.

5 Output Feedback Fuzzy SMC of Discrete-Time Nonlinear Systems

5.1 INTRODUCTION

Over the past decades, some valuable fuzzy output feedback SMC results on nonlinear systems have been reported [94, 95]. Specifically, the authors in [94] proposed some fuzzy adaptive output feedback SMC results for nonlinear systems with mismatched uncertainties, and the bound of external disturbances was estimated by an adaptive law. In [95], an adaptive output feedback SMC design approach was proposed for delta operator fuzzy systems. Nevertheless, it is noticed that the output feedback fuzzy SMC results in [94, 95] were basically derived based on T-S fuzzy systems with local linear models and through a CQLF, which may be conservative.

This chapter focuses on the fuzzy output feedback SMC design for discrete-time nonlinear 1-D systems based on T-S fuzzy affine models. Through adopting a state-input augmentation scheme, the dynamical properties of the sliding motion are characterized by a descriptor system. New results on asymptotic stability analysis for the sliding motion are proposed based on CQLF and PQLFs, respectively, and the sliding surface gain can be computed in a convex optimization setup. Then some new output feedback fuzzy sliding mode controller and dynamic sliding mode controller design approaches are proposed to force the resulting closed-loop system states into a neighbourhood of the designed sliding surface locally for a prescribed sliding region. It is also worthy of pointing out that the designed dynamic sliding mode controller can also relax the assumption that each local affine model shares one common control input matrix, and the control input matrices are allowed to contain parameter uncertainties.

5.2 OUTPUT FEEDBACK SMC DESIGN VIA T-S FUZZY AFFINE MODELS

5.2.1 MODEL DESCRIPTION: CASE I

Consider the first and second equation of the discrete-time T-S fuzzy affine system described as in (1.2) with a common control input matrix B, which is characterized by r fuzzy **IF-THEN** rules as follows.

Plant Rule $\mathscr{R}^l$**: IF** $\theta_1(k)$ **is** $\mathscr{F}_1^l$ **and** $\cdots$ **and** $\theta_\varphi(k)$ **is** $\mathscr{F}_\varphi^l$**, THEN**

$$\begin{cases} x(k+1) = (A_l + \Delta A_l)x(k) + a_l + \Delta a_l + B\left(u(k) + f(x(k),k)\right) \\ y(k) = Cx(k),\ l \in \mathscr{L} := \{1,\dots,r\}, \end{cases} \tag{5.1}$$

DOI: 10.1201/9781032695716-5

where B is the common control input matrix with full-column rank; $C = C_l, l \in \mathscr{L}$ is the common measurement output matrix; ΔA_l and Δa_l represent the parameter uncertainties in the l-th local model satisfying the condition in (1.3) as $\begin{bmatrix} \Delta A_l & \Delta a_l \end{bmatrix} = H_{l1}\Delta_l(k) \begin{bmatrix} H_{l2} & H_{l3} \end{bmatrix}, l \in \{1,\ldots,r\}$, where H_{l1}, H_{l2} and H_{l3} are known real-valued matrices and $\Delta_l^{\mathrm{T}}(k)\Delta_l(k) \leq \mathbf{I}$. $f(x(k),k) \in \Re^{n_u}$ represents the nonlinear disturbance, and it is assumed that $f(x(k),k)$ satisfies

$$\|f(x(k),k)\| \leq \gamma\|x(k)\| + \lambda(y(k),k) \tag{5.2}$$

where $\gamma > 0$ and $\lambda(y(k),k) : [\Re^{n_y} \times \Re^{+}] \rightarrow \Re^{+}$ stands for a known scalar-valued function.

Similar to Chapter 2, utilizing a singleton fuzzifier, product inference and center-average defuzzifier, the system (5.1) can be inferred as

$$\begin{cases} x(k+1) = (A(\mu) + \Delta A(\mu))x(k) + a(\mu) + \Delta a(\mu) + B(u(k) + f(x(k),k)) \\ y(k) = Cx(k) \end{cases} \tag{5.3}$$

where

$$\begin{cases} A(\mu) = \sum\limits_{l=1}^{r} \mu_l A_l, \Delta A(\mu) = \sum\limits_{l=1}^{r} \mu_l \Delta A_l, \\ a(\mu) = \sum\limits_{l=1}^{r} \mu_l a_l, \Delta a(\mu) = \sum\limits_{l=1}^{r} \mu_l \Delta a_l. \end{cases} \tag{5.4}$$

In this chapter, some new output feedback SMC synthesis methods for discrete-time uncertain T-S fuzzy affine system will be proposed via a descriptor-system-based method, and it will be shown that the resulting closed-loop system states can be driven into a neighbourhood of the sliding surface locally for a prescribed sliding patch.

5.2.2 SLIDING SURFACE DESIGN

For the system (5.1) or equivalently (5.3), design a sliding surface relying on the measurement output as

$$s(k) = Ky(k), \tag{5.5}$$

where $K \in \Re^{n_u \times n_y}$ is the sliding surface gain to be determined.

Notice that the control input matrix B in the T-S fuzzy affine system (5.1) is with full column rank, and consequently there always exists a full-row rank matrix $M \in \Re^{(n_x - n_u) \times n_x}$ satisfying $MB = 0$. Thus, a nonsingular transformation matrix T can be constructed as

$$T = \begin{bmatrix} M \\ B^{\mathrm{T}} \end{bmatrix}. \tag{5.6}$$

Based on (5.6), it can be found that $MT^{-1} = \begin{bmatrix} \mathbf{I} & \mathbf{0}_{(n_x - n_u) \times n_u} \end{bmatrix}$. Through a state transformation with matrix T, the system dynamics in (5.3) are formulated as

$$\begin{aligned} Tx(k+1) = {} & T\left(A(\mu) + \Delta A(\mu)\right)x(k) + T\left(a(\mu) + \Delta a(\mu)\right) \\ & + TB\left(u(k) + f(x(k),k)\right). \end{aligned} \tag{5.7}$$

The premise variable space can be decomposed by a similar line in Chapter 2. The following set $\mathscr{N}(i) := \{s|\mu_s[\theta(k)] > 0,\ \theta(k) \in S_i,\ s \in \mathscr{L},\ i \in \mathscr{I}\}$ is also introduced to denote the indices for local models used in the interpolation in each region S_i.

Based on the set $\mathscr{N}(i)$, the system (5.7) is formulated as

$$\begin{aligned} Tx(k+1) = T\left(\mathbf{A}_i + \Delta\mathbf{A}_i\right)x(k) + T(\mathbf{a}_i + \Delta\mathbf{a}_i) \\ +TB\left(u(k) + f(x(k),k)\right),\ \theta(k) \in S_i, i \in \mathscr{I}, \end{aligned} \tag{5.8}$$

where

$$\begin{cases} \mathbf{A}_i := \sum\limits_{s\in\mathscr{N}(i)} \mu_s A_s, \Delta\mathbf{A}_i := \sum\limits_{s\in\mathscr{N}(i)} \mu_s \Delta A_s, \\ \mathbf{a}_i := \sum\limits_{s\in\mathscr{N}(i)} \mu_s a_s, \Delta\mathbf{a}_i := \sum\limits_{s\in\mathscr{N}(i)} \mu_s \Delta a_s \end{cases} \tag{5.9}$$

with $\sum_{s\in\mathscr{N}(i)} \mu_s[\theta(k)] = 1, 0 < \mu_s[\theta(k)] \leq 1$.

With $MB = 0$, one has

$$\begin{cases} Mx(k+1) = M\left(\mathbf{A}_i + \Delta\mathbf{A}_i\right)x(k) + M(\mathbf{a}_i + \Delta\mathbf{a}_i) \\ B^{\mathrm{T}}x(k+1) = B^{\mathrm{T}}\left(\mathbf{A}_i + \Delta\mathbf{A}_i\right)x(k) + B^{\mathrm{T}}(\mathbf{a}_i + \Delta\mathbf{a}_i) \\ \qquad +B^{\mathrm{T}}B(u(k) + f(x(k),k)),\ \theta(k) \in S_i, i \in \mathscr{I}, \end{cases} \tag{5.10}$$

where $B^{\mathrm{T}}B$ is nonsingular. It is easy to see that the first equation in (5.10) describes the sliding mode dynamics of the system (5.3) or (5.8). Noticing that the sliding surface function satisfies $s(k) = 0$, the holonomic sliding mode dynamical equation is formulated as

$$Ex(k+1) = (\mathscr{A}_i + \Delta\mathscr{A}_i)x(k) + \mathrm{a}_i + \Delta\mathrm{a}_i,\ \theta(k) \in S_i, i \in \mathscr{I}, \tag{5.11}$$

where

$$\begin{cases} E = \begin{bmatrix} M \\ 0 \end{bmatrix}, \\ \mathscr{A}_i + \Delta\mathscr{A}_i = \begin{bmatrix} M(\mathbf{A}_i + \Delta\mathbf{A}_i) \\ KC \end{bmatrix}, \mathrm{a}_i + \Delta\mathrm{a}_i = \begin{bmatrix} M(\mathbf{a}_i + \Delta\mathbf{a}_i) \\ \mathbf{0} \end{bmatrix}. \end{cases} \tag{5.12}$$

Via a state transformation $z(k) = Tx(k)$, together with the fact that $MT^{-1} = \begin{bmatrix} \mathbf{I} & \mathbf{0}_{(n_x-n_u)\times n_u} \end{bmatrix}$, the system in (5.11) can be derived as,

$$\tilde{E}z(k+1) = (\tilde{\mathscr{A}}_i + \Delta\tilde{\mathscr{A}}_i)z(k) + \tilde{\mathrm{a}}_i + \Delta\tilde{\mathrm{a}}_i,\ \tilde{\theta}(k) \in \tilde{S}_i, i \in \mathscr{I}, \tag{5.13}$$

where $\tilde{\theta}(k)$ and $\tilde{S}_i$, respectively, refer to the new premise variables and partitioned space with the transformation matrix T, and

$$\begin{cases} \tilde{E} = \begin{bmatrix} \mathbf{I}_{n_x-n_u} & \mathbf{0} \\ \mathbf{0} & \mathbf{0}_{n_u} \end{bmatrix}, \tilde{\mathscr{A}}_i + \Delta\tilde{\mathscr{A}}_i = \begin{bmatrix} M(\mathbf{A}_i + \Delta\mathbf{A}_i)T_1 & M(\mathbf{A}_i + \Delta\mathbf{A}_i)T_2 \\ KCT_1 & KCT_2 \end{bmatrix}, \\ \tilde{\mathrm{a}}_i + \Delta\tilde{\mathrm{a}}_i = \begin{bmatrix} M(\mathbf{a}_i + \Delta\mathbf{a}_i) \\ \mathbf{0} \end{bmatrix}, T^{-1} = \begin{bmatrix} T_1 & T_2 \end{bmatrix} \end{cases} \tag{5.14}$$

and $T_1 \in \Re^{n_x\times(n_x-n_u)}$, $T_2 \in \Re^{n_x\times n_u}$ are known constant matrices.

Combining the designed sliding surface (5.5) with the first equation in system (5.10), the sliding motion is fully described by the descriptor system (5.13). The existence of the proposed sliding surface is ensured by the admissibility of the system (5.13). In the following, we will firstly propose sufficient conditions for the existence of sliding surface (5.5), and then develop two output feedback SMC synthesis methods.

5.2.3 STABILITY ANALYSIS OF SLIDING MOTION

New results on the asymptotic stability analysis of the sliding motion will be given based on PQLFs in this section. Then two output feedback SMC design approaches will be proposed to guarantee that the system states can be forced into a neighbourhood the sliding surface locally.

Firstly, we will analyze the asymptotic stability of the sliding motion through PQLFs in a convex optimization framework. For future use, a new set $\Omega := \{(i,j)|\tilde{\theta}(k) \in \tilde{S}_i, \tilde{\theta}(k+1) \in \tilde{S}_j, i,j \in \mathscr{I}\}$ is introduced in order to characterize region transitions.

Similar to Chapter 2, it is assumed in this chapter that an ellipsoid $\mathscr{E}_i := \{z| \|Q_i z + q_i\| \leq 1\}$ is used to outer approximate each polyhedral region $\tilde{S}_i$ in (5.13), and the relationship given as in (2.83) holds for each region.

The sliding motion stability analysis result is presented as follows.

Theorem 5.1. The sliding mode dynamical system (5.13) is asymptotically stable if there exist matrices $0 < P_{i1} = P_{i1}^T \in \Re^{(n_x-n_u)\times(n_x-n_u)}$, $P_{i2} \in \Re^{n_u\times(n_x-n_u)}$, $P_{i3} \in \Re^{n_u\times n_u}$, $i \in \mathscr{I}$, and $\hat{K} \in \Re^{n_u\times n_y}$, $\{G_{i1(1)}, G_{i2(1)}\} \in \Re^{(n_x-n_u)\times(n_x-n_u)}$, $\{G_{i1(2)}, G_{i2(2)}\} \in \Re^{n_u\times(n_x-n_u)}$, $G_{(3)} \in \Re^{n_u\times n_u}$, and scalars $\{\varepsilon_i > 0, i \in \mathscr{I}\}$, $\{\rho_i < 0, i \in \mathscr{I}_1\}$, such that the conditions in (5.15)-(5.16) hold,

$$\begin{bmatrix} \Sigma_{11} + \varepsilon_i T^{-\mathrm{T}} H_{s2}^{\mathrm{T}} H_{s2} T^{-1} & \star & \star \\ \Sigma_{21} - G_{i2}^{\mathrm{T}} & \Sigma_{22} & \star \\ \Sigma_{31} & \Sigma_{32} & -\varepsilon_i \mathbf{I} \end{bmatrix} < 0, i \in \mathscr{I}_0, s \in \mathscr{N}(i), \tag{5.15}$$

$$\begin{bmatrix} \rho_i(q_i^{\mathrm{T}} q_i - 1) + \varepsilon_i H_{s3}^{\mathrm{T}} H_{s3} & \star & \star & \star \\ \rho_i Q_i^{\mathrm{T}} q_i + \Sigma_{10} + \varepsilon_i T^{-\mathrm{T}} H_{s2}^{\mathrm{T}} H_{s3} & \rho_i Q_i^{\mathrm{T}} Q_i + \Sigma_{11} + \varepsilon_i T^{-\mathrm{T}} H_{s2}^{\mathrm{T}} H_{s2} T^{-1} & \star & \star \\ \Sigma_{20} & \Sigma_{21} - G_{i2}^{\mathrm{T}} & \Sigma_{22} & \star \\ \mathbf{0} & \Sigma_{31} & \Sigma_{32} & -\varepsilon_i \mathbf{I} \end{bmatrix} < 0, i \in \mathscr{I}_1, s \in \mathscr{N}(i), \tag{5.16}$$

where

$$\begin{cases} \Sigma_{11} = \mathrm{Sym}\left\{\begin{bmatrix} G_{i2(1)} \\ G_{i2(2)} \end{bmatrix} MA_sT^{-1} + \begin{bmatrix} \mathbf{0} \\ \rho_2\mathbf{I} \end{bmatrix}\hat{K}CT^{-1}\right\} - \tilde{E}^{\mathrm{T}}P_i\tilde{E}, \\ \Sigma_{21} = \begin{bmatrix} G_{i1(1)} \\ G_{i1(2)} \end{bmatrix} MA_sT^{-1} + \begin{bmatrix} J \\ \rho_1\mathbf{I} \end{bmatrix}\hat{K}CT^{-1}, \\ \Sigma_{22} = P_j - G_{i1} - G_{i1}^{\mathrm{T}}, \\ \Sigma_{31} = H_{s1}^{\mathrm{T}}M^{\mathrm{T}}\begin{bmatrix} G_{i2(1)}^{\mathrm{T}} & G_{i2(2)}^{\mathrm{T}} \end{bmatrix}, \Sigma_{32} = H_{s1}^{\mathrm{T}}M^{\mathrm{T}}\begin{bmatrix} G_{i1(1)}^{\mathrm{T}} & G_{i1(2)}^{\mathrm{T}} \end{bmatrix}, \\ \Sigma_{10} = \begin{bmatrix} G_{i2(1)} \\ G_{i2(2)} \end{bmatrix} Ma_s, \Sigma_{20} = \begin{bmatrix} G_{i1(1)} \\ G_{i1(2)} \end{bmatrix} Ma_s, P_i = \begin{bmatrix} P_{i1} & \star \\ P_{i2} & P_{i3} \end{bmatrix}, \\ T^{-1} = \begin{bmatrix} T_1 & T_2 \end{bmatrix}, G_{i1} = \begin{bmatrix} G_{i1(1)} & JG_{(3)} \\ G_{i1(2)} & \rho_1 G_{(3)} \end{bmatrix}, G_{i2} = \begin{bmatrix} G_{i2(1)} & \mathbf{0} \\ G_{i2(2)} & \rho_2 G_{(3)} \end{bmatrix}, \end{cases} \tag{5.17}$$

in which $J \in \Re^{(n_x-n_u)\times n_u}$ is an arbitrary matrix, ρ_1, ρ_2 are constant scalars, and $T_1 \in \Re^{n_x\times(n_x-n_u)}$, $T_2 \in \Re^{n_x\times n_u}$ are known constant matrices given in (5.14).

Moreover, the gain of the sliding surface (5.5) can be attained by

$$K = G_{(3)}^{-1}\hat{K}. \tag{5.18}$$

Proof. Note that the condition (5.15) for $i \in \mathscr{I}_0$ is a special case of the condition (5.16) for $i \in \mathscr{I}_1$. Without loss of generality, we merely prove the more complex case for $i \in \mathscr{I}_1$. Consider a PQLF as,

$$V(k) = z^{\mathrm{T}}(k)\tilde{E}^{\mathrm{T}}P_i\tilde{E}z(k), \tag{5.19}$$

where

$$P_i = \begin{bmatrix} P_{i1} & \star \\ P_{i2} & P_{i3} \end{bmatrix}, i \in \mathscr{I}, \tag{5.20}$$

and $P_{i3} \in \Re^{n_u\times n_u}$, $P_{i2} \in \Re^{n_u\times(n_x-n_u)}$, $0 < P_{i1} = P_{i1}^T \in \Re^{(n_x-n_u)\times(n_x-n_u)}$.

Based on (5.19), the following inequality ensures the asymptotic stability of the sliding mode dynamical system (5.13),

$$\Delta V(k) = \left(\tilde{E}z(k+1)\right)^{\mathrm{T}} P_j\left(\tilde{E}z(k+1)\right) - z^{\mathrm{T}}(k)\tilde{E}^{\mathrm{T}}P_i\tilde{E}z(k) < 0, (i,j) \in \Omega. \tag{5.21}$$

Define

$$\vartheta(k) = \begin{bmatrix} 1 & z^{\mathrm{T}}(k) & z^{\mathrm{T}}(k+1)\tilde{E}^{\mathrm{T}} \end{bmatrix}^{\mathrm{T}}, \tag{5.22}$$

and then the inequality (5.21) can be rewritten as,

$$\vartheta^{\mathrm{T}}(k)\begin{bmatrix} 0 & \star & \star \\ \mathbf{0} & -\tilde{E}^{\mathrm{T}}P_i\tilde{E} & \star \\ \mathbf{0} & \mathbf{0} & P_j \end{bmatrix}\vartheta(k) < 0, (i,j) \in \Omega. \tag{5.23}$$

Utilizing the S-procedure based on the partition information described as in (2.83), for $\rho_i < 0$, $i \in \mathscr{I}_1$, the following inequality implies (5.23)

$$\vartheta^{\mathrm{T}}(k)\begin{bmatrix} \rho_i(q_i^{\mathrm{T}}q_i-1) & \star & \star \\ \rho_i Q_i^{\mathrm{T}} q_i & -\tilde{E}^{\mathrm{T}}P_i\tilde{E}+\rho_i Q_i^{\mathrm{T}}Q_i & \star \\ \mathbf{0} & \mathbf{0} & P_j \end{bmatrix}\vartheta(k)<0, i\in\mathscr{I}_1,(i,j)\in\Omega. \tag{5.24}$$

Based on system (5.13), the following equality holds for any matrix $\mathscr{G}_i$, $i \in \mathscr{I}$,

$$\begin{aligned} 0 &= 2\mathscr{G}_i\left\{\tilde{\mathrm{a}}_i+\Delta\tilde{\mathrm{a}}_i+(\tilde{\mathscr{A}}_i+\Delta\tilde{\mathscr{A}}_i)z(k)-\tilde{E}z(k+1)\right\} \\ &= 2\mathscr{G}_i\begin{bmatrix} \tilde{\mathrm{a}}_i+\Delta\tilde{\mathrm{a}}_i & \tilde{\mathscr{A}}_i+\Delta\tilde{\mathscr{A}}_i & -\mathbf{I}\end{bmatrix}\vartheta(k). \end{aligned} \tag{5.25}$$

Combining (5.24) with (5.25), one has,

$$\vartheta^{\mathrm{T}}(k)\left\{\begin{bmatrix} \rho_i(q_i^{\mathrm{T}}q_i-1) & \star & \star \\ \rho_i Q_i^{\mathrm{T}} q_i & \rho_i Q_i^{\mathrm{T}}Q_i-\tilde{E}^{\mathrm{T}}P_i\tilde{E} & \star \\ \mathbf{0} & \mathbf{0} & P_j \end{bmatrix}+\mathtt{Sym}\{\mathscr{G}_i\mathscr{A}_i\}\right\}\vartheta(k)<0,$$
$$i\in\mathscr{I}_1,(i,j)\in\Omega, \tag{5.26}$$

where $\mathscr{A}_i = \begin{bmatrix} \tilde{\mathrm{a}}_i+\Delta\tilde{\mathrm{a}}_i & \tilde{\mathscr{A}}_i+\Delta\tilde{\mathscr{A}}_i & -\mathbf{I}\end{bmatrix}$.

In order to guarantee $\Delta V(k) < 0$, it follows from (5.26) that

$$\begin{bmatrix} \rho_i(q_i^{\mathrm{T}}q_i-1) & \star & \star \\ \rho_i Q_i^{\mathrm{T}} q_i & \rho_i Q_i^{\mathrm{T}}Q_i-\tilde{E}^{\mathrm{T}}P_i\tilde{E} & \star \\ \mathbf{0} & \mathbf{0} & P_j \end{bmatrix}+\mathtt{Sym}\{\mathscr{G}_i\mathscr{A}_i\}<0,$$
$$i\in\mathscr{I}_1,(i,j)\in\Omega. \tag{5.27}$$

Extracting the MFs, one has

$$\begin{bmatrix} \rho_i(q_i^{\mathrm{T}}q_i-1) & \star & \star \\ \rho_i Q_i^{\mathrm{T}} q_i & \rho_i Q_i^{\mathrm{T}}Q_i-\tilde{E}^{\mathrm{T}}P_i\tilde{E} & \star \\ \mathbf{0} & \mathbf{0} & P_j \end{bmatrix}+\mathtt{Sym}\{\mathscr{G}_i\mathbb{A}_s\}<0,$$
$$s\in\mathscr{N}(i),i\in\mathscr{I}_1,(i,j)\in\Omega, \tag{5.28}$$

where

$$\begin{cases} \mathbb{A}_s = \begin{bmatrix} \tilde{a}_s+\Delta\tilde{a}_s & \tilde{A}_s+\Delta\tilde{A}_s & -\mathbf{I}\end{bmatrix}, \\ \tilde{A}_s+\Delta\tilde{A}_s = \begin{bmatrix} M(A_s+\Delta A_s) \\ KC \end{bmatrix}T^{-1}, \tilde{a}_s+\Delta\tilde{a}_i = \begin{bmatrix} M(a_s+\Delta a_s) \\ \mathbf{0} \end{bmatrix}. \end{cases} \tag{5.29}$$

Note that the sliding surface gain does not appear in the first row of $\tilde{a}_s+\Delta\tilde{a}_s$ and $\tilde{A}_s+\Delta\tilde{A}_s$. For numerical tractability, $\mathscr{G}_i$ can be specified as $\mathscr{G}_i = \begin{bmatrix} \mathbf{0} & G_{i2}^{\mathrm{T}} & G_{i1}^{\mathrm{T}} \end{bmatrix}^{\mathrm{T}}$, and the slack variable matrices G_{i1} and G_{i2} are specified as,

$$G_{i1} = \begin{bmatrix} G_{i1(1)} & JG_{(3)} \\ G_{i1(2)} & \rho_1 G_{(3)} \end{bmatrix}, G_{i2} = \begin{bmatrix} G_{i2(1)} & \mathbf{0} \\ G_{i2(2)} & \rho_2 G_{(3)} \end{bmatrix}, \tag{5.30}$$

where ρ_1 and ρ_2 are scalar parameters, and $J \in \Re^{(n_x-n_u)\times n_u}$ is an arbitrary matrix.

Define

$$\hat{K} = G_{(3)}K. \tag{5.31}$$

Based on (5.30)–(5.31), inequality (5.28) can be derived as,

$$\begin{bmatrix} \rho_i(q_i^{\mathrm{T}} q_i - 1) & \star & \star \\ \rho_i Q_i^{\mathrm{T}} q_i + \Gamma_{10} & \Gamma_{11} & \star \\ \Gamma_{20} & \Gamma_{21} - G_{i2}^{\mathrm{T}} & P_j - G_{i1} - G_{i1}^{\mathrm{T}} \end{bmatrix} < 0, \quad s \in \mathscr{N}(i), i \in \mathscr{I}_1, (i,j) \in \Omega, \tag{5.32}$$

where

$$\begin{cases} \Gamma_{10} = \begin{bmatrix} G_{i2(1)} \\ G_{i2(2)} \end{bmatrix} M(a_s + \Delta a_s), \\ \Gamma_{11} = \rho_i Q_i^{\mathrm{T}} Q_i - \tilde{E}^{\mathrm{T}} P_i \tilde{E} \\ \qquad + \mathtt{Sym}\left\{ \begin{bmatrix} G_{i2(1)} \\ G_{i2(2)} \end{bmatrix} M(A_s + \Delta A_s) T^{-1} + \begin{bmatrix} \mathbf{0} \\ \rho_2 \mathbf{I} \end{bmatrix} \hat{K} C T^{-1} \right\}, \\ \Gamma_{20} = \begin{bmatrix} G_{i1(1)} \\ G_{i1(2)} \end{bmatrix} M(a_s + \Delta a_s), \\ \Gamma_{21} = \begin{bmatrix} G_{i1(1)} \\ G_{i1(2)} \end{bmatrix} M(A_s + \Delta A_s) T^{-1} + \begin{bmatrix} J \\ \rho_1 \mathbf{I} \end{bmatrix} \hat{K} C T^{-1}. \end{cases} \tag{5.33}$$

Adopting Lemma A1 to tackle the parameter uncertainties presented in (1.3), then inequality (5.16) implies inequality (5.32) for a set of scalars $\varepsilon_i > 0$, $i \in \mathscr{I}$ based on Schur complement.

In addition, the condition in (5.16) indicates that $G_{(3)} + G_{(3)}^{\mathrm{T}} - P_{j3} > 0$, which implies that $G_{(3)}$ is invertible. The sliding surface gain can be attained via (5.18) consequently. The proof is then accomplished. ■

Remark 5.2.1. Notice that the state transformation matrix T in (5.6) is also not unique, which has been sufficiently discussed and proved as in Remark 2.2.2. It is also noted that when $P_{i1} = P_{j1} \equiv P_1 \in \Re^{(n_x - n_u)\times(n_x - n_u)}$, $P_{i2} = P_{j2} \equiv P_2 \in \Re^{n_u \times (n_x - n_u)}$, and $P_{i3} = P_{j3} \equiv P_3 \in \Re^{n_u \times n_u}$, the corresponding stability analysis results in Theorem 5.1 can be easily derived based on a CQLF, which tend to be conservative as the space partition information is not fully considered.

In the following, two output feedback sliding mode controllers will be designed to drive the state trajectories of the closed-loop system into a neighbourhood of the sliding surface (5.5) locally.

5.3 OUTPUT FEEDBACK SLIDING MODE CONTROLLER DESIGN

Before designing the sliding mode controller, introduce a sliding patch Ψ firstly,

$$\Psi = \{x(k) \in \Re^{n_x} : \|x(k)\| \leq \sigma\} \tag{5.34}$$

where $\sigma > 0$ is a known positive scalar.

With the sliding surface gain obtained by Theorem 5.1, two output feedback sliding mode controller design methods will be proposed in the following subsection.

5.3.1 SLIDING MODE CONTROLLER DESIGN

Theorem 5.2. For the T-S fuzzy affine system (5.1) or equivalently (5.3) with the sliding surface given in (5.5), the resulting closed-loop system states can be forced into a neighbourhood the sliding surface (5.5) for the sliding patch Ψ by the following output feedback sliding mode controller,

$$u(k) = -(KCB)^{-1}\{\eta s(k)+(\|KCB\|\cdot\lambda(y(k),k)+\zeta(k))\,\mathrm{sgn}(s(k))\}, \quad (5.35)$$

where

$$\begin{aligned}\zeta(k) \geq \sigma \sum_{l=1}^{r} \mu_l \{\|KCA_l\|+\|KCH_{l1}\|\cdot\|H_{l2}\|+\gamma\|KCB\|\} \\ +\sum_{l=1}^{r} \mu_l \{\|KCa_l\|+\|KCH_{l1}\|\cdot\|H_{l3}\|\}\end{aligned} \quad (5.36)$$

with given small positive scalars η and σ.

Proof. Construct the subsequent Lyapunov function

$$V(k)=\frac{1}{2}s^{\mathrm{T}}(k)s(k). \quad (5.37)$$

The incremental $\Delta V(k)$ is

$$\begin{aligned}\Delta V(k) &= V(k+1)-V(k) \\ &= s^{\mathrm{T}}(k)\Delta s(k)+\Pi,\end{aligned} \quad (5.38)$$

where $\Delta s(k)=s(k+1)-s(k)$, $\Pi=\frac{1}{2}\Delta s^{\mathrm{T}}(k)\Delta s(k)$.

Based on the controller (5.35) and considering the matching condition in (5.2), then (5.38) can be formulated as

$$\begin{aligned}\Delta V(k) &= s^{\mathrm{T}}(k)\{KCx(k+1)-s(k)\}+\Pi \\ &= s^{\mathrm{T}}(k)\left\{\sum_{l=1}^{r}\mu_l KC[(A_l+\Delta A_l)x(k)+a_l+\Delta a_l+B(u(k)+f(x,k))]-s(k)\right\} \\ &\quad +\Pi \\ &= s^{\mathrm{T}}(k)\left\{\sum_{l=1}^{r}\mu_l[KC(A_l+\Delta A_l)x(k)+KC(a_l+\Delta a_l)+KCBf(x,k)-\eta s(k)\right. \\ &\qquad \left. -(\|KCB\|\cdot\lambda(y,k)+\zeta(k))\,\mathrm{sgn}(s(k))]-s(k)\right\}+\Pi \\ &\leq \sigma\sum_{l=1}^{r}\mu_l\{\|KCA_l\|+\|KC\Delta A_l\|+\gamma\|KCB\|\}\|s(k)\| \\ &\quad +\sum_{l=1}^{r}\mu_l\{\|KCa_l\|+\|KC\Delta a_l\|\}\|s(k)\|-(\eta+1)\|s(k)\|^2-\zeta(k)\|s(k)\|+\Pi \\ &\leq \sigma\sum_{l=1}^{r}\mu_l\{\|KCA_l\|+\|KCH_{l1}\|\cdot\|H_{l2}\|+\gamma\|KCB\|\}\|s(k)\| \\ &\quad +\sum_{l=1}^{r}\mu_l\{\|KCa_l\|+\|KCH_{l1}\|\cdot\|H_{l3}\|\}\|s(k)\| \\ &\quad -(\eta+1)\|s(k)\|^2-\zeta(k)\|s(k)\|+\Pi.\end{aligned} \quad (5.39)$$

Noticing that

$$\begin{aligned}\zeta(k) \geq \sigma \sum_{l=1}^{r} \mu_l \{\|KCA_l\| + \|KCH_{l1}\| \cdot \|H_{l2}\| + \gamma\|KCB\|\} \\ + \sum_{l=1}^{r} \mu_l \{\|KCa_l\| + \|KCH_{l1}\| \cdot \|H_{l3}\|\},\end{aligned} \tag{5.40}$$

then one has

$$\Delta V(k) \leq -(\eta+1)\|s(k)\|^2 + \Pi. \tag{5.41}$$

As the sliding patch Ψ is introduced in (5.34), one has $\|x(k)\| \leq \sigma$. From this perspective, it is noted that $\|\Delta s(k)\| \leq 2\|KC\|\sigma$, and then we have $\Pi \leq 2\|KC\|^2\sigma^2$. $\Delta V(k) < 0$ can be ensured by tuning the parameter η. Then it can be shown that the states of the fuzzy affine system (5.1) can be forced into a neighbourhood of the sliding surface by the output feedback sliding mode controller (5.35) in the region Ψ. The proof is accomplished. ■

It is worthy of mentioning that an assumption that $\|x(k)\| \leq \sigma$ in (5.34) for is introduced for Theorem 5.2 in fact. The sliding mode controller (5.35) can only guarantee the local reachability of the sliding surface. In order to relax this assumption, in the following, an adaptive output feedback SMC method will be proposed for the case when σ is not available.

5.3.2 ADAPTIVE SLIDING MODE CONTROLLER DESIGN

Theorem 5.3. For the system (5.1) or equivalently (5.3) with the sliding surface given in (5.5), the resulting closed-loop system states can be driven into a neighbourhood the sliding surface in the region Ψ by the following output feedback adaptive sliding mode controller

$$u(k) = -(KCB)^{-1}\left\{\eta s(k) + \left[\|KCB\| \cdot \lambda(y(k),k) + \hat{\zeta}(k)\right]\operatorname{sgn}(s(k))\right\}, \tag{5.42}$$

where

$$\begin{aligned}\hat{\zeta}(k) \geq \hat{\sigma}(k) \sum_{l=1}^{r} \mu_l \{\|KCA_l\| + \|KCH_{l1}\| \cdot \|H_{l2}\| + \gamma\|KCB\|\} \\ + \sum_{l=1}^{r} \mu_l \{\|KCa_l\| + \|KCH_{l1}\| \cdot \|H_{l3}\|\}\end{aligned} \tag{5.43}$$

and $\hat{\sigma}(k)$ is generated by the following adaptive law,

$$\begin{aligned}\Delta\hat{\sigma}(k) &= \hat{\sigma}(k+1) - \hat{\sigma}(k) \\ &= \varpi \sum_{l=1}^{r} \mu_l \{\|KCA_l\| + \|KCH_{l1}\| \cdot \|H_{l2}\| + \gamma\|KCB\|\} \|s(k)\|\end{aligned} \tag{5.44}$$

where $\eta > 0$ and $\varpi > 0$ are two given positive scalars, and $\hat{\sigma}(k)$ is the estimation of σ.

Proof. Construct the subsequent Lyapunov function

$$V(k)=\frac{1}{2}\left[s^{\mathrm{T}}(k)s(k)+\frac{1}{\varpi}\sigma_e^2(k)\right], \tag{5.45}$$

where $\sigma_e(k)=\sigma-\hat{\sigma}(k)$.

The incremental $\Delta V(k)$ is

$$\begin{aligned}\Delta V(k)&=V(k+1)-V(k)\\&=s^{\mathrm{T}}(k)\Delta s(k)+\frac{1}{\varpi}\sigma_e(k)\Delta\sigma_e(k)+\Phi,\end{aligned} \tag{5.46}$$

where $\Delta\sigma_e(k)=\sigma_e(k+1)-\sigma_e(k)$, $\Delta s(k)=s(k+1)-s(k)$, and $\Phi=\frac{1}{2\varpi}\Delta\sigma_e^2(k)+\frac{1}{2}\Delta s^{\mathrm{T}}(k)\Delta s(k)$.

Define $\Delta\hat{\sigma}(k)=\hat{\sigma}(k+1)-\hat{\sigma}(k)$. Noticing $\Delta\sigma_e(k)=-\Delta\hat{\sigma}(k)$ and considering the matching condition in (5.2), then based on the controller (5.42), (5.46) is formulated as

$$\begin{aligned}\Delta V(k)&=s^{\mathrm{T}}(k)\left\{KCx(k+1)-s(k)\right\}-\frac{1}{\varpi}\sigma_e(k)\Delta\hat{\sigma}(k)+\Phi\\
&=s^{\mathrm{T}}(k)\left\{\sum_{l=1}^{r}\mu_l KC\left[(A_l+\Delta A_l)x(k)+a_l\right.\right.\\
&\quad\left.\left.+\Delta a_l+B(u(k)+f(x,k))\right]-s(k)\right\}-\frac{1}{\varpi}\sigma_e(k)\Delta\hat{\sigma}(k)+\Phi\\
&=s^{\mathrm{T}}(k)\left\{\sum_{l=1}^{r}\mu_l\left[KC(A_l+\Delta A_l)x(k)+KC(a_l+\Delta a_l)+KCBf(x,k)\right.\right.\\
&\quad\left.\left.-\eta s(k)-\left(\|KCB\|\cdot\lambda(y,k)+\hat{\zeta}(k)\right)\mathrm{sgn}(s(k))\right]-s(k)\right\}-\frac{1}{\varpi}\sigma_e(k)\Delta\hat{\sigma}(k)+\Phi\\
&\le\sigma\sum_{l=1}^{r}\mu_l\left\{\|KCA_l\|+\|KCH_{l1}\|\cdot\|H_{l2}\|+\gamma\|KCB\|\right\}\|s(k)\|\\
&\quad+\sum_{l=1}^{r}\mu_l\left\{\|KCa_l\|+\|KCH_{l1}\|\cdot\|H_{l3}\|\right\}\|s(k)\|-\hat{\zeta}(k)\|s(k)\|-(\eta+1)\|s(k)\|^2\\
&\quad-(\sigma-\hat{\sigma}(k))\sum_{l=1}^{r}\mu_l\left\{\|KCA_l\|+\|KCH_{l1}\|\cdot\|H_{l2}\|+\gamma\|KCB\|\right\}\|s(k)\|+\Phi\\
&=\hat{\sigma}(k)\sum_{l=1}^{r}\mu_l\left\{\|KCA_l\|+\|KCH_{l1}\|\cdot\|H_{l2}\|+\gamma\|KCB\|\right\}\|s(k)\|\\
&\quad+\sum_{l=1}^{r}\mu_l\left\{\|KCa_l\|+\|KCH_{l1}\|\cdot\|H_{l3}\|\right\}\|s(k)\|-\hat{\zeta}(k)\|s(k)\|\\
&\quad-(\eta+1)\|s(k)\|^2+\Phi.\end{aligned} \tag{5.47}$$

Noticing that

$$\begin{aligned}\hat{\zeta}(k)\ge\hat{\sigma}(k)&\sum_{l=1}^{r}\mu_l\left\{\|KCA_l\|+\|KCH_{l1}\|\cdot\|H_{l2}\|+\gamma\|KCB\|\right\}\\&+\sum_{l=1}^{r}\mu_l\left\{\|KCa_l\|+\|KCH_{l1}\|\cdot\|H_{l3}\|\right\},\end{aligned} \tag{5.48}$$

then one has

$$\Delta V(k) \leq -(\eta+1)\|s(k)\|^2 + \Phi. \tag{5.49}$$

Following a similar process in the proof of Theorem 5.2 and based on (5.44), $\Delta V(k) < 0$ can be guaranteed via properly regulating the parameter η. Thus, the states of fuzzy affine system (5.1) can be driven into a neighbourhood of the sliding surface (5.5) by the output feedback adaptive sliding mode controller (5.42) in the region Ψ. The proof is accomplished. ■

Note that the proposed output feedback adaptive sliding mode controller (5.42) can only guarantee the local convergence of the sliding surface. The parameter ϖ can be tuned to determine the rate of the convergence and the precision of the estimation $\hat{\sigma}(k)$.

5.3.3 ILLUSTRATIVE EXAMPLES

Example 5.1. Consider a simplified pitch model of the helicopter as in Example 4.1. In this example, the helicopter model is characterized by the following equation

$$\begin{cases} \dot{\mathbf{x}}_1 = \mathbf{x}_2 \\ \dot{\mathbf{x}}_2 = \dfrac{1}{I_{yy}}\left[-mI_{cgx}g\cos(\mathbf{x}_1) - mI_{cgz}g\sin(\mathbf{x}_1) - F_{vM}\mathbf{x}_2 + u\right], \end{cases} \tag{5.50}$$

where $\mathbf{x}_1$ and $\mathbf{x}_2$ are the pitch angle and pitch rate, respectively. The nominal plant parameters in this example can be found in Table 4.1.

Following a same coordinate change and modelling method as in Example 4.1 and discretizzing the helicopter system with sampling period $T = 0.05$s, and then the system (5.50) can be described by the following discrete-time T-S fuzzy affine model with three rules,

Plant Rule $\mathscr{R}^l$: IF $x_1(k)$ is $\mathscr{F}_1^l$, THEN

$$\begin{cases} x(k+1) = (A_l + \Delta A_l)x(k) + a_l + \Delta a_l + B(u(k) + f(x(k),k)) \\ y(k) = Cx(k),\ l \in \{1,2,3\}, \end{cases} \tag{5.51}$$

where the fuzzy sets are given as: "$\mathscr{F}_1^1$" is "about $-\frac{\pi}{5}$", and "$\mathscr{F}_1^2$" is "about 0", and "$\mathscr{F}_1^3$" is "about $\frac{\pi}{5}$", and the system matrices are given as in (4.46), and the measurement output matrix $C = \begin{bmatrix} 0.1 & 1 \end{bmatrix}$. The parameter uncertainties ΔA_l and Δa_l are given in the form of (1.3) with matrices H_{11}, H_{12}, and H_{13} given as in (4.47).

Figure 5.1 shows the MFs in this example with $d_1 = \frac{\pi}{5}$ and $d_2 = \frac{3\pi}{5}$. It can be observed from Figure 5.1 that the premise variable space is decomposed into three regions as in (3.66) with $n_x = 2$.

Based on the approach proposed in section 5.2, the transformation matrix T can be chosen as $T = \begin{bmatrix} 1 & 0 \\ 0 & 1.7651 \end{bmatrix}$, and then $z_1(k) = x_1(k)$. These three regions can be accurately characterized by degenerate ellipsoids as in (2.16) with Q_i and q_i given as in (4.48).

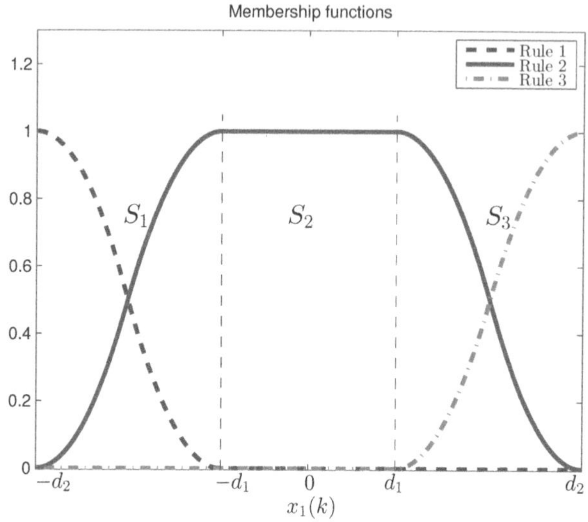

Figure 5.1 Membership functions of Example 5.1-Example 5.4.

The purpose of this example is to design a sliding surface (5.5) and an output feedback sliding mode controller (5.35) such that the closed-loop system states can be driven into a neighbourhood of the sliding surface (5.5) locally and the sliding motion is asymptotically stable. *Note that the SMC design approaches given in [94, 95] cannot be utilized for the discrete-time T-S fuzzy affine systems in this example.* Applying Theorem 5.1 with $\rho_1 = 1$, $\rho_2 = 0$ and $J = -10$, the sliding surface gain can be obtained as $K = 1.0320$.

To show the effectiveness of the proposed control scheme, simulations are conducted with different initial conditions. Under initial condition $x_0 = \begin{bmatrix} \frac{2\pi}{5} & 0 \end{bmatrix}^{\mathrm{T}}$ and $f(x(k),k) = \begin{bmatrix} 0.2 & 0.3 \end{bmatrix} x(k) + 0.3\sin(0.1k)y(k)$, via using the output feedback sliding mode controller in (5.35) with $\sigma = 0.05$, $\gamma = 0.45$, and $\eta = 1.12$, the closed-loop system states and control input are shown in Figure 5.2 and Figure 5.3, respectively. The sliding surface is shown in Figure 5.4.

To show the effectiveness of the proposed output feedback adaptive sliding mode controller, another numerical example will be considered.

Example 5.2. Consider a discrete-time T-S fuzzy affine system (5.1) with three local models as,

$$\left\{ \begin{aligned} A_1 &= \begin{bmatrix} 0.2 & -0.1 & 0 \\ 0.1 & 0.2 & 0 \\ 0.1 & 0.2 & 1.5 \end{bmatrix}, A_2 = \begin{bmatrix} 0.1 & 0.9 & 0.2 \\ 0.1 & 0.3 & 0.1 \\ 0.2 & 0.1 & 1.1 \end{bmatrix}, A_3 = \begin{bmatrix} 0.3 & 0.1 & 0.2 \\ 0.2 & 0.3 & 0.5 \\ 0.2 & -0.2 & 1.2 \end{bmatrix}, \\ a_1 &= \begin{bmatrix} 0 \\ 0 \\ 0.5 \end{bmatrix}, a_2 = \begin{bmatrix} 0 \\ 0 \\ 0 \end{bmatrix}, a_3 = \begin{bmatrix} 0 \\ 0 \\ -0.5 \end{bmatrix}, B = \begin{bmatrix} 0 \\ 0 \\ 2 \end{bmatrix}, C = \begin{bmatrix} 1 & 0 & 1 \end{bmatrix} \end{aligned} \right. \tag{5.52}$$

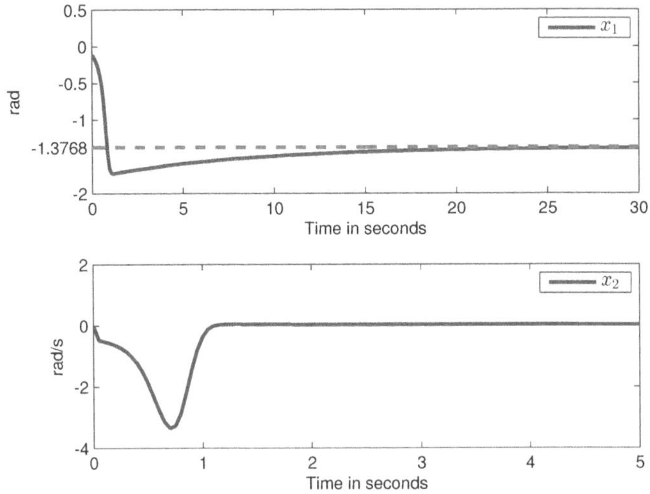

Figure 5.2 The closed-loop system states with sliding mode controller (5.35) in Example 5.1.

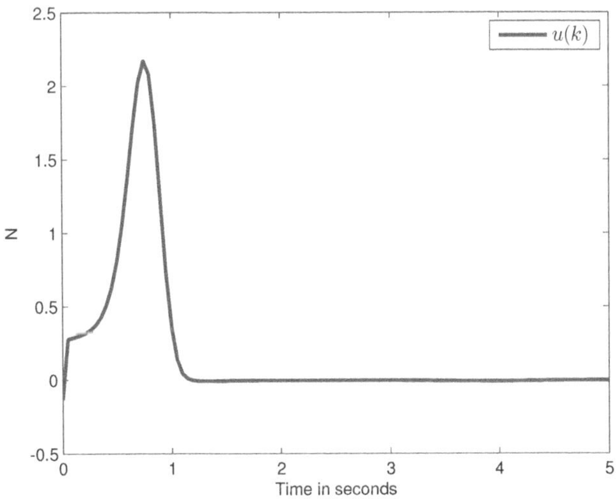

Figure 5.3 The control input with sliding mode controller (5.35) in Example 5.1.

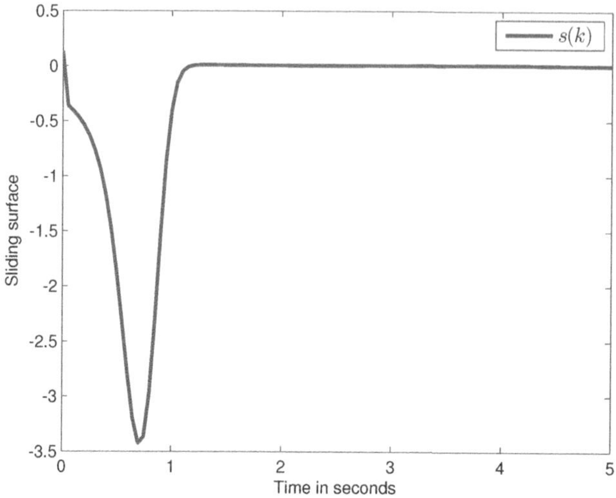

Figure 5.4 The sliding surface with sliding mode controller (5.35) in Example 5.1.

and the parameter uncertainties ΔA_l and Δa_l are given in the form of (1.3) with

$$\begin{cases} H_{11} = \begin{bmatrix} 0.1 \\ 0.1 \\ 0.1 \end{bmatrix}, H_{21} = \begin{bmatrix} 0.1 \\ 0.15 \\ 0.1 \end{bmatrix}, H_{31} = \begin{bmatrix} 0.15 \\ 0.1 \\ 0.1 \end{bmatrix}, \\ H_{12} = \begin{bmatrix} 0.1 & 0.1 & 0.1 \end{bmatrix}, H_{22} = \begin{bmatrix} 0.12 & 0.1 & 0.1 \end{bmatrix}, H_{32} = \begin{bmatrix} 0.1 & 0.11 & 0.1 \end{bmatrix}, \\ H_{13} = 0.1, H_{23} = 0.08, H_{33} = 0.1. \end{cases} \tag{5.53}$$

The MFs in this example are also shown in Figure 5.1 with $d_1 = 50$ and $d_2 = 300$. The premise variable space is decomposed into three regions as in (3.66) with $n_x = 3$. The transformation matrix T is chosen as $T = \begin{bmatrix} 1 & 0 & 0 \\ 0 & 1 & 0 \\ 0 & 0 & 2 \end{bmatrix}$, and it can be seen that $z_1(k) = x_1(k)$. The characterizing matrices can be obtained as $Q_1 = Q_3 = \begin{bmatrix} \frac{2}{d_2 - d_1} & 0 & 0 \end{bmatrix}$, $Q_2 = \begin{bmatrix} \frac{1}{d_1} & 0 & 0 \end{bmatrix}$, $q_1 = \frac{d_1 + d_2}{d_2 - d_1}$, $q_2 = 0$, and $q_3 = \frac{d_1 + d_2}{d_1 - d_2}$.

By applying the PQLFs-based scheme in Theorem 5.1 with $\rho_1 = 5$, $\rho_2 = -15$, and $J = \begin{bmatrix} -10 & 0 \end{bmatrix}$, one attains feasible solutions with the sliding surface gain $K = 0.6024$.

When the parameter σ is available and given as $\sigma = 0.01$, under initial condition $x_0 = \begin{bmatrix} 3 & -1.2 & 0.6 \end{bmatrix}^{\mathrm{T}}$ and the nonlinear disturbance $f(x(k), k) = \begin{bmatrix} 0.1 & 0.1 & 0.1 \end{bmatrix} x(k) + 0.8 \sin(k) y(k)$, use the output feedback sliding mode controller (5.35) with $\eta = 0.6$ and $\gamma = 0.2$, and then the closed-loop system states and control input are shown in Figure 5.5 and Figure 5.6, respectively.

If the parameter σ is not available, the sliding mode controller (5.35) and the output feedback SMC design methods proposed in [102] are not applicable, and the

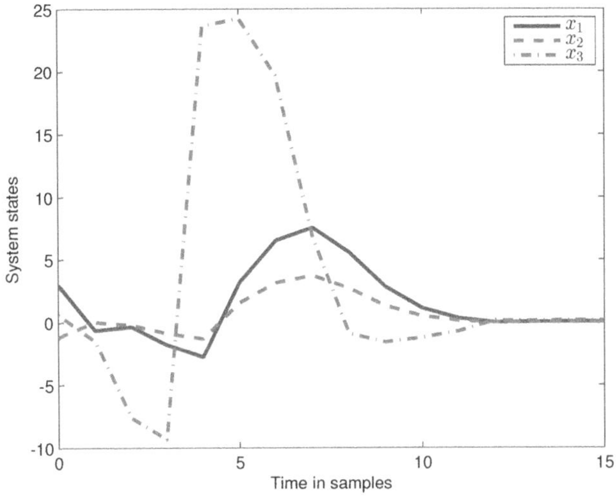

Figure 5.5 The closed-loop system states with sliding mode controller (5.35) in Example 5.2.

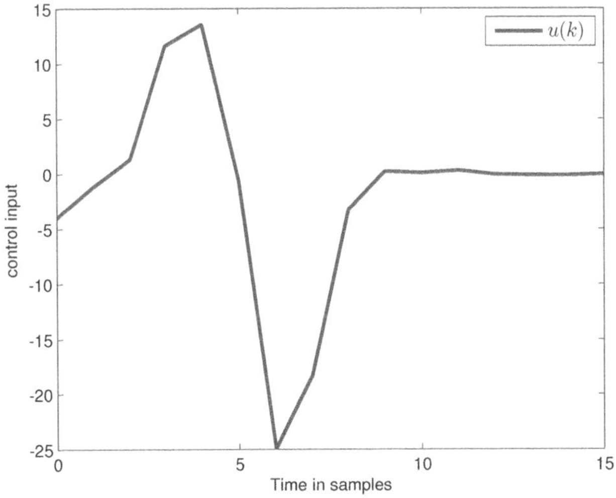

Figure 5.6 The control input with sliding mode controller (5.35) in Example 5.2.

output feedback adaptive sliding mode controller in (5.42) can be utilized instead. With the same initial condition x_0 and disturbance $f(x(k),k)$, apply the output feedback adaptive sliding mode controller (5.42) with $\eta = 1.2$, $\gamma = 0.2$, and $\varpi = 0.01$, and the system states and control input are shown in Figure 5.7 and Figure 5.8, respectively.

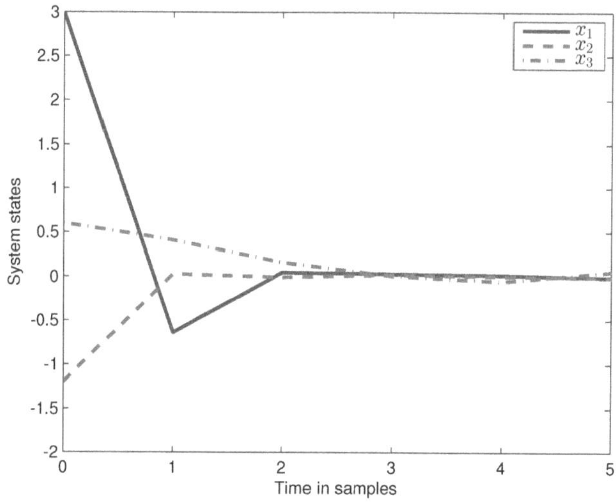

Figure 5.7 The closed-loop system states with adaptive sliding mode controller (5.42) in Example 5.2.

The simulation results clearly illustrate the effectiveness of the proposed two output feedback sliding mode controller design methods.

5.4 OUTPUT FEEDBACK DYNAMIC SMC DESIGN VIA T-S FUZZY AFFINE MODELS

When each local model of T-S fuzzy systems does not share a common control input matrix, the above-mentioned output feedback SMC design methods are not applicable. In the following, an output feedback dynamic sliding mode controller design approach will be proposed, and the asymptotic stability and robust $\mathscr{H}_\infty$ performance of the sliding motion will be also analyzed under the framework of PQLFs.

5.4.1 MODEL DESCRIPTION: CASE II

Consider the following discrete-time uncertain T-S fuzzy affine system which is characterized by r fuzzy **IF-THEN** rules as

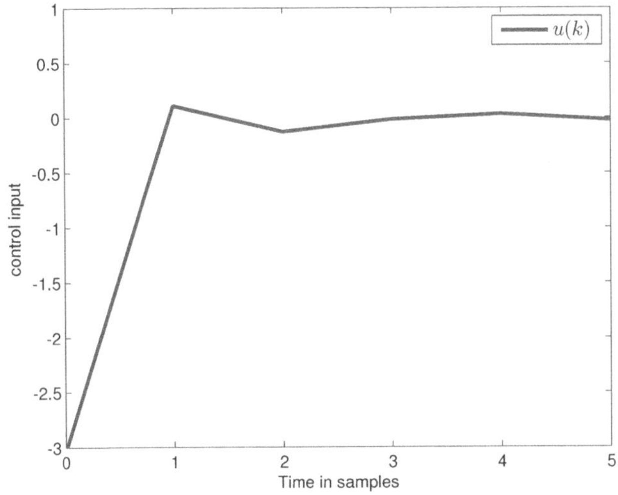

Figure 5.8 The control input with adaptive sliding mode controller (5.42) in Example 5.2.

Plant Rule $\mathscr{R}^l$: IF $\theta_1(k)$ is $\mathscr{F}_1^l$ and $\cdots$ and $\theta_\varphi(k)$ is $\mathscr{F}_\varphi^l$, THEN

$$\begin{cases} x(k+1) = (A_l + \Delta A_l)x(k) + a_l + \Delta a_l + B_l u(k) + D_l w(k) \\ y(k) = Cx(k) \\ z(k) = L_l x(k) + N_l u(k),\ l \in \mathscr{L} := \{1, 2, \ldots, r\}, \end{cases} \tag{5.54}$$

where $\mathscr{R}^l$ is the l-th fuzzy inference rule; $\mathscr{F}_\phi^l(\phi = 1, \ldots, \varphi)$ are fuzzy sets; r is the number of inference rules; $\theta(k) := [\theta_1(k), \ldots, \theta_\varphi(k)]$ represent the measurable premise variables; $x(k) \in \Re^{n_x}$ is the system state; $u(k) \in \Re^{n_u}$ is the control input; $w(k) \in \Re^{n_w}$ is the external disturbance belonging to $l_2[0, \infty)$; $y(k) \in \Re^{n_z}$ is the measurement output; $z(k) \in \Re^{n_z}$ is the regulated output; $C = C_l$, $l \in \mathscr{L}$ is the common measurement output matrix; ΔA_l and Δa_l are the parameter uncertainties satisfying the same condition described as in (5.1).

Similar to Chapter 2, the global T-S fuzzy affine model can be formulated into the following,

$$\begin{cases} x(k+1) = (A(\mu) + \Delta A(\mu))x(k) + a(\mu) + \Delta a(\mu) + B(\mu)u(k) + D(\mu)w(k) \\ z(k) = L(\mu)x(k) + N(\mu)u(k) \end{cases} \tag{5.55}$$

where

$$\begin{cases} A(\mu) = \sum_{l=1}^{r} \mu_l A_l, \Delta A(\mu) = \sum_{l=1}^{r} \mu_l \Delta A_l, a(\mu) = \sum_{l=1}^{r} \mu_l a_l, \Delta a(\mu) = \sum_{l=1}^{r} \mu_l \Delta a_l, \\ B(\mu) = \sum_{l=1}^{r} \mu_l B_l, D(\mu) = \sum_{l=1}^{r} \mu_l D_l, L(\mu) = \sum_{l=1}^{r} \mu_l L_l, N(\mu) = \sum_{l=1}^{r} \mu_l N_l. \end{cases} \tag{5.56}$$

The premise variable space can be decomposed by a similar line in Chapter 2. The following set $\mathscr{N}(i) := \{m | \mu_m[\theta(k)] > 0,\ m \in \mathscr{L},\ \theta(k) \in S_i,\ i \in \mathscr{I}\}$ is introduced to describe the indices for local models utilized within the interpolation in each region S_i.

Based on set $\mathscr{N}(i)$, the system (5.55) can be formulated as

$$\begin{cases} x(k+1) = (\mathscr{A}_i + \Delta\mathscr{A}_i)x(k) + \mathrm{a}_i + \Delta\mathrm{a}_i + \mathscr{B}_i u(k) + \mathscr{D}_i w(k) \\ z(k) = \mathscr{L}_i x(k) + \mathscr{N}_i u(k), \theta(k) \in S_i, i \in \mathscr{I}, \end{cases} \tag{5.57}$$

where

$$\begin{cases} \mathscr{A}_i := \sum_{m \in \mathscr{N}(i)} \mu_m A_m, \Delta\mathscr{A}_i := \sum_{m \in \mathscr{N}(i)} \mu_m \Delta A_m, \mathrm{a}_i := \sum_{m \in \mathscr{N}(i)} \mu_m a_m, \\ \Delta\mathrm{a}_i := \sum_{m \in \mathscr{N}(i)} \mu_m \Delta a_m, \mathscr{B}_i := \sum_{m \in \mathscr{N}(i)} \mu_m B_m, \mathscr{D}_i := \sum_{m \in \mathscr{N}(i)} \mu_m D_m, \\ \mathscr{L}_i := \sum_{m \in \mathscr{N}(i)} \mu_m L_m, \mathscr{N}_i := \sum_{m \in \mathscr{N}(i)} \mu_m N_m \end{cases} \tag{5.58}$$

with $0 < \mu_m[\theta(k)] \leq 1$, $\sum_{m \in \mathscr{N}(i)} \mu_m[\theta(k)] = 1$.

In the following, an output feedback dynamic sliding mode controller design approach will be proposed for the discrete-time T-S fuzzy affine system (5.54) to guarantee the reachability of the sliding surface. Sufficient conditions for asymptotic stability and robust $\mathscr{H}_\infty$ performance analysis of the sliding motion will be proposed based on PQLFs. For a prescribed level of disturbance attenuation $\gamma > 0$, asymptotical stability of the sliding motion can be ensured when $w(k) \equiv 0$, and for nonzero $w(k) \in l_2[0, \infty)$, it is satisfied that $\sum_{k=0}^{\infty} z^{\mathrm{T}}(k)z(k) < \gamma^2 \sum_{k=0}^{\infty} w^{\mathrm{T}}(k)w(k)$ under zero initial condition $x(k) = 0$.

5.4.2 SLIDING SURFACE DESIGN

For the discrete-time T-S fuzzy affine system (5.54), design the sliding surface based on the measurement output and control input as

$$s(k) = Ky(k) + u(k) = 0, \tag{5.59}$$

where $K \in \Re^{n_u \times n_y}$ is the sliding surface gain to be determined.

With (5.59), it can be easily obtained that

$$u(k) = -Ky(k) = -KCx(k). \tag{5.60}$$

Then the sliding motion can be described by the following descriptor system based on system (5.57) and (5.60),

$$\begin{cases} x(k+1) = (\mathscr{A}_i + \Delta\mathscr{A}_i)x(k) + \mathrm{a}_i + \Delta\mathrm{a}_i + \mathscr{B}_i u(k) + \mathscr{D}_i w(k) \\ 0 \times u(k+1) = -KCx(k) - u(k) \\ z(k) = \mathscr{L}_i x(k) + \mathscr{N}_i u(k), \theta(k) \in S_i, i \in \mathscr{I}. \end{cases} \tag{5.61}$$

Rewrite the system (5.61) into a compact form as

$$\begin{cases} E\bar{x}(k+1) = (\bar{\mathscr{A}_i} + \Delta\bar{\mathscr{A}_i})\bar{x}(k) + \bar{\mathrm{a}}_i + \Delta\bar{\mathrm{a}}_i + \bar{\mathscr{D}}_i w(k) \\ z(k) = \bar{\mathscr{L}}_i \bar{x}(k), i \in \mathscr{I}, \end{cases} \tag{5.62}$$

where

$$\begin{cases} E = \begin{bmatrix} \mathbf{I}_{n_x} & \mathbf{0} \\ \mathbf{0} & \mathbf{0} \end{bmatrix}, \bar{x}(k) = \begin{bmatrix} x(k) \\ u(k) \end{bmatrix}, \bar{\mathscr{A}_i} = \begin{bmatrix} \mathscr{A}_i & \mathscr{B}_i \\ -K & -\mathbf{I} \end{bmatrix}, \Delta\bar{\mathscr{A}_i} = \begin{bmatrix} \Delta\mathscr{A}_i & \mathbf{0} \\ \mathbf{0} & \mathbf{0} \end{bmatrix}, \\ \bar{\mathrm{a}}_i = \begin{bmatrix} \mathrm{a}_i \\ \mathbf{0} \end{bmatrix}, \Delta\bar{\mathrm{a}}_i = \begin{bmatrix} \Delta\mathrm{a}_i \\ \mathbf{0} \end{bmatrix}, \bar{\mathscr{D}}_i = \begin{bmatrix} \mathscr{D}_i \\ \mathbf{0} \end{bmatrix}, \bar{\mathscr{L}}_i = \begin{bmatrix} \mathscr{L}_i & \mathscr{N}_i \end{bmatrix}. \end{cases} \tag{5.63}$$

By combining the first equation of system (5.57) with the sliding surface (5.59), the dynamical properties of the sliding motion can be fully described by the descriptor system (5.62). Along a similar line in Chapter 2, it is assumed that one can use an ellipsoid $\mathscr{E}_i := \{x| \|Q_i x + q_i\| \leq 1\}$ outer approximate each polyhedral region S_i in (5.61), and the relationship given as in (2.83) holds for each region.

For future use, a set $\Omega := \{(i,j)|x(k) \in S_i, x(k+1) \in S_j, i, j \in \mathscr{I}\}$ is introduced to describe region transitions similar to Section 4.4.

5.4.3 STABILITY AND ROBUST PERFORMANCE ANALYSIS OF SLIDING MOTION

Note that the CQLF can be generally viewed as a special case of the PQLFs, thus, the stability and robust performance analysis of the sliding motion will be conducted based on PQLFs in this subsection. The readers can derive the stability and robust performance analysis results easily based on a CQLF by using the method discussed as in Remark 5.2.1.

Theorem 5.4. The sliding mode dynamical system (5.62) is asymptotically stable with robust $\mathscr{H}_\infty$ performance γ, if there exist matrices $0 < P_i = P_i^{\mathrm{T}} \in \Re^{(n_x+n_u)\times(n_x+n_u)}$, $\{Y_{i1}, W_{i1}\} \in \Re^{n_x \times n_x}$, $\{Y_{i2}, W_{i2}\} \in \Re^{n_u \times n_x}$, $i \in \mathscr{I}$, $Y_0 \in \Re^{n_u \times n_u}$, $\bar{K} \in \Re^{n_u \times n_y}$, and scalars $\{\varepsilon_i > 0, i \in \mathscr{I}\}$, $\{\kappa_i < 0, i \in \mathscr{I}_1\}$ exist, such that the following conditions hold,

$$\begin{bmatrix} \Psi_{11} & \star & \star & \star \\ \Psi_{21} - W_i^{\mathrm{T}} & P_j - Y_i - Y_i^{\mathrm{T}} & \star & \star \\ \Psi_{31} & \Psi_{32} & -\gamma^2\mathbf{I} & \star \\ \Psi_{41} & \Psi_{42} & \mathbf{0} & -\varepsilon_i\mathbf{I} \end{bmatrix} < 0, \tag{5.64}$$
$$i \in \mathscr{I}_0, m \in \mathscr{N}(i),$$

$$\begin{bmatrix} \kappa_i(q_i^{\mathrm{T}}q_i-1)+\varepsilon_i H_{m3}^{\mathrm{T}}H_{m3} & \star & \star & \star & \star \\ \kappa_i\bar{Q}_i^{\mathrm{T}}q_i+\Psi_{10} & \rho_i\bar{Q}_i^{\mathrm{T}}\bar{Q}_i+\Psi_{11} & \star & \star & \star \\ \begin{bmatrix} Y_{i1} \\ Y_{i2} \end{bmatrix} a_m & \Psi_{21}-W_i^{\mathrm{T}} & P_j-Y_i-Y_i^{\mathrm{T}} & \star & \star \\ \mathbf{0} & \Psi_{31} & \Psi_{32} & -\gamma^2\mathbf{I} & \star \\ \mathbf{0} & \Psi_{41} & \Psi_{42} & \mathbf{0} & -\varepsilon_i\mathbf{I} \end{bmatrix} < 0,$$
$$i \in \mathscr{I}_1, m \in \mathscr{N}(i), \tag{5.65}$$

where

$$\begin{cases} \Psi_{11} = \mathrm{Sym}\left\{ \begin{bmatrix} W_{i1} \\ W_{i2} \end{bmatrix} \begin{bmatrix} A_m & B_m \end{bmatrix} - \begin{bmatrix} J_2 \\ \rho_2\mathbf{I} \end{bmatrix} \begin{bmatrix} \bar{K}C & Y_0 \end{bmatrix} \right\} + \varepsilon_i \begin{bmatrix} H_{m2}^{\mathrm{T}} \\ \mathbf{0} \end{bmatrix} (\star) \\ \qquad -E^{\mathrm{T}}P_iE+\bar{L}_m^{\mathrm{T}}\bar{L}_m, \\ \Psi_{21} = \begin{bmatrix} Y_{i1} \\ Y_{i2} \end{bmatrix} \begin{bmatrix} A_m & B_m \end{bmatrix} - \begin{bmatrix} J_1 \\ \rho_1\mathbf{I} \end{bmatrix} \begin{bmatrix} \bar{K}C & Y_0 \end{bmatrix}, \\ \Psi_{31} = D_m^{\mathrm{T}} \begin{bmatrix} W_{i1}^{\mathrm{T}} & W_{i2}^{\mathrm{T}} \end{bmatrix}, \Psi_{32} = D_m^{\mathrm{T}} \begin{bmatrix} Y_{i1}^{\mathrm{T}} & Y_{i2}^{\mathrm{T}} \end{bmatrix}, \\ \Psi_{41} = H_{m1}^{\mathrm{T}} \begin{bmatrix} W_{i1}^{\mathrm{T}} & W_{i2}^{\mathrm{T}} \end{bmatrix}, \Psi_{42} = H_{m1}^{\mathrm{T}} \begin{bmatrix} Y_{i1}^{\mathrm{T}} & Y_{i2}^{\mathrm{T}} \end{bmatrix}, \\ \Psi_{10} = \begin{bmatrix} W_{i1} \\ W_{i2} \end{bmatrix} a_m + \varepsilon_i \begin{bmatrix} H_{m2}^{\mathrm{T}} \\ \mathbf{0} \end{bmatrix} H_{m3}, \\ Y_i = \begin{bmatrix} Y_{i1} & J_1Y_0 \\ Y_{i2} & \rho_1Y_0 \end{bmatrix}, W_i = \begin{bmatrix} W_{i1} & J_2Y_0 \\ W_{i2} & \rho_2Y_0 \end{bmatrix}, \\ \bar{L}_m = \begin{bmatrix} L_m & N_m \end{bmatrix}, \bar{Q}_i = \begin{bmatrix} Q_i & \mathbf{0}_{1\times n_u} \end{bmatrix} \end{cases} \tag{5.66}$$

and $\{J_1, J_2\} \in \Re^{n_x\times n_u}$ are arbitrary matrices, and $\rho_1 \neq 0$, ρ_2 are constant scalars.

The gain of the sliding surface (5.59) can be computed via

$$K = Y_0^{-1}\bar{K}. \tag{5.67}$$

Proof. It is worthy of pointing out that the condition (5.64) for $i \in \mathscr{I}_0$ is a special case of the condition (5.65) for $i \in \mathscr{I}_1$ in fact. Without loss of generality, only the more complex case for $i \in \mathscr{I}_1$ will be proved in the following. Consider the following PQLF as

$$V(k) = \bar{x}^{\mathrm{T}}(k)E^{\mathrm{T}}P_iE\bar{x}(k), i \in \mathscr{I}, \tag{5.68}$$

where $0 < P_i = P_i^{\mathrm{T}} \in \Re^{(n_x+n_u)\times(n_x+n_u)}$.

Based on the PQLF in (5.68), the sliding mode dynamical system (5.62) is asymptotically stable with robust $\mathscr{H}_\infty$ performance γ, if the following condition is satisfied,

$$V(k+1)-V(k)+z^{\mathrm{T}}(k)z(k)-\gamma^2 w^{\mathrm{T}}(k)w(k) < 0. \tag{5.69}$$

The inequality (5.69) can be rewritten as,

$$\begin{bmatrix} 1 \\ \bar{x}(k) \\ E\bar{x}(k+1) \\ w(k) \end{bmatrix}^{\mathrm{T}} \begin{bmatrix} 0 & \star & \star & \star \\ \mathbf{0} & \bar{\mathscr{L}}_i^{\mathrm{T}}\bar{\mathscr{L}}_i - E^{\mathrm{T}}P_iE & \star & \star \\ \mathbf{0} & \mathbf{0} & P_j & \star \\ \mathbf{0} & \mathbf{0} & \mathbf{0} & -\gamma^2\mathbf{I} \end{bmatrix} \begin{bmatrix} 1 \\ \bar{x}(k) \\ E\bar{x}(k+1) \\ w(k) \end{bmatrix} < 0,$$
$$(i,j) \in \Omega. \tag{5.70}$$

Noticing the partition information described as in (2.83) and using the S-procedure, the following inequality implies (5.70) for $\rho_i < 0$, $i \in \mathscr{I}_1$,

$$\xi^{\mathrm{T}}(k)\begin{bmatrix} \kappa_i(q_i^{\mathrm{T}}q_i-1) & \star & \star & \star \\ \kappa_i\bar{Q}_i^{\mathrm{T}}q_i & \kappa_i\bar{Q}_i^{\mathrm{T}}\bar{Q}_i+\bar{\mathscr{L}}_i^{\mathrm{T}}\bar{\mathscr{L}}_i-E^{\mathrm{T}}P_iE & \star & \star \\ \mathbf{0} & \mathbf{0} & P_j & \star \\ \mathbf{0} & \mathbf{0} & \mathbf{0} & -\gamma^2\mathbf{I} \end{bmatrix}\xi(k)<0, \quad i \in \mathscr{I}_1,(i,j)\in\Omega, \tag{5.71}$$

where

$$\begin{cases} \xi(k)=\begin{bmatrix} 1 & \bar{x}^{\mathrm{T}}(k) & E\bar{x}^{\mathrm{T}}(k+1) & w^{\mathrm{T}}(k) \end{bmatrix}^{\mathrm{T}}, \\ \bar{Q}_i=\begin{bmatrix} Q_i & \mathbf{0}_{1\times n_u} \end{bmatrix}. \end{cases} \tag{5.72}$$

Based on the sliding mode dynamical system (5.62), and for arbitrary matrices $\mathscr{G}_i$, $i \in \mathscr{I}$, it is satisfied that

$$\begin{aligned} 0 &= 2\mathscr{G}_i\left\{\bar{\mathrm{a}}_i+\Delta\bar{\mathrm{a}}_i+(\bar{\mathscr{A}}_i+\Delta\bar{\mathscr{A}}_i)\bar{x}(k)+\bar{\mathscr{D}}_iw(k)-E\bar{x}(k+1)\right\} \\ &= 2\mathscr{G}_i\begin{bmatrix} \bar{\mathrm{a}}_i+\Delta\bar{\mathrm{a}}_i & \bar{\mathscr{A}}_i+\Delta\bar{\mathscr{A}}_i & -\mathbf{I} & \bar{\mathscr{D}}_i \end{bmatrix}\xi(k). \end{aligned} \tag{5.73}$$

Combining (5.71) with (5.73), yields

$$\xi^{\mathrm{T}}(k)\left(\Pi+\mathtt{Sym}\{\mathscr{G}_i\mathscr{A}_i\}\right)\xi(k)<0, i \in \mathscr{I}_1,(i,j)\in\Omega, \tag{5.74}$$

where $\mathscr{A}_i=\begin{bmatrix} \bar{\mathrm{a}}_i+\Delta\bar{\mathrm{a}}_i & \bar{\mathscr{A}}_i+\Delta\bar{\mathscr{A}}_i & -\mathbf{I} & \bar{\mathscr{D}}_i \end{bmatrix}$.

It follows from (5.74) that the following inequality implies (5.69),

$$\Pi+\mathtt{Sym}\{\mathscr{G}_i\mathscr{A}_i\}<0, i \in \mathscr{I}_1,(i,j)\in\Omega. \tag{5.75}$$

Extracting the MFs in (5.75), yields

$$\begin{bmatrix} \kappa_i(q_i^{\mathrm{T}}q_i-1) & \star & \star & \star \\ \kappa_i\bar{Q}_i^{\mathrm{T}}q_i & \kappa_i\bar{Q}_i^{\mathrm{T}}\bar{Q}_i+\bar{L}_m^{\mathrm{T}}\bar{L}_m-E^{\mathrm{T}}P_iE & \star & \star \\ \mathbf{0} & \mathbf{0} & P_j & \star \\ \mathbf{0} & \mathbf{0} & \mathbf{0} & -\gamma^2\mathbf{I} \end{bmatrix}+\mathtt{Sym}\{\mathscr{G}_n\mathscr{A}_m\}<0, \quad m\in\mathscr{N}(i), i \in \mathscr{I}_1,(i,j)\in\Omega, \tag{5.76}$$

where

$$\begin{cases} \mathscr{A}_m=\begin{bmatrix} \bar{a}_m+\Delta\bar{a}_m & \bar{A}_m+\Delta\bar{A}_m & -\mathbf{I} & \bar{D}_m \end{bmatrix}, \\ \bar{A}_m+\Delta\bar{A}_m=\begin{bmatrix} A_m+\Delta A_m & B_m \\ -K & -\mathbf{I} \end{bmatrix}, \\ \bar{a}_m+\Delta\bar{a}_m=\begin{bmatrix} a_m+\Delta a_m \\ \mathbf{0} \end{bmatrix}, \bar{L}_m=\begin{bmatrix} L_m & N_m \end{bmatrix}. \end{cases} \tag{5.77}$$

Note that the first row of $\bar{a}_m+\Delta\bar{a}_m$ and $\bar{A}_m+\Delta\bar{A}_m$ does not contain the sliding surface gain K. For numerical convexification, specify $\mathscr{G}_i=\begin{bmatrix} \mathbf{0} & W_i^{\mathrm{T}} & Y_i^{\mathrm{T}} & \mathbf{0} \end{bmatrix}^{\mathrm{T}}$, where the slack variable matrices Y_i and W_i are given as

$$Y_i=\begin{bmatrix} Y_{i1} & J_1Y_0 \\ Y_{i2} & \rho_1Y_0 \end{bmatrix}, W_i=\begin{bmatrix} W_{i1} & J_2Y_0 \\ W_{i2} & \rho_2Y_0 \end{bmatrix}, \tag{5.78}$$

where $\rho_1 \neq 0$ and $\rho_2 \neq 0$ are constant, and $\{J_1, J_2\} \in \Re^{n_x \times n_u}$ are arbitrary matrices.
Define

$$\bar{K} = Y_0 K. \tag{5.79}$$

Based on (5.78)–(5.79), one can derive the inequality (5.76) as,

$$\begin{bmatrix} \kappa_i(q_i^{\mathrm{T}} q_i - 1) & \star & \star & \star \\ \kappa_i \bar{Q}_i^{\mathrm{T}} q_i + \Lambda_{10} & \Lambda_{11} & \star & \star \\ \Lambda_{20} & \Lambda_{21} - W_i^{\mathrm{T}} & P_j - Y_i - Y_i^{\mathrm{T}} & \star \\ \mathbf{0} & \Lambda_{31} & \Lambda_{32} & -\gamma^2 \mathbf{I} \end{bmatrix} < 0, \tag{5.80}$$
$$m \in \mathscr{N}(i), i \in \mathscr{I}_1, (i,j) \in \Omega,$$

where

$$\begin{cases} \Lambda_{11} = \mathtt{Sym}\left\{ \begin{bmatrix} W_{i1} \\ W_{i2} \end{bmatrix} \begin{bmatrix} A_m + \Delta A_m & B_m \end{bmatrix} - \begin{bmatrix} J_2 \\ \rho_2 \mathbf{I} \end{bmatrix} \begin{bmatrix} \bar{K}C & Y_0 \end{bmatrix} \right\} \\ \quad + \kappa_i \bar{Q}_i^{\mathrm{T}} \bar{Q}_i - E^{\mathrm{T}} P_i E + \bar{L}_m^{\mathrm{T}} \bar{L}_m, \\ \Lambda_{21} = \begin{bmatrix} Y_{i1} \\ Y_{i2} \end{bmatrix} \begin{bmatrix} A_m + \Delta A_m & B_m \end{bmatrix} - \begin{bmatrix} J_1 \\ \rho_1 \mathbf{I} \end{bmatrix} \begin{bmatrix} \bar{K}C & Y_0 \end{bmatrix}, \\ \Lambda_{31} = D_m^{\mathrm{T}} \begin{bmatrix} W_{i1}^{\mathrm{T}} & W_{i2}^{\mathrm{T}} \end{bmatrix}, \Lambda_{32} = D_m^{\mathrm{T}} \begin{bmatrix} Y_{i1}^{\mathrm{T}} & Y_{i2}^{\mathrm{T}} \end{bmatrix}, \\ \Lambda_{10} = \begin{bmatrix} W_{i1} \\ W_{i2} \end{bmatrix} (a_m + \Delta a_m), \Lambda_{20} = \begin{bmatrix} V_{n1} \\ V_{n2} \end{bmatrix} (a_m + \Delta a_m). \end{cases} \tag{5.81}$$

Utilizing Lemma A1 to deal with the parameter uncertainties in (5.80), then for scalars $\varepsilon_i > 0$, the inequality (5.65) indicates (5.80) by Schur complement.

In addition, the conditions in (5.64)–(5.65) imply that $\rho_1 Y_0 + \rho_1 Y_0^{\mathrm{T}} - P_{j3} > 0$, which means that Y_0 is nonsingular due to $\rho_1 \neq 0$. Thus, the sliding surface gain can be computed by (5.67). This ends the proof. ∎

Due to the application of the descriptor piecewise affine system method and introduction of auxiliary matrix multipliers, the stability and robust $\mathscr{H}_\infty$ performance analysis results in Theorem 5.4 are proposed in terms of linear matrix inequalities, which can be solved efficiently by commercially available software. It is noted that analysis and design conservatism can be also further reduced due to the increased freedom introduced by the slack variables.

5.5 OUTPUT FEEDBACK DYNAMIC SLIDING MODE CONTROLLER DESIGN

Before the dynamic sliding mode controller design, similar to subchapter 5.3, the following sliding patch Ω is introduced as

$$\Omega := \{x(k) \in \Re^{n_x} : \|x(k)\| \leq \lambda\}. \tag{5.82}$$

5.5.1 DYNAMIC SLIDING MODE CONTROLLER DESIGN

Theorem 5.5. In view of the discrete-time T-S fuzzy affine system (5.54) with the sliding surface given in (5.59), the following output feedback dynamic sliding mode controller can drive the resultant closed-loop system states into a neighbourhood of the sliding surface (5.59) in the region Ω locally,

$$u(k+1)=u(k)-\sum_{l=1}^{r}\mu_l KC(a_l+B_l u(k))-(\eta+\vartheta(k))\text{sgn}(s(k)), \tag{5.83}$$

where $\eta>0$ is a given positive constant scalar, and

$$\begin{aligned}\vartheta(k)=\sum_{l=1}^{r}\mu_l\{(&\|KCA_l\|+\|KCH_{l1}\|\cdot\|H_{l2}\|+\|KC\|)\lambda+\|KCH_{l1}\|\cdot\|H_{l3}\|\\&+\|KCD_l\|\varpi(k)\},\end{aligned} \tag{5.84}$$

where $\varpi(k)$ represents the uniform upper-bound of external disturbance $w(k)$.

Proof. Construct the following Lyapunov function as,

$$V(k)=\frac{1}{2}s^{\mathrm{T}}(k)s(k). \tag{5.85}$$

The incremental $\Delta V(k)$ is obtained as

$$\begin{aligned}\Delta V(k)&=V(k+1)-V(k)\\&=s^{\mathrm{T}}(k)\Delta s(k)+\Upsilon,\end{aligned} \tag{5.86}$$

where $\Delta s(k)=s(k+1)-s(k)$, $s(k+1)=Ky(k+1)+u(k+1)$, $\Upsilon=\frac{1}{2}\Delta s^{\mathrm{T}}(k)\Delta s(k)$.

Noticing the sliding patch Ω given as in (5.82) and considering the output feedback dynamic sliding mode controller (5.83), yields,

$$\begin{aligned}\Delta V(k)&=s^{\mathrm{T}}(k)\{KCx(k+1)+u(k+1)-KCx(k)-u(k)\}+\Gamma\\&=s^{\mathrm{T}}(k)\left\{\sum_{l=1}^{r}\mu_l KC[(A_l+\Delta A_l)x(k)+a_l+\Delta a_l+B_l u(k)+D_l w(k)]\right.\\&\qquad\left.-KCx(k)+\Delta u(k)\right\}+\Upsilon\\&=s^{\mathrm{T}}(k)\sum_{l=1}^{r}\mu_l KC\{(A_l+\Delta A_l-\mathbf{I})x(k)+\Delta a_l+D_l w(k)\}-\vartheta(k)\|s(k)\|\\&\quad-\eta\|s(k)\|+\Upsilon\\&\le\|s(k)\|\sum_{l=1}^{r}\mu_l\{(\|KCA_l\|+\|KC\Delta A_l\|+\|KC\|)\lambda+\|KC\Delta a_l\|\\&\qquad+\|KCD_l\|\varpi(k)\}-\vartheta(k)\|s(k)\|-\eta\|s(k)\|+\Upsilon\\&\le\sum_{l=1}^{r}\mu_l\{(\|KCA_l\|+\|KCH_{l1}\|\|H_{l2}\|+\|KC\|)\lambda+\|KCH_{l1}\|\|H_{l3}\|\\&\qquad+\|KD_l\|\varpi(k)\}\|s(k)\|-\vartheta(k)\|s(k)\|-\eta\|s(k)\|+\Upsilon,\end{aligned} \tag{5.87}$$

where $\Delta u(k)=u(k+1)-u(k)$.

Notice that

$$\vartheta(k) = \sum_{l=1}^{r} \mu_l \{(\|KCA_l\| + \|KCH_{l1}\|\|H_{l2}\| + \|KC\|)\lambda + \|KCH_{l1}\|\|H_{l3}\| + \|KCD_l\|\varpi(k)\}, \tag{5.88}$$

and then one has

$$\Delta V(k) \leq -\eta \|s(k)\| + \Upsilon. \tag{5.89}$$

Notice that

$$\begin{aligned} \Delta s(k) &= KCx(k+1) - KCx(k) + u(k+1) - u(k) \\ &\leq \sum_{l=1}^{r} \mu_l \{(\|KCA_l\| + \|KCH_{l1}\|\|H_{l2}\| + \|KC\|)\lambda \\ &\qquad + \|KCH_{l1}\|\|H_{l3}\| + \|KCD_l\|\varpi(k)\} \\ &\qquad - \vartheta(k)\mathrm{sgn}(s(k)) - \eta\,\mathrm{sgn}(s(k)). \end{aligned} \tag{5.90}$$

Then the upper-bound of $s(k)$ can be obtained as

$$\|\Delta s(k)\| \leq \eta + 2\vartheta(k), \tag{5.91}$$

which means that $\|\Delta s(k)\|$ is bounded.

By tuning the parameter η appropriately, $\Delta V(k) < 0$ can be ensured. Thus, the system states of the discrete-time T-S fuzzy system (5.54) can be driven into a neighbourhood of the sliding surface (5.59) and maintained hereafter by the output feedback dynamic sliding mode controller (5.83). The proof is ended. ∎

5.5.2 ILLUSTRATIVE EXAMPLES

In this subsection, three simulation examples will be given to illustrate the effectiveness of the developed output feedback dynamic sliding mode controller design scheme.

Example 5.3. Consider a simplified pitch model of the helicopter, whose dynamics are described as in (5.50), and the nominal plant parameters are shown as in Table 4.1. Conduct a same coordinate change and modelling method as in Example 4.1 and discretize the helicopter system with sampling period $T = 0.05$s. The measurement output $y(k)$ and the regulated output $z(k)$ of this exmaple are given as $y(k) = x_2(k)$ and $z(k) = 0.1x_1(k) + 0.2x_2(k) + 0.1u(k)$, respectively. Then the helicopter (5.50) can be described by the following discrete-time T-S fuzzy affine model with three rules as,

Plant Rule $\mathscr{R}^l$: IF $x_1(k)$ is $\mathscr{F}_1^l$, THEN

$$\begin{cases} x(k+1) = (A_l + \Delta A_l)x(k) + a_l + \Delta a_l + B_l u(k) + D_l w(k) \\ y(k) = Cx(k) \\ z(k) = L_l x(k) + N_l u(k),\ l \in \mathscr{L} := \{1,2,3\}, \end{cases} \tag{5.92}$$

where the system matrices are given as in (4.46), and $C = \begin{bmatrix} 0 & 1 \end{bmatrix}$, $L_l = \begin{bmatrix} 0.5 & 0.2 \end{bmatrix}$, $N_l = 0.1$. The parameter uncertainties ΔA_l and Δa_l are given in the form of (1.3) with matrices H_{11}, H_{12}, and H_{13} given as in (4.47).

The normalized MFs of this example is shown as in Figure 5.1 with $d_1 = \frac{\pi}{5}$ and $d_2 = \frac{3\pi}{5}$. The premise variable space is divided into three regions described as in (3.66) with $n_x = 2$, which can be accurately characterized by degenerate ellipsoids as in (2.82) with $Q_1 = Q_3 = \begin{bmatrix} \frac{2}{d_2 - d_1} & 0 \end{bmatrix}$, $Q_2 = \begin{bmatrix} \frac{1}{d_1} & 0 \end{bmatrix}$, $q_1 = \frac{d_1 + d_2}{d_2 - d_1}$, $q_2 = 0$, and $q_3 = \frac{d_1 + d_2}{d_1 - d_2}$.

The purpose of this example is to design a sliding surface (5.59) and an output feedback dynamic sliding mode controller (5.83) to drive the closed-loop system states into a neighbourhood of the sliding surface (5.59) locally and guarantee the asymptotical stability of the sliding motion with robust $\mathscr{H}_\infty$ performance γ. Use Theorem 5.4 with $\rho_1 = 1$, $\rho_2 = 0$, $J_1 = \begin{bmatrix} 3 & 1 \end{bmatrix}^{\mathrm{T}}$, and $J_2 = \begin{bmatrix} 1 & 0 \end{bmatrix}^{\mathrm{T}}$, one can obtain the sliding surface gain $K = 0.7180$ with robust $\mathscr{H}_\infty$ performance index $\gamma_{\min} = 2.0873$.

To show the effectiveness of the proposed output feedback dynamic SMC design scheme, simulations are conducted under different initial conditions. With initial condition $x_0 = \begin{bmatrix} \frac{\pi}{4} & 0 \end{bmatrix}^{\mathrm{T}}$ and $w(k) = 10\mathrm{e}^{-0.8k}\sin(0.05k)$, applying the output feedback dynamic sliding mode controller (5.83) with $\eta = 0.1$ and $\lambda = 0.41$, the resulting closed-loop system states and control input are shown in Figure 5.9 and Figure 5.10, respectively. The sliding surface is shown in Figure 5.11. Under zero initial condition, the $\mathscr{H}_\infty$ performance is presented in Figure 5.12. The ratio $\frac{\sqrt{\sum_{k=0}^{\lambda} z^{\mathrm{T}}(k)z(k)}}{\sqrt{\sum_{k=0}^{\lambda} w^{\mathrm{T}}(k)w(k)}}$ is about 0.63, which is lower than the minimum disturbance attenuation level $\gamma_{\min} = 2.0873$.

To show the effectiveness of the proposed output feedback dynamic sliding mode controller, another numerical example will be considered.

Example 5.4. Consider a T-S fuzzy affine system in the form of (5.54) with three rules, and the system matrices are

$$\begin{cases} \begin{bmatrix} A_1 \mid A_2 \mid A_3 \end{bmatrix} = \left[\begin{array}{cc|cc|cc} 1.2 & 0.2 & 1.0 & 0.1 & 1.1 & 0.1 \\ -0.4 & 0.9 & -0.2 & 0.8 & 0.1 & 0.6 \end{array}\right], \\ \begin{bmatrix} a_1 \mid a_2 \mid a_3 \end{bmatrix} = \left[\begin{array}{c|c|c} 0 & 0 & 0 \\ 0 & -0.2 & 0.2 \end{array}\right], \\ \begin{bmatrix} B_1 \mid B_2 \mid B_3 \end{bmatrix} = \left[\begin{array}{c|c|c} 1.5 & 1.0 & 1.5 \\ 1.0 & 2.0 & 2.0 \end{array}\right], D_l = \begin{bmatrix} 0 \\ 1.0 \end{bmatrix}, C = \begin{bmatrix} 1.0 & 0 \end{bmatrix}, \\ L_l = \begin{bmatrix} 0.1 & 0.2 \end{bmatrix}, N_l = 0.1, \ l = \{1,2,3\} \end{cases} \tag{5.93}$$

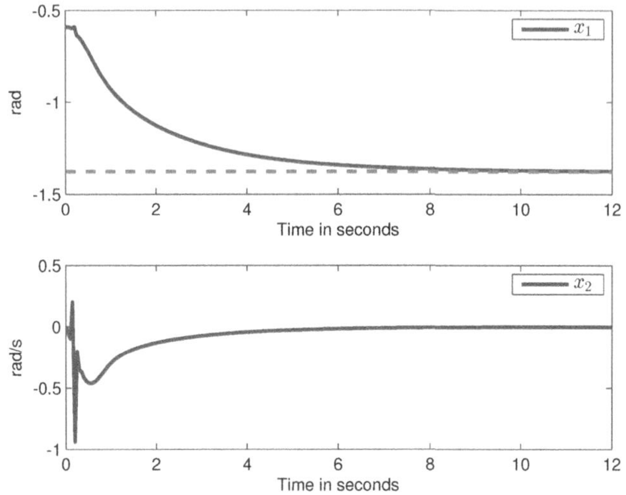

Figure 5.9 The closed-loop system states with dynamic sliding mode controller (5.83) in Example 5.3.

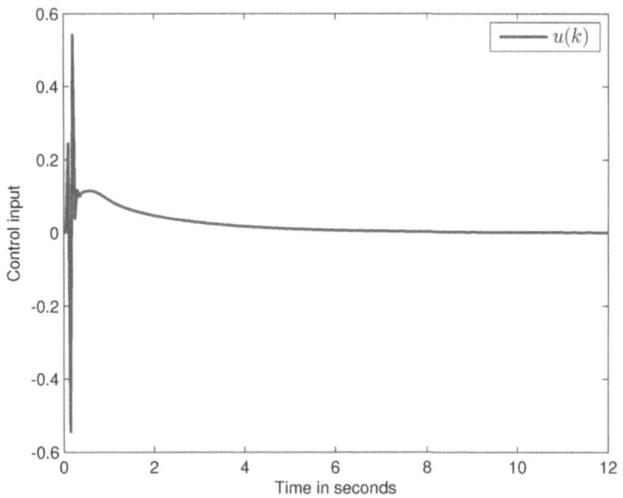

Figure 5.10 The control input with dynamic sliding mode controller (5.83) in Example 5.3.

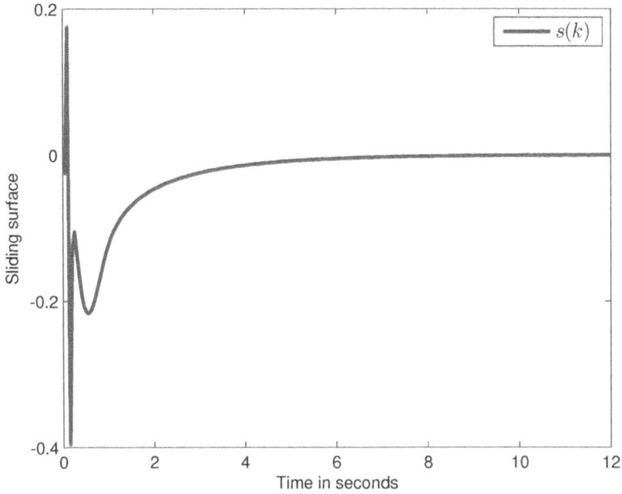

Figure 5.11 The sliding surface with dynamic sliding mode controller (5.83) in Example 5.3.

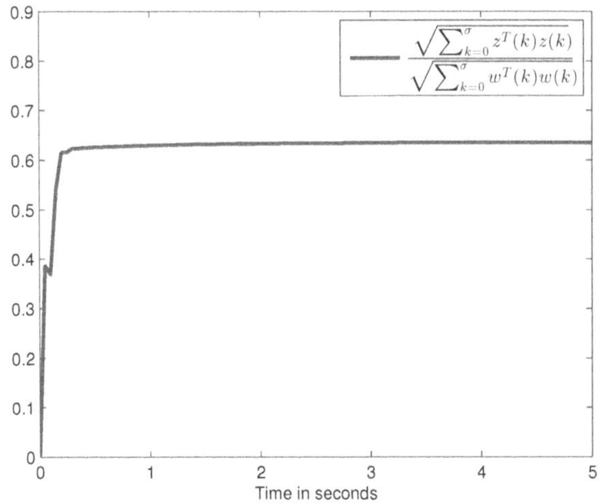

Figure 5.12 Response of the ratio $\frac{\sqrt{\Sigma_{k=0}^{\sigma} z^{\mathrm{T}}(k)z(k)}}{\sqrt{\Sigma_{k=0}^{\sigma} w^{\mathrm{T}}(k)w(k)}}$ in Example 5.3.

and the parameter uncertainties ΔA_l, Δa_l and ΔB_l are given in the form of (1.3) with

$$\begin{cases} H_{l1} = \begin{bmatrix} 0.01 \\ 0.01 \end{bmatrix}, H_{l2} = \begin{bmatrix} 0.01 & 0.01 \end{bmatrix}, \\ H_{l3} = 0.1, l = \{1,2,3\}. \end{cases} \tag{5.94}$$

The MFs of this example are also shown as in Figure 5.1 with $d_1 = 50$ and $d_2 = 450$. It can be seen that the control input matrices B_l in this example are not common for each local model. The aforementioned SMC design method in Theorem 5.2 and adaptive SMC design method in Theorem 5.3 are both not applicable. By applying the PQLFs-based method presented in Theorem 5.4 with $\rho_1 = 1$, $\rho_2 = 0$, $J_1 = \begin{bmatrix} 4 & 1 \end{bmatrix}^{\mathrm{T}}$, and $J_2 = \begin{bmatrix} 1 & 0 \end{bmatrix}^{\mathrm{T}}$, one indeed obtains feasible solutions, and the sliding surface gain is $K = 0.4069$ with robust $\mathscr{H}_\infty$ performance $\gamma_{\min} = 1.4892$.

Simulations are conducted to verify the validity of the proposed output feedback dynamic SMC design approach. Under initial condition $x_0 = \begin{bmatrix} 0.2 & -0.1 \end{bmatrix}^{\mathrm{T}}$ and $w(k) = 5\mathrm{e}^{-0.2k}\sin(0.1k)$, use the output feedback dynamic sliding mode controller (5.83) with $\eta = 0.01$ and $\lambda = 0.23$, the closed-loop system states and control input are shown in Figure 5.13 and Figure 5.14, respectively. Figure 5.15 shows the sliding surface. Under zero initial condition, the $\mathscr{H}_\infty$ performance is presented in Figure 5.16. The ratio $\frac{\sqrt{\sum_{k=0}^{\lambda} z^{\mathrm{T}}(k)z(k)}}{\sqrt{\sum_{k=0}^{\lambda} w^{\mathrm{T}}(k)w(k)}}$ is about 0.73, which is lower than the minimum disturbance attenuation level $\gamma_{\min} = 1.4892$.

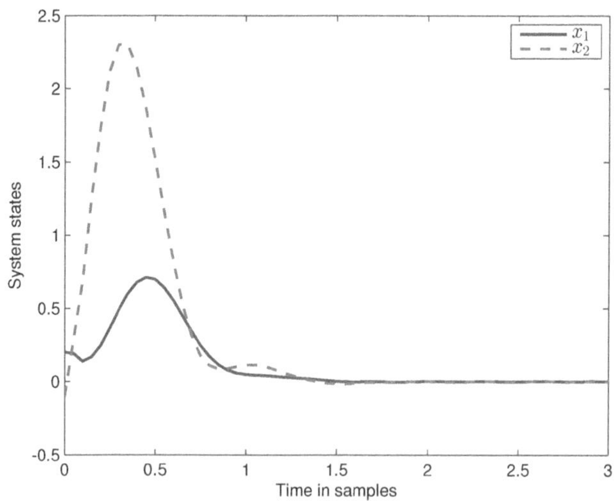

Figure 5.13 The closed-loop system states with dynamic sliding mode controller (5.83) in Example 5.4.

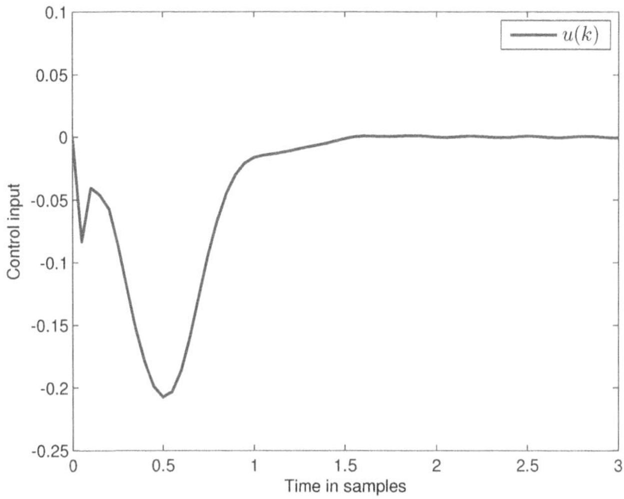

Figure 5.14 The control input with dynamic sliding mode controller (5.83) in Example 5.4.

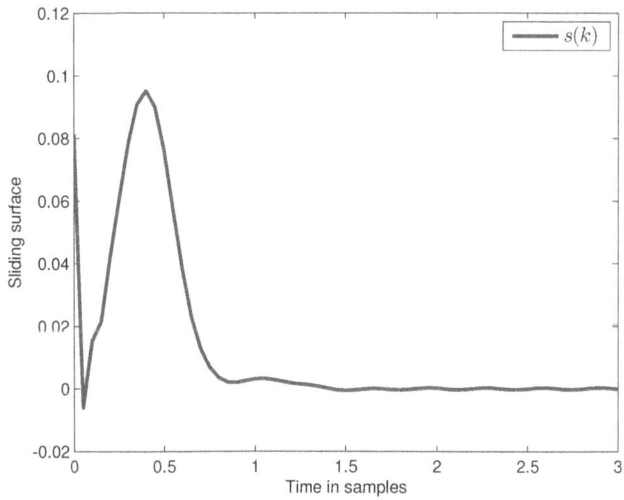

Figure 5.15 The sliding surface with dynamic sliding mode controller (5.83) in Example 5.4.

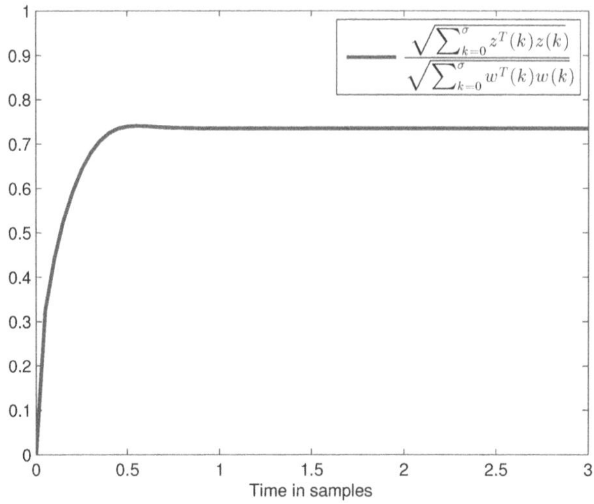

Figure 5.16 Response of the ratio $\frac{\sqrt{\sum_{k=0}^{\sigma} z^{\mathrm{T}}(k)z(k)}}{\sqrt{\sum_{k=0}^{\sigma} w^{\mathrm{T}}(k)w(k)}}$ in Example 5.4.

To further show the effectiveness of the proposed state feedback SMC design methods, in the following, simulation on another nonlinear continuous stirred tank reactor (CSTR) system will be conducted to show the effectiveness of the proposed output feedback dynamic SMC design method.

Example 5.5. Consider a nonlinear continuous stirred tank reactor (CSTR) system, whose dynamics are described as follows,

$$\begin{cases} \dot{T} = M_0(T_n - T) + M_1 k_0 \mathrm{e}^{-\frac{M_2}{T}} C_A - M_3 T + M_3 \mathbf{u} \\ \dot{C}_A = M_0(C_0 - C_A) - M_0 \mathrm{e}^{-\frac{M_2}{T}} C_A \end{cases} \tag{5.95}$$

and the parameters are given as in Example 2.1.

The measurement output $y(t)$ and the regulated output $z(t)$ of the CSTR system are given as $y(t) = x_1(t)$ and $z(t) = 0.1x_1(t) + 0.2x_2(t) + 0.1u(t)$, respectively. Follow a same fuzzy modelling method in Example 4.5 and discretize the CSTR system with sampling period $T = 0.05$min. Consider the external disturbance, and then the CSTR system (5.95) can be described by the following discrete-time T-S fuzzy affine model as,

Plant Rule $\mathscr{R}^l$: IF $x_1(k)$ is $\mathscr{F}_1^l$, THEN

$$\begin{cases} x(k+1) = (A_l + \Delta A_l)x(k) + a_l + \Delta a_l + B_l u(k) + D_l w(k) \\ y(k) = Cx(k) \\ z(k) = L_l x(k) + N_l u(k), \ l \in \{1, 2, 3\}, \end{cases} \tag{5.96}$$

where the system matrices are

$$\begin{cases} A_1 = \begin{bmatrix} 0.9512 & 1.4545 \\ -0.0005 & 0.9430 \end{bmatrix}, A_2 = \begin{bmatrix} 1.2190 & 10.4595 \\ -0.0018 & 0.9001 \end{bmatrix}, \\ A_3 = \begin{bmatrix} 1.3818 & 42.1122 \\ -0.0026 & 0.7487 \end{bmatrix}, a_1 = \begin{bmatrix} -1.7943 \\ 0.0084 \end{bmatrix}, a_2 = \begin{bmatrix} 0 \\ 0 \end{bmatrix}, \\ a_3 = \begin{bmatrix} 4.7483 \\ -0.0224 \end{bmatrix}, B_l = \begin{bmatrix} 0.1046 \\ 0 \end{bmatrix}, D_l = \begin{bmatrix} 0.05 \\ 0 \end{bmatrix}, \\ C = \begin{bmatrix} 1 & 0 \end{bmatrix}, L_l = \begin{bmatrix} 1 & 0.1 \end{bmatrix}, N_l = 0.1, \ l \in \{1,2,3\}, \end{cases} \tag{5.97}$$

and the parameter uncertainties ΔA_l and Δa_l are given in the form of (1.3) with

$$\begin{cases} H_{l1} = \begin{bmatrix} 0 & 0.01 \end{bmatrix}^{\mathrm{T}}, H_{l2} = \begin{bmatrix} 0.01 & 0.01 \end{bmatrix}, \\ H_{l3} = 0.01, \ l \in \{1,2,3\}. \end{cases} \tag{5.98}$$

Figure 5.17 shows the fuzzy MFs of Example 5.5 with $d_1 = -50$, $d_2 = -25.6$, $d_3 = 20.65$ and $d_4 = 50$, and it is found from Figure 5.5 that where S_2 is a fuzzy region with $\mathcal{N}(2) = \{1,2,3\}$, and S_1 and S_3 are crisp regions with $\mathcal{N}(1) = \{1\}$ and $\mathcal{N}(3) = \{3\}$. These three regions can be precisely described by degenerate ellipsoids as in (2.82) with $Q_1 = \begin{bmatrix} \frac{2}{d_2 - d_1} & 0 \end{bmatrix}$, $Q_2 = \begin{bmatrix} \frac{2}{d_3 - d_2} & 0 \end{bmatrix}$, $Q_3 = \begin{bmatrix} \frac{2}{d_4 - d_3} & 0 \end{bmatrix}$ $q_1 = \frac{d_1 + d_2}{d_1 - d_2}$, $q_2 = \frac{d_3 + d_2}{d_2 - d_3}$, $q_3 = \frac{d_4 + d_3}{d_3 - d_4}$.

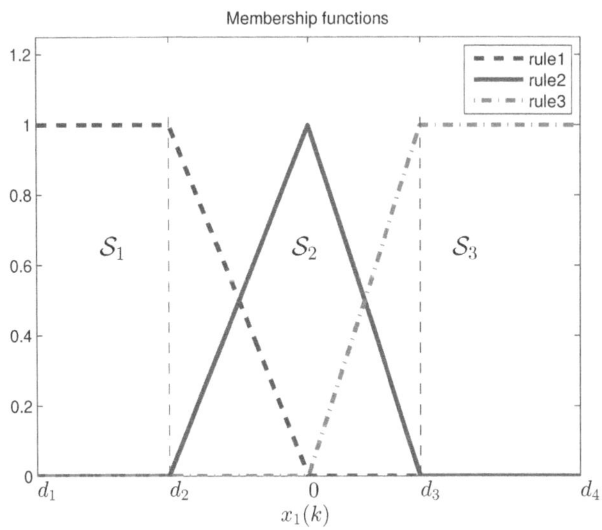

Figure 5.17 Membership functions of Example 5.5.

By applying the PQLFs-based method presented in Theorem 5.4 with $\rho_1 = 1$, $\rho_2 = 0$, $J_1 = \begin{bmatrix} 3 & 1 \end{bmatrix}^T$, and $J_2 = \begin{bmatrix} 1 & 0 \end{bmatrix}^T$, one can obtain the sliding surface gain $K = 12.6949$ with robust $\mathscr{H}_\infty$ performance index $\gamma_{\min} = 0.7892$.

With initial condition $x_0 = \begin{bmatrix} -15 & 0.5 \end{bmatrix}^T$ and external disturbance $w(k) = 60e^{-0.2k}\sin(0.2k)$, utilizing the output feedback dynamic sliding mode controller (5.83) with $\eta = 0.1$ and $\lambda = 0.2$, the closed-loop system states and the control input are shown in Figure 5.18 and Figure 5.19, respectively.

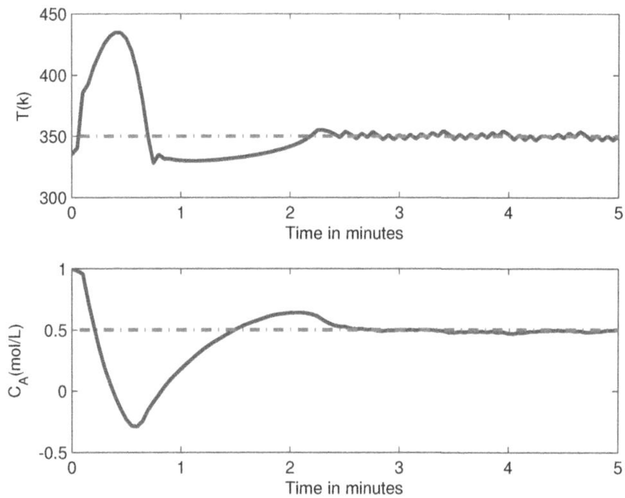

Figure 5.18 The closed-loop system states with dynamic sliding mode controller (5.83) in Example 5.5.

The simulation results clearly show that effectiveness of the proposed state feedback dynamic sliding mode controller design method.

5.6 CONCLUSION

This chapter has investigated the output feedback SMC issue for a class of uncertain nonlinear systems through T-S fuzzy affine models. Through utilizing CQLF-based and PQLFs-based methods, sufficient conditions for asymptotic stability analysis of the sliding motion are derived in terms of LMIs. Some new fuzzy output feedback sliding mode controller design approaches are proposed to force the system states into a vicinity of the sliding surface locally. Illustrative simulation examples are shown to verify the effectiveness of the proposed approaches.

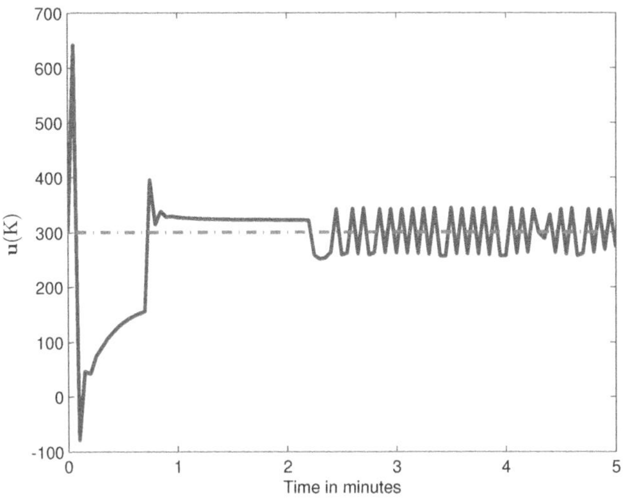

Figure 5.19 The control input with dynamic sliding mode controller (5.83) in Example 5.5.

6 Fuzzy Integral SMC of Nonlinear Singularly Perturbed Systems

6.1 INTRODUCTION

Note that based on T-S fuzzy affine models and piecewise quadratic Lyapunov functions, some new results on fuzzy sliding mode control problem for nonlinear 1-D systems are proposed in Chapter 2 - Chapter 5. On another research front, recognized as a typical class of complex physical systems, singularly perturbed systems (SPSs) contain multiple time-scale dynamical properties, which widely exist in practical situations, such as electronic circuits, industrial processes, marine control systems, motor control systems and so on [104]. Nevertheless, the inherent ill-conditioned dynamical properties existing in the singularly perturbed systems also bring some stiffness and unwieldiness into stability analysis and controller synthesis for SPSs. In recent years, singular perturbation theory has been verified as an efficient technique to circumvent the above-mentioned problems [49, 105]. Via using a small singular perturbation parameter ε, the original SPS is partitioned into decoupled slow subsystems and fast subsystems, and a so-called model-reduction method is utilized to tackle these subsystems hereafter. In view of this perspective, many research interests have been concentrated on SPSs and a substantial body of remarkable results have been reported [106, 107, 108, 109, 110, 111].

Due to the appealing advantages of T-S fuzzy models, the SMC issue for nonlinear SPSs has become another vigorous research topic in past few years. It is also noted that during the reaching phase, SMC may be sensitive to matched or mismatched disturbances. Thus integral sliding mode control (ISMC) is proposed, which can guarantee the robustness of the closed-loop systems to disturbances and uncertainties. The authors in [112] studied the integral SMC problem for T-S fuzzy SPSs with matched external disturbances. Based on neural networks, an adaptive sliding mode controller was designed for T-S fuzzy SPSs in [114]. Nevertheless, it is noted that in [112] and [114], all local models of the fuzzy SPSs have one common control input matrix. Furthermore, the control input matrices in [112] and [114] were not allowed to contain parameter uncertainties. However, few efforts have been devoted to release the above-mentioned restrictions, which inspires the study in this chapter.

This chapter focuses on the integral SMC problem for nonlinear SPSs subject to mismatched external disturbances by T-S fuzzy models. Some new stability and robust performance analysis results for sliding motion are proposed under a convex optimization framework. In sequel, a dynamic sliding mode controller is synthesized to drive the system states onto the sliding surface in finite time.

DOI: 10.1201/9781032695716-6

6.2 MODEL DESCRIPTION AND PROBLEM FORMULATION

Consider a T-S fuzzy singularly perturbed (SP) system characterized r by fuzzy **IF-THEN** rules as,

Plant Rule $\mathscr{R}^l$: IF $\vartheta_1(t)$ is $\mathscr{F}_1^l$ and $\cdots$ and $\vartheta_\varphi(t)$ is $\mathscr{F}_\varphi^l$, THEN

$$\begin{cases} E(\varepsilon)\dot{x}(t) = (A_l + \Delta A_l)x(t) + (B_l + \Delta B_l)u(t) + D_l w(t) \\ z(t) = L_l x(t) + N_l u(t),\ l \in \mathscr{L} = \{1,\ldots,r\}, \end{cases} \tag{6.1}$$

where $\mathscr{F}_\phi^l (\phi = 1,\ldots,\varphi)$ are fuzzy sets; $\mathscr{R}^l$ is the l-th fuzzy inference rule; $\vartheta(t) := [\vartheta_1(t),\ldots,\vartheta_\varphi(t)]$ are measurable premise variables; r is the number of inference rules; $x(t) \in \Re^{n_x}$ is the system states including slow time-scale states $x_1(t) \in \Re^{n_{x1}}$ and fast time-scale states $x_2(t) \in \Re^{n_{x2}}$, and $n_x = n_{x1} + n_{x2}$; $E(\varepsilon) = \text{diag}\{\mathbf{I}_{n_{x1}}, \varepsilon\mathbf{I}_{n_{x2}}\}$ and $0 < \varepsilon \leq 1$ is the singularly perturbed parameter; $w(t) \in \Re^{n_w}$ is the external disturbance and $w(t) \in L_2[0,\infty)$; $u(t) \in \Re^{n_u}$ is the control input; $z(t) \in \Re^{n_z}$ is the regulated output; $(A_l, B_l, D_l, L_l, N_l)$ is the l-th local model of system (6.1); ΔA_l and ΔB_l are the uncertainty terms satisfying

$$\begin{bmatrix} \Delta A_l & \Delta B_l \end{bmatrix} = H_{l1}\Delta_l(t)\begin{bmatrix} H_{l2} & H_{l3} \end{bmatrix}, l \in \mathscr{L} \tag{6.2}$$

with known real-valued matrices H_{l1}, H_{l2}, and H_{l3}. $\Delta_l(t) \in \Re^{s_1 \times s_2}$ are unknown matrix functions satisfying

$$\Delta_l^{\mathrm{T}}(t)\Delta_l(t) \leq \mathbf{I}_{s_2}. \tag{6.3}$$

Let $\mu_l[\vartheta(t)]$ denote the normalized membership function (MF),

$$\mu_l[\vartheta(t)] := \frac{\prod_{\phi=1}^{\varphi} \mu_{l\phi}[\vartheta_\phi(t)]}{\sum_{v=1}^{r}\prod_{\phi=1}^{\varphi} \mu_{v\phi}[\vartheta_\phi(t)]} \geq 0,\ \sum_{l=1}^{r} \mu_l[\vartheta(t)] = 1 \tag{6.4}$$

with $\mu_{l\phi}[\vartheta_\phi(t)]$ beging the grade of membership of $\vartheta_\phi(t)$ belonging to $\mathscr{F}_\phi^l$. For brevity, define $\mu_l := \mu_l[\vartheta(t)]$.

Based on (6.4), the system (6.1) can be inferred as

$$\begin{cases} E(\varepsilon)\dot{x}(t) = (A(\mu) + \Delta A(\mu))x(t) + (B(\mu) + \Delta B(\mu))u(t) \\ \qquad\qquad + D(\mu)w(t) \\ z(t) = L(\mu)x(t) + N(\mu)u(t) \end{cases} \tag{6.5}$$

where

$$\begin{cases} A(\mu) = \sum_{l=1}^{r} \mu_l A_l, \Delta A(\mu) = \sum_{l=1}^{r} \mu_l \Delta A_l, \\ B(\mu) = \sum_{l=1}^{r} \mu_l B_l, \Delta B(\mu) = \sum_{l=1}^{r} \mu_l \Delta B_l, \\ D(\mu) = \sum_{l=1}^{r} \mu_l D_l, L(\mu) = \sum_{l=1}^{r} \mu_l L_l, N(\mu) = \sum_{l=1}^{r} \mu_l N_l. \end{cases} \tag{6.6}$$

This chapter attempts to synthesize a new dynamic sliding mode controller for T-S fuzzy SPS (6.5) to drive the system states to the designed integral sliding surface in finite time. For a prescribed robust $\mathscr{H}_\infty$ performance $\gamma > 0$, the stability of the sliding motion can be ensured when $w(t) \equiv 0$. It is also satisfied that

$$\int_{t_r}^{\infty} z^{\mathrm{T}}(t)z(t)dt < \gamma^2 \int_{t_r}^{\infty} w^{\mathrm{T}}(t)w(t)dt \tag{6.7}$$

with zero initial condition for states under nonzero $w(t) \in L_2[0, \infty)$, and t_r represents the time when the system states reach the sliding surface.

6.3 DYNAMIC INTEGRAL SMC DESIGN

In this section, an integral sliding surface will be designed based on the system states and control input, and then a new dynamic sliding mode controller synthesis method will be proposed to drive the closed-loop system states onto the sliding surface in finite time. The stability and robust performance is also analyzed by using some optimal convexification techniques, and the sliding surface gain can be calculated simultaneously.

6.3.1 INTEGRAL SLIDING SURFACE DESIGN

For the T-S fuzzy SPS (6.1), design an integral sliding surface as follows,

$$\begin{aligned} s(t) = {} & C_x E(\varepsilon)[x(t) - x(0)] + C_u[u(t) - u(0)] \\ & - \int_0^t \sum_{l=1}^{r} \mu_l \cdot C_x [A_l x(\tau) + B_l u(\tau)] d\tau \\ & - \int_0^t \sum_{l=1}^{r} \mu_l \cdot C_u [K_{l1} x(\tau) + K_{l2} u(\tau)] d\tau, \end{aligned} \tag{6.8}$$

where $\{K_{l1}, C_x\} \in \Re^{n_u \times n_x}$, $\{K_{l2}, C_u\} \in \Re^{n_u \times n_u}$, and C_u is selected to be nonsingular. Specifically, $K_{l1} \in \Re^{n_u \times n_x}$ and $K_{l2} \in \Re^{n_u \times n_u}$ are designed such that matrices $\begin{bmatrix} A_l & B_l \\ K_{l1} & K_{l2} \end{bmatrix}$, $l \in \mathscr{L}$ are Hurwitz.

6.3.2 DYNAMIC SLIDING MODE CONTROLLER DESIGN

The integral dynamic sliding mode control approach is proposed by the following Theorem, which can guarantee the finite time reachability of the sliding surface (6.8).

Theorem 6.1. Consider the fuzzy SPS (6.1) or equivalently (6.5) with the sliding surface given as in (6.8). The trajectories of the closed-loop system states can be driven onto the designed sliding surface (6.8) in finite time by the following dynamic sliding mode controller

$$\dot{u}(t) = \sum_{l=1}^{r} \mu_l \{K_{l1} x(t) + K_{l2} u(t)\} - (\eta + \zeta(t)) C_u^{-1} \mathrm{sgn}(s(t)), \tag{6.9}$$

where $\eta > 0$ is a given positive constant, and K_{l1}, K_{l2} are defined in (6.8), and

$$\zeta(t) = \sum_{l=1}^{r} \mu_l \|C_x\| \{\|H_{l1}\| (\|H_{l2}\| \cdot \|x(t)\| + \|H_{l3}\| \cdot \|u(t)\|) + \|D_l\| \varpi(t)\} \tag{6.10}$$

with $\varpi(t)$ representing the uniform upper bound of $w(t)$.

Proof. Construct the Lyapunov function as,

$$V(t) = \frac{1}{2}s^{\mathrm{T}}(t)s(t). \tag{6.11}$$

Based on the dynamic sliding mode controller (6.9), the time-derivative of $V(t)$ is

$$\begin{aligned}
\dot{V}(t) = s^{\mathrm{T}}(t)\sum_{l=1}^{r}\mu_l C_x\{(A_l+\Delta A_l)x(t)+(B_l+\Delta B_l)\,u(t)+D_l w(t)\}\\
-s^{\mathrm{T}}(t)\sum_{l=1}^{r}\mu_l C_x\{A_l x(t)+B_l u(t)\}\\
-s^{\mathrm{T}}(t)\sum_{l=1}^{r}\mu_l C_u\{K_{l1}x(t)+K_{l2}u(t)\}\\
+s^{\mathrm{T}}(t)\sum_{l=1}^{r}\mu_l C_u\{K_{l1}x(t)+K_{l2}u(t)\}-\eta\|s(t)\|-\zeta(t)\|s(t)\|\\
= s^{\mathrm{T}}(t)\sum_{l=1}^{r}\mu_l C_x\{\Delta A_l x(t)+\Delta B_l u(t)+D_l w(t)\}-\eta\|s(t)\|-\zeta(t)\|s(t)\|\\
\le \sum_{l=1}^{r}\mu_l\|s(t)\|\cdot\|C_x\|\{\|H_{l1}\|(\|H_{l2}\|\cdot\|x(t)\|+\|H_{l3}\|\cdot\|u(t)\|)+\|D_l\|\varpi(t)\}\\
-\zeta(t)\|s(t)\|-\eta\|s(t)\|.
\end{aligned} \tag{6.12}$$

Noticing (6.10), it follows from (6.12) that

$$\dot{V}(t) \le -\eta\|s(t)\|, \tag{6.13}$$

which implies that the sliding surface can be reached in finite time. The proof is then accomplished. ■

Note that the designed sliding mode controller (6.9) is in a dynamic form, and one does not need to make any singular valuable decomposition (SVD) to B_l for design purpose. Furthermore, the dynamic sliding mode controller (6.9) can also handle mismatched parameter uncertainties and external disturbances, which distinguishes our results from the existing SMC design methods in [112] and [114].

Define $T_1 = \begin{bmatrix} \mathbf{I}_{n_x} & \mathbf{0} \end{bmatrix}^{\mathrm{T}}$ and $T_2 = \begin{bmatrix} \mathbf{0} & \mathbf{I}_{n_u} \end{bmatrix}^{\mathrm{T}}$. Applying the sliding mode controller (6.9) to fuzzy singularly perturbed system (6.5), then the closed-loop system is obtained as,

$$\begin{cases}
\bar{E}(\varepsilon)\dot{\bar{x}}(t) = \left[T_1(\bar{A}(\mu)+\Delta\bar{A}(\mu))+T_2\bar{K}(\mu)\right]\bar{x}(t)\\
\qquad\qquad +T_1 D(\mu)w(t)-T_2(\eta+\zeta(t))C_u^{-1}\mathrm{sgn}(s(t))\\
z(t) = \bar{L}(\mu)\bar{x}(t)
\end{cases} \tag{6.14}$$

where

$$\begin{cases}
\bar{E}(\varepsilon) = \mathrm{diag}\{E(\varepsilon),\mathbf{I}_{n_u}\},\bar{x}(t) = \begin{bmatrix} x^{\mathrm{T}}(t) & u^{\mathrm{T}}(t) \end{bmatrix}^{\mathrm{T}},\\
\bar{A}(\mu) = \begin{bmatrix} A(\mu) & B(\mu) \end{bmatrix},\bar{L}(\mu) = \begin{bmatrix} L(\mu) & N(\mu) \end{bmatrix},\\
\Delta\bar{A}(\mu) = \begin{bmatrix} \Delta A(\mu) & \Delta B(\mu) \end{bmatrix},\\
\bar{K}(\mu) = \sum_{l=1}^{r}\mu_l\bar{K}_l,\bar{K}_l = \begin{bmatrix} K_{l1} & K_{l2} \end{bmatrix}.
\end{cases} \tag{6.15}$$

Then we can formulate the integral sliding surface (6.8) into the compact form as

$$s(t)=\bar{C}\left\{\bar{E}(\varepsilon)\left[\bar{x}(t)-\bar{x}(0)\right]-\int_0^t\left[T_1\bar{A}(\mu)+T_2\bar{K}(\mu)\right]\bar{x}(\tau)d\tau\right\} \tag{6.16}$$

with $\bar{C}=\left[\begin{array}{cc} C_x & C_u \end{array}\right]$.

The term $\bar{E}(\varepsilon)\bar{x}(0)+\int_0^t\left[T_1\bar{A}(\mu)+T_2\bar{K}(\mu)\right]\bar{x}(\tau)d\tau$ in (6.16) actually characterizes the dynamical property of the following fuzzy SPS,

$$\bar{E}(\varepsilon)\dot{\bar{x}}(t)=\left[T_1\bar{A}(\mu)+T_2\bar{K}(\mu)\right]\bar{x}(t). \tag{6.17}$$

Note that the system (6.17) is also viewed as the nominal closed-loop system. After the sliding surface (6.8) or equivalently (6.16) is reached and maintained afterward, the real resulting closed-loop system states coincide those of the system (6.17). That is to say, the real resulting closed-loop system behaves as the system (6.17) does.

Remark 6.3.1. Based on the SMC theory [85], it follows that $s(t)=0$ and $\dot{s}(t)=0$ after the sliding surface is reached. Via designing the parameters of system (6.17) appropriately, one can achieve satisfactory performance of the sliding motion.

Theorem 6.2. The system (6.17) is asymptotically stable, if matrices $0<X_{11}=X_{11}^{\mathrm{T}}\in\Re^{n_{x1}\times n_{x1}}$, $0<X_{22}=X_{22}^{\mathrm{T}}\in\Re^{n_{x2}\times n_{x2}}$, $0<X_{33}=X_{33}^{\mathrm{T}}\in\Re^{n_u\times n_u}$, $X_{21}\in\Re^{n_{x2}\times n_{x1}}$, $X_{23}\in\Re^{n_{x2}\times n_u}$, $X_{31}\in\Re^{n_u\times n_{x1}}$, and $\tilde{K}_l\in\Re^{n_u\times(n_x+n_u)}$ exist, such that the following conditions hold,

$$\mathtt{Sym}\left\{T_1\bar{A}_lX+T_2\tilde{K}_l\right\}<0, l\in\mathscr{L}, \tag{6.18}$$

where

$$\bar{A}_l=\left[\begin{array}{cc} A_l & B_l \end{array}\right], X=\left[\begin{array}{ccc} X_{11} & \mathbf{0} & X_{31}^{\mathrm{T}} \\ X_{21} & X_{22} & X_{23} \\ X_{31} & \mathbf{0} & X_{33} \end{array}\right]. \tag{6.19}$$

In addition, the controller gains can be attained by

$$\bar{K}_l=\tilde{K}_lX^{-1}. \tag{6.20}$$

Proof. Consider a matrix P as

$$P=\left[\begin{array}{ccc} P_{11} & \mathbf{0} & P_{31}^{\mathrm{T}} \\ P_{21} & P_{22} & P_{23} \\ P_{31} & \mathbf{0} & P_{33} \end{array}\right] \tag{6.21}$$

with $0<P_{11}=P_{11}^{\mathrm{T}}\in\Re^{n_{x1}\times n_{x1}}$, $0<P_{22}=P_{22}^{\mathrm{T}}\in\Re^{n_{x2}\times n_{x2}}$, $0<P_{33}=P_{33}^{\mathrm{T}}\in\Re^{n_u\times n_u}$, $P_{21}\in\Re^{n_{x2}\times n_{x1}}$, $P_{23}\in\Re^{n_{x2}\times n_u}$, $P_{31}\in\Re^{n_u\times n_{x1}}$.

Based on the special structure of P and following a similar line in [49], there exists a positive scalar $\delta_1^*>0$, such that $\forall\varepsilon\in(0,\ \delta_1^*]$, matrix $P(\varepsilon)=\left[\begin{array}{ccc} P_{11} & \varepsilon P_{21}^{\mathrm{T}} & P_{31}^{\mathrm{T}} \\ P_{21} & P_{22} & P_{23} \\ P_{31} & \varepsilon P_{23}^{\mathrm{T}} & P_{33} \end{array}\right]$ satisfies $\bar{E}^{\mathrm{T}}(\varepsilon)P(\varepsilon)=P^{\mathrm{T}}(\varepsilon)\bar{E}(\varepsilon)>0$.

Construct the following Lyapunov function

$$V(t) = \bar{x}^{\mathrm{T}}(t)\bar{E}^{\mathrm{T}}(\varepsilon)P(\varepsilon)\bar{x}(t). \tag{6.22}$$

The system (6.17) is asymptotically stable, if

$$\dot{V}(t) = 2\bar{x}^{\mathrm{T}}(t)\left(T_1\bar{A}(\mu)+T_2\bar{K}(\mu)\right)^{\mathrm{T}} P(\varepsilon)\bar{x}(t) < 0. \tag{6.23}$$

By continuity [49], one can see that there exists $\delta_2^* > 0$, such that for $\forall \varepsilon \in (0,\ \delta_2^*]$,

$$\dot{V}(t) = 2\bar{x}^{\mathrm{T}}(t)\left\{\left(T_1\bar{A}(\mu)+T_2\bar{K}(\mu)\right)^{\mathrm{T}} P + o(\varepsilon)\right\}\bar{x}(t) < 0, \tag{6.24}$$

where $o(\varepsilon)$ represents the higher order of ε.

Denote $\delta_3^* = \min\{\delta_1^*, \delta_2^*\}$. Obviously, $\forall \varepsilon \in (0,\ \delta_3^*]$, the following inequality implies (6.24),

$$\mathtt{Sym}\left\{P^{\mathrm{T}}\left[T_1\bar{A}(\mu)+T_2\bar{K}(\mu)\right]\right\} < 0. \tag{6.25}$$

Extracting the membership functions, one has

$$\mathtt{Sym}\left\{P^{\mathrm{T}}\left[T_1\bar{A}_l+T_2\bar{K}_l\right]\right\} < 0. \tag{6.26}$$

Denote $X = P^{-1}$ with $X = \begin{bmatrix} X_{11} & \mathbf{0} & X_{31}^{\mathrm{T}} \\ X_{21} & X_{22} & X_{23} \\ X_{31} & \mathbf{0} & X_{33} \end{bmatrix}$, where $0 < X_{11} = X_{11}^{\mathrm{T}} \in \Re^{n_{x1}\times n_{x1}}$, $0 < X_{22} = X_{22}^{\mathrm{T}} \in \Re^{n_{x2}\times n_{x2}}$, $0 < X_{33} = X_{33}^{\mathrm{T}} \in \Re^{n_u\times n_u}$, $X_{21} \in \Re^{n_{x2}\times n_{x1}}$, $X_{23} \in \Re^{n_{x2}\times n_u}$, $X_{31} \in \Re^{n_u\times n_{x1}}$.

Define

$$\tilde{K}_l = \bar{K}_l X. \tag{6.27}$$

Pre- and post-multiply (6.26) by X^{T} and X, respectively, then (6.18) implies (6.26). The proof is then completed. ■

6.3.3 STABILITY AND ROBUST PERFORMANCE ANALYSIS OF SLIDING MOTION

After the sliding surface is reached, $\dot{s}(t) = 0$ should be satisfied to maintain the sliding motion. To be more specific, it is shown that

$$\begin{aligned}\dot{s}(t) = {} & \sum_{l=1}^{r} \mu_l C_x\left\{(A_l+\Delta A_l)x(t)+(B_l+\Delta B_l)u(t)+D_l w(t)\right\}+C_u\dot{u}(t) \\ & -\sum_{l=1}^{r} \mu_l\left\{C_x\left(A_l x(t)+B_l u(t)\right)+C_u\left(K_{l1}x(t)+K_{l2}u(t)\right)\right\} = 0,\end{aligned} \tag{6.28}$$

where K_{l1} and K_{l2} can be attained via (6.20).

With the nonsingular property of C_u and (6.28), one has

$$\dot{u}(t) = \sum_{l=1}^{r} \mu_l\left\{(K_{l1}x(t)+K_{l2}u(t)) - C_{ux}\left(\Delta A_l x(t)+\Delta B_l u(t)+D_l w(t)\right)\right\} \tag{6.29}$$

where $C_{ux} = C_u^{-1}C_x$.

Note that the controller in (6.29) is characterized as the *equivalent dynamic controller in the sliding motion* in this chapter. Then by combining the equivalent sliding mode controller (6.29) with T-S fuzzy SPS (6.5), the sliding motion dynamics are embedded into the following singularly perturbed sliding mode dynamical system

$$\begin{cases} \bar{E}(\varepsilon)\dot{\bar{x}}(t) = \left[T_1\bar{A}(\mu) + T_2\bar{K}(\mu) + (T_1 - T_2C_{ux}) \times \Delta\bar{A}(\mu)\right]\bar{x}(t) \\ \qquad +(T_1 - T_2C_{ux})D(\mu)w(t) \\ z(t) = \bar{L}(\mu)\bar{x}(t) \end{cases} \tag{6.30}$$

where $\bar{A}(\mu)$, $\Delta\bar{A}(\mu)$, $\bar{D}(\mu)$ and $\bar{L}(\mu)$ are given in (6.15).

The asymptotic stability and robust $\mathscr{H}_\infty$ performance analysis results on the system (6.30) will be proposed by the following theorem.

Theorem 6.3. Given a constant parameter $\gamma > 0$, the singularly perturbed sliding mode dynamical system (6.30) is asymptotically stable, if scalars $\{\kappa_l > 0, l \in \mathscr{L}\}$, and matrices $0 < W = W^{\mathrm{T}} \in \Re^{(n_x+n_u)\times(n_x+n_u)}$, $0 < Q_{11} = Q_{11}^{\mathrm{T}} \in \Re^{n_{x1}\times n_{x1}}$, $0 < Q_{22} = Q_{22}^{\mathrm{T}} \in \Re^{n_{x2}\times n_{x2}}$, $0 < Q_{33} = Q_{33}^{\mathrm{T}} \in \Re^{n_u\times n_u}$, $Q_{21} \in \Re^{n_{x2}\times n_{x1}}$, $Q_{23} \in \Re^{n_{x2}\times n_u}$, and $Q_{31} \in \Re^{n_u\times n_{x1}}$ exist, such that the following conditions hold,

$$\begin{bmatrix} \Pi_1 & \star \\ \Pi_2 & -\mathrm{diag}\{T_2^{\mathrm{T}}WT_2, T_1^{\mathrm{T}}WT_1, \kappa_l\mathbf{I}\} \end{bmatrix} < 0,\ l \in \mathscr{L}, \tag{6.31}$$

where $\bar{K}_l$ can be obtained via (6.20), and

$$\begin{cases} \Pi_1 = \begin{bmatrix} \mathtt{Sym}\{Q^{\mathrm{T}}T_1\bar{A}_l + Q^{\mathrm{T}}T_2\bar{K}_l\} + \bar{L}_l^{\mathrm{T}}\bar{L}_l + \kappa_l\bar{H}_{l2}^{\mathrm{T}}\bar{H}_{l2} & \star & \star \\ D_l^{\mathrm{T}}T_1^{\mathrm{T}}Q & -\gamma^2\mathbf{I} & \star \\ \mathbf{0} & \mathbf{0} & -T_1^{\mathrm{T}}WT_1 \end{bmatrix}, \\ \Pi_2 = \begin{bmatrix} \sqrt{2}T_2^{\mathrm{T}}Q & \mathbf{0} & \mathbf{0} \\ \mathbf{0} & T_1^{\mathrm{T}}WT_1D_l & \mathbf{0} \\ H_{l1}^{\mathrm{T}}T_1^{\mathrm{T}}Q & \mathbf{0} & H_{l1}^{\mathrm{T}}T_1^{\mathrm{T}}WT_1 \end{bmatrix}, Q = \begin{bmatrix} Q_{11} & \mathbf{0} & Q_{31}^{\mathrm{T}} \\ Q_{21} & Q_{22} & Q_{23} \\ Q_{31} & \mathbf{0} & Q_{33} \end{bmatrix}, \\ \bar{A}_l = \begin{bmatrix} A_l & B_l \end{bmatrix}, \bar{H}_{l2} = \begin{bmatrix} H_{l2} & H_{l3} \end{bmatrix}, \bar{L}_l = \begin{bmatrix} L_l & N_l \end{bmatrix}. \end{cases} \tag{6.32}$$

Furthermore, the sliding surface gain can be attained by

$$\bar{C} = T_2^{\mathrm{T}}W. \tag{6.33}$$

Proof. Define a matrix $Q = \begin{bmatrix} Q_{11} & \mathbf{0} & Q_{31}^{\mathrm{T}} \\ Q_{21} & Q_{22} & Q_{23} \\ Q_{31} & \mathbf{0} & Q_{33} \end{bmatrix}$. Similar to [49], there exists a positive scalar $\delta_4^* > 0$, such that $\forall\varepsilon \in (0,\ \delta_4^*]$, matrix $Q(\varepsilon) = \begin{bmatrix} Q_{11} & \varepsilon Q_{21}^{\mathrm{T}} & Q_{31}^{\mathrm{T}} \\ Q_{21} & Q_{22} & Q_{23} \\ Q_{31} & \varepsilon Q_{23}^{\mathrm{T}} & Q_{33} \end{bmatrix}$ satisfies $Q^{\mathrm{T}}(\varepsilon)\bar{E}(\varepsilon) = \bar{E}^{\mathrm{T}}(\varepsilon)Q(\varepsilon) > 0$.

Construct the Lyapunov function

$$V(t) = \bar{x}^{\mathrm{T}}(t)\bar{E}^{\mathrm{T}}(\varepsilon)Q(\varepsilon)\bar{x}(t). \tag{6.34}$$

Based on Lyapunov function given in (6.34), the asymptotic stability of the singularly perturbed sliding mode dynamical system (6.30) can be ensured with robust $\mathscr{H}_\infty$ performance γ, if

$$\dot{V}(t) + z^{\mathrm{T}}(t)z(t) - \gamma^2 w^{\mathrm{T}}(t)w(t) < 0. \tag{6.35}$$

Based on (6.30), inequality (6.35) is formulated as

$$\begin{aligned}&2\bar{x}^{\mathrm{T}}(t)\left\{T_1\bar{A}(\mu) + T_2\bar{K}(\mu) + (T_1 - T_2C_{ux})\Delta\bar{A}(\mu)\right\}^{\mathrm{T}} Q(\varepsilon)\bar{x}(t) + 2\bar{x}^{\mathrm{T}}(t)\\ &\times Q^{\mathrm{T}}(\varepsilon)(T_1 - T_2C_{ux})D(\mu)w(t) + \bar{x}^{\mathrm{T}}(t)\bar{L}^{\mathrm{T}}(\mu)\bar{L}(\mu)\bar{x}(t) - \gamma^2 w^{\mathrm{T}}(t)w(t) < 0.\end{aligned} \tag{6.36}$$

By continuity [49], there exists a constant $\delta_5^* > 0$, and $\forall \varepsilon \in (0,\ \delta_5^*]$, the following inequality implies (6.36),

$$\begin{aligned}&2\bar{x}^{\mathrm{T}}(t)\left\{\left[T_1\bar{A}(\mu) + T_2\bar{K}(\mu) + (T_1 - T_2C_{ux})\Delta\bar{A}(\mu)\right]^{\mathrm{T}} Q + \bar{L}^{\mathrm{T}}(\mu)\bar{L}(\mu)\right\}\bar{x}(t)\\ &+ 2\bar{x}^{\mathrm{T}}(t)Q^{\mathrm{T}}(T_1 - T_2C_{ux})D(\mu)w(t) - \gamma^2 w^{\mathrm{T}}(t)w(t) + o(\varepsilon)\cdot\|x(t)\|^2 < 0,\end{aligned} \tag{6.37}$$

where $o(\varepsilon)$ represents the higher order of ε.

Choose the sliding surface gain as $\bar{C} = T_2^{\mathrm{T}}W$, where $0 < W = W^{\mathrm{T}} \in \Re^{(n_x+n_u)\times(n_x+n_u)}$. It is easy to see that a matrix $M > 0$ always exists such that $W = M^{\mathrm{T}}M$. Then one has

$$\begin{aligned}&-2\bar{x}^{\mathrm{T}}(t)Q^{\mathrm{T}}T_2C_{ux}\Delta\bar{A}(\mu)\bar{x}(t)\\ &= -2\bar{x}^{\mathrm{T}}(t)Q^{\mathrm{T}}T_2(T_2^{\mathrm{T}}WT_2)^{-1}T_2^{\mathrm{T}}M^{\mathrm{T}}MT_1\Delta\bar{A}(\mu)\bar{x}(t)\\ &\leq \bar{x}^{\mathrm{T}}(t)\left\{Q^{\mathrm{T}}T_2(T_2^{\mathrm{T}}WT_2)^{-1}T_2^{\mathrm{T}}M^{\mathrm{T}}MT_2(T_2^{\mathrm{T}}WT_2)^{-1}T_2^{\mathrm{T}}Q\right.\\ &\qquad\left. + \Delta\bar{A}^{\mathrm{T}}(\mu)T_1^{\mathrm{T}}M^{\mathrm{T}}MT_1\Delta\bar{A}(\mu)\right\}\bar{x}(t)\\ &= \bar{x}^{\mathrm{T}}(t)\left\{Q^{\mathrm{T}}T_2(T_2^{\mathrm{T}}WT_2)^{-1}T_2^{\mathrm{T}}Q + \Delta\bar{A}^{\mathrm{T}}(\mu)T_1^{\mathrm{T}}WT_1\Delta\bar{A}(\mu)\right\}\bar{x}(t).\end{aligned} \tag{6.38}$$

Similarly, we also have

$$\begin{aligned}&-2\bar{x}^{\mathrm{T}}(t)Q^{\mathrm{T}}T_2C_{ux}D(\mu)w(t)\\ &\leq \bar{x}^{\mathrm{T}}(t)Q^{\mathrm{T}}T_2(T_2^{\mathrm{T}}WT_2)^{-1}T_2^{\mathrm{T}}Q\bar{x}(t) + w^{\mathrm{T}}(t)D^{\mathrm{T}}(\mu)T_1^{\mathrm{T}}WT_1D(\mu)w(t).\end{aligned} \tag{6.39}$$

Denote $\delta_6^* = \min\{\delta_4^*, \delta_5^*\}$, and then $\forall \varepsilon \in (0,\ \delta_6^*]$, the following inequality implies (6.37) based on (6.38)–(6.39),

$$\xi^{\mathrm{T}}(t)\begin{bmatrix}\Gamma & \star\\ D^{\mathrm{T}}(\mu)T_1^{\mathrm{T}}Q & -\gamma^2\mathbf{I} + D^{\mathrm{T}}(\mu)T_1^{\mathrm{T}}WT_1D(\mu)\end{bmatrix}\xi(t) < 0, \tag{6.40}$$

where

$$\begin{cases}\xi(t) = \begin{bmatrix}\bar{x}^{\mathrm{T}}(t) & w^{\mathrm{T}}(t)\end{bmatrix}^{\mathrm{T}},\\ \Gamma = \mathtt{Sym}\{Q^{\mathrm{T}}T_1(\bar{A}(\mu) + \Delta\bar{A}(\mu)) + Q^{\mathrm{T}}T_2\bar{K}(\mu)\}\\ \qquad + 2Q^{\mathrm{T}}T_2(T_2^{\mathrm{T}}WT_2)^{-1}T_2^{\mathrm{T}}Q + \Delta\bar{A}^{\mathrm{T}}(\mu)T_1^{\mathrm{T}}WT_1\Delta\bar{A}(\mu) + \bar{L}^{\mathrm{T}}(\mu)\bar{L}(\mu).\end{cases} \tag{6.41}$$

Then $\dot{V}(t) < 0$, if

$$\begin{bmatrix}\Gamma & \star\\ D^{\mathrm{T}}(\mu)T_1^{\mathrm{T}}Q & -\gamma^2\mathbf{I} + D^{\mathrm{T}}(\mu)T_1^{\mathrm{T}}WT_1D(\mu)\end{bmatrix} < 0. \tag{6.42}$$

Extracting the membership functions and utilizing Schur complement, the following inequality implies (6.42)

$$\begin{bmatrix} \Phi & \star & \star & & \\ D_l^{\mathrm{T}} T_1^{\mathrm{T}} Q & -\gamma^2 \mathbf{I} & \star & \star & \star \\ T_1^{\mathrm{T}} W T_1 \Delta \bar{A}_l & \mathbf{0} & -T_1^{\mathrm{T}} W T_1 & \star & \star \\ \sqrt{2} T_2^{\mathrm{T}} Q & \mathbf{0} & \mathbf{0} & \star & \star \\ \mathbf{0} & T_1^{\mathrm{T}} W T_1 D_l & \mathbf{0} & \mathbf{0} & -T_1^{\mathrm{T}} W T_1 \end{bmatrix} < 0, \tag{6.43}$$

where

$$\Phi = \mathtt{Sym}\{Q^{\mathrm{T}} T_1 (\bar{A}_l + \Delta \bar{A}_l) + Q^{\mathrm{T}} T_2 \bar{K}_l\} + \bar{L}_l^{\mathrm{T}} \bar{L}_l. \tag{6.44}$$

Employing Lemma A1 to deal with the parameter uncertainties shown in (6.2), for $\kappa_l > 0$, $l \in \mathscr{L}$, (6.31) implies (6.43). The proof is completed. ∎

6.4 ILLUSTRATIVE EXAMPLES

In this section, two illustrative simulation examples will be given to show the effectiveness of the proposed dynamic SMC design method.

Example 6.1. As shown in Figure 6.1, consider a nonlinear circuit containing nonlinear resistor and parasitic capacitor [112].

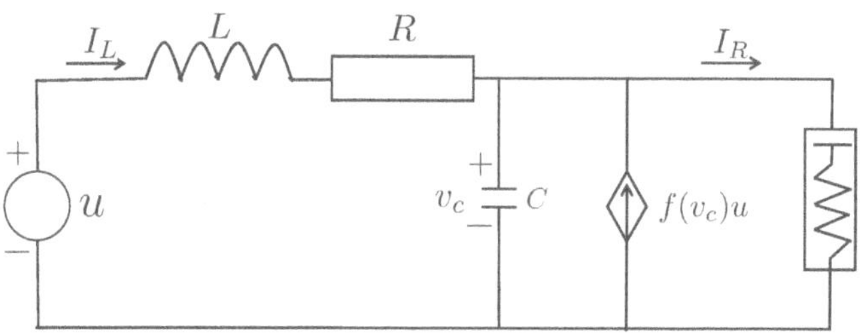

Figure 6.1 Nonlinear circuit with nonlinear resistor and parasitic capacitor.

The v_c-I_R characteristic of the nonlinear resistor is $I_R = 0.2v_c \cdot \cos^2(0.5v_c)$. Based on the Kirchoff voltage and current law, then the circuit system is described by

$$\begin{cases} L\dot{I}_L(t) = -I_L R - v_c + u(t) + 0.1w(t) \\ \varepsilon \dot{v}_c(t) = I_L - 0.2v_c \cos^2(0.5v_c) + f(v_c) \cdot u(t) + 0.1w(t) \end{cases} \tag{6.45}$$

where $\varepsilon = C$, $f(v_c) = 0.5 + \sin(0.25v_c^2)$, $L = 0.1\mathrm{H}$, $R = 0.2\Omega$, and $u(t)$ denotes the constant power source which is also the control input, and $w(t)$ represents the external nonlinear disturbance. Denote $x_1(t) = L \cdot I_L(t)$ and $x_2(t) = v_c(t)$, respectively, as the magnetic linkage of the inductance L and the voltage of the capacitor C. The regulated output is given as $z(t) = 0.1x_1(t)$. Linearize the nonlinear circuit system (6.45) at $(0,0)^{\mathrm{T}}$, $\left(\pm\frac{2\pi}{3},0\right)^{\mathrm{T}}$, and then we have the following T-S fuzzy singularly perturbed model,

Plant Rule $\mathscr{R}^l$: IF $x_2(t)$ is $\mathscr{F}_1^l$, THEN

$$\begin{cases} E(\varepsilon)\dot{x}(t) = (A_l + \Delta A_l)x(t) + (B_l + \Delta B_l)u(t) + D_l w(t) \\ z(t) = L_l x(t) + N_l u(t),\ l \in \{1,2\}, \end{cases} \tag{6.46}$$

where

$$\begin{cases} E(\varepsilon) = \begin{bmatrix} 1 & 0 \\ 0 & \varepsilon \end{bmatrix}, A_1 = \begin{bmatrix} -2 & -1 \\ 10 & -0.2 \end{bmatrix}, A_2 = \begin{bmatrix} -2 & -1 \\ 10 & -0.05 \end{bmatrix}, \\ B_1 = \begin{bmatrix} 1 \\ 0.5 \end{bmatrix}, B_2 = \begin{bmatrix} 1 \\ 1.3897 \end{bmatrix}, D_l = \begin{bmatrix} 0.1 \\ 0.1 \end{bmatrix}, \\ L_l = \begin{bmatrix} 0.1 & 0 \end{bmatrix}, N_l = 0.3,\ l \in \{1,2\}, \end{cases} \tag{6.47}$$

and the parameter uncertainties ΔA_l and ΔB_l are given in the form of (6.2) with

$$H_{l1} = \begin{bmatrix} 0 \\ 0.1 \end{bmatrix}, H_{l2} = \begin{bmatrix} 0 & 0.1 \end{bmatrix}, H_{l3} = 0.1,\ l \in \{1,2\}. \tag{6.48}$$

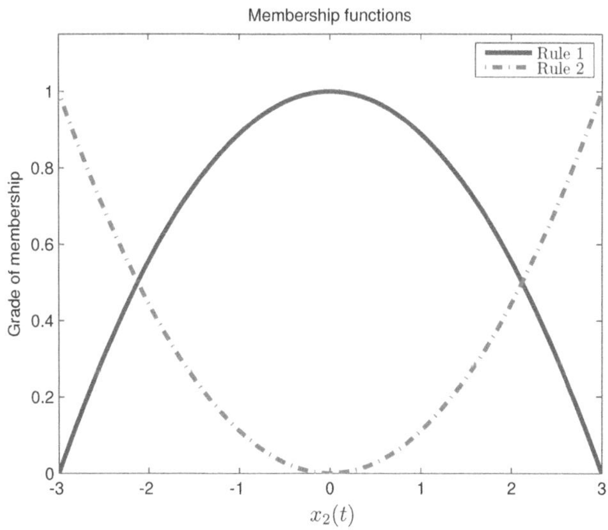

Figure 6.2 Membership functions of Example 6.1.

The MFs are chosen as $\mu_1(t) = 1 - \frac{x_2^2(t)}{9}$ and $\mu_2(t) = 1 - \mu_1(t)$, which are shown in Figure 6.2. *As shown in (6.47), the control input matrices B_l in each local model* *are different and have parameter uncertainties, and the exogenous nonlinear disturbance is in a mismatched form, the fuzzy SMC approaches presented in [112] and [114] are not applicable to this example.* Fortunately, for $\gamma = 0.7$, feasible solutions can be obtained via using Theorem 6.2 and Theorem 6.3, and the controller and the sliding surface gains are

$$\begin{cases} \bar{K}_1 = \begin{bmatrix} -9.4759 & 0.2559 & -1.2734 \end{bmatrix}, \\ \bar{K}_2 = \begin{bmatrix} -8.7127 & -0.4644 & -1.2111 \end{bmatrix}, \\ \bar{C} = \begin{bmatrix} 5.4264 & 5.4065 & 22.0585 \end{bmatrix}. \end{cases} \tag{6.49}$$

To show the effectiveness of the proposed method, simulations are conducted under different initial conditions with the singularly perturbed parameter $\varepsilon = 0.01$ in this example. With initial condition $x_0 = \begin{bmatrix} 0 & -0.8 \end{bmatrix}^{\mathrm{T}}$ and external disturbance $w(t) = 170\mathrm{e}^{-t}\sin(2t)$, Figure 6.3 shows the open-loop system states. Using the dynamic sliding mode controller in (6.9) with $\eta = 80$, Figure 6.4 and Figure 6.5, respectively, show the closed-loop system states and the control input. The sliding surface is presented in Figure 6.6.

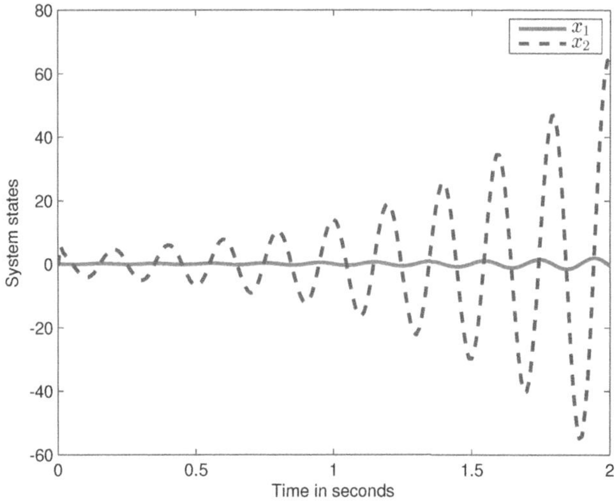

Figure 6.3 Responses of the open-loop system states in Example 6.1.

To further show the effectiveness of the proposed dynamic sliding mode controller, simulation on an inverted pendulum will be conducted.

Example 6.2. Consider an inverted pendulum controlled by a motor by a gear train [113], and the dynamics of the inverted pendulum are described as

$$\begin{cases} \dot{x}_1(t) = x_2(t) \\ \dot{x}_2(t) = \dfrac{g}{l}\sin(x_1(t)) + \dfrac{NK_m}{ml^2}x_3(t) \\ L_a\dot{x}_3(t) = -K_bNx_2(t) - R_ax_3(t) + u(t) + w(t) \\ z(t) = 0.1x_1(t) + 0.3u(t) \end{cases} \tag{6.50}$$

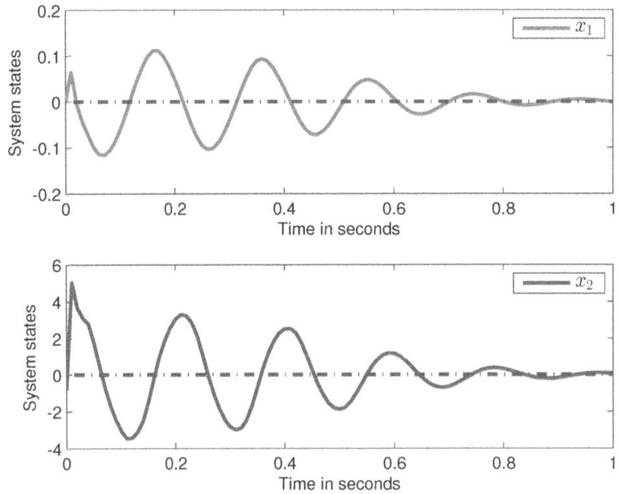

Figure 6.4 Responses of the closed-loop system states in Example 6.1.

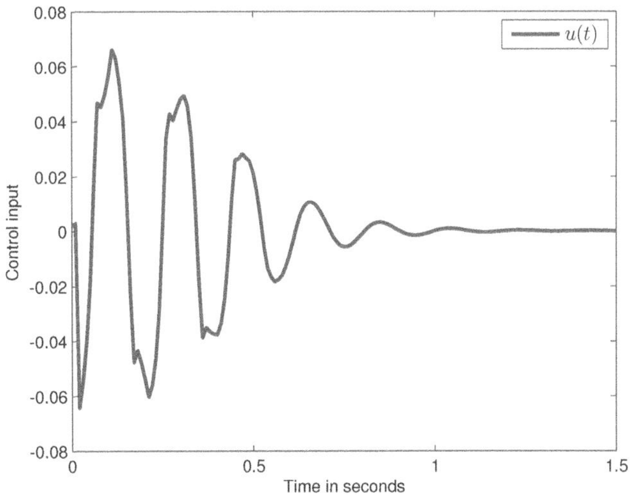

Figure 6.5 The control input in Example 6.1.

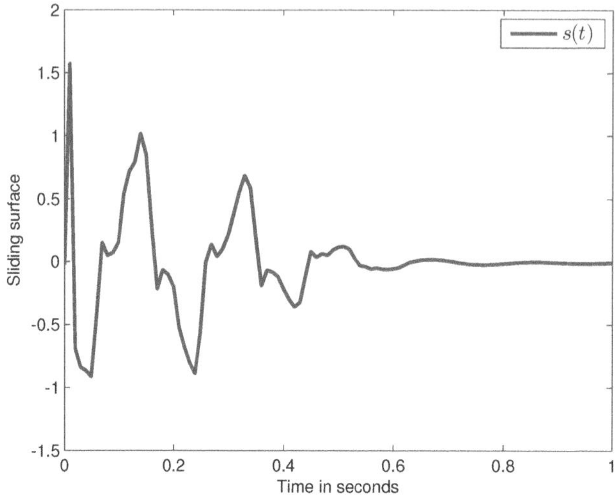

Figure 6.6 The sliding surface in Example 6.1.

where $x_1(t)$ and $x_2(t)$ are the angle of inverted pendulum from the vertical and the angular velocity, respectively; $x_3(t) = I_a(t)$ is the current; $w(t)$ is the external disturbance; K_m is the motor torque constant; K_b is the back emf constant; N is the gear ratio. The parameters for the inverted pendulum (6.50) are given as $g = 9.8\text{m/s}^2$, $l = 1\text{m}$, $m = 1\text{kg}$, $N = 10$, $K_m = 0.1\text{Nm/A}$, $K_b = 0.1\text{Vs/rad}$, $R_a = 1\Omega$, and $L_a = \varepsilon\text{mH}$.

Linearize the nonlinear system (6.50) at three operating points $\begin{pmatrix} 0 & 0 & 0 \end{pmatrix}^{\mathrm{T}}$ and $\begin{pmatrix} \pm\frac{\pi}{2} & 0 & 0 \end{pmatrix}^{\mathrm{T}}$. Then the system (6.50) can be described by a T-S fuzzy singularly perturbed system model with two rules as,

Plant Rule $\mathscr{R}^l$: IF $x_1(t)$ is $\mathscr{F}_1^l$, THEN

$$\begin{cases} E(\varepsilon)\dot{x}(t) = (A_l + \Delta A_l)x(t) + (B_l + \Delta B_l)u(t) + D_l w(t) \\ z(t) = L_l x(t) + N_l u(t), \ l \in \{1,2\}, \end{cases} \tag{6.51}$$

where

$$\begin{cases} E(\varepsilon) = \begin{bmatrix} 1 & 0 & 0 \\ 0 & 1 & 0 \\ 0 & 0 & \varepsilon \end{bmatrix}, A_1 = \begin{bmatrix} 0 & 1 & 0 \\ 9.8 & 0 & 1 \\ 0 & -1 & -1 \end{bmatrix}, A_2 = \begin{bmatrix} 0 & 1 & 0 \\ 0 & 0 & 1 \\ 0 & -1 & -1 \end{bmatrix}, \\ B_1 = \begin{bmatrix} 0 \\ 0 \\ 1 \end{bmatrix}, D_l = \begin{bmatrix} 0 \\ 0 \\ 1 \end{bmatrix}, L_l = \begin{bmatrix} 0.1 & 0 & 0 \end{bmatrix}, N_l = 0.3, l \in \{1,2\}, \end{cases} \tag{6.52}$$

and the parameter uncertainties ΔA_l and ΔB_l are given in the form of (6.2) with

$$H_{l1} = \begin{bmatrix} 0 \\ 0.1 \\ 0 \end{bmatrix}, H_{l2} = \begin{bmatrix} 0.1 & 0 & 0 \end{bmatrix}, \ l \in \{1,2\}. \tag{6.53}$$

The MFs are chosen as $\mu_1(t) = 1 - \frac{|x_1(t)|}{\pi}$ and $\mu_2(t) = 1 - \mu_1(t)$. Given $\gamma = 0.5$, feasible solutions can be obtained by using Theorem 6.2 and Theorem 6.3, and the controller and the sliding surface gains are

$$\begin{cases} \bar{K}_1 = \begin{bmatrix} -193.1280 & -70.5092 & -2.0931 & -8.8828 \end{bmatrix}, \\ \bar{K}_2 = \begin{bmatrix} -344.2366 & -122.8752 & -2.0931 & -16.6627 \end{bmatrix}, \\ \bar{C} = \begin{bmatrix} 42.4468 & 41.8715 & -18.9125 & 119.9031 \end{bmatrix}. \end{cases} \tag{6.54}$$

To show the effectiveness of the proposed method, simulations are conducted under different initial conditions with the singularly perturbed parameter $\varepsilon = 0.1$ in this example. With initial condition $x_0 = \begin{bmatrix} \pi/4 & 0 & 0 \end{bmatrix}^{\mathrm{T}}$ and external disturbance $w(t) = 10\mathrm{e}^{-t}\sin(2t)$, Using the dynamic sliding mode controller in (6.9) with $\eta = 0.02$, Figure 6.7 and Figure 6.8, respectively, show the closed-loop system states and the control input. Under zero initial condition, the robust $\mathscr{H}_\infty$ performance is presented in Figure 6.9. The ratio $\frac{\sqrt{\int_0^{t_f} z^T(t)z(t)dt}}{\sqrt{\int_0^{t_f} w^T(t)w(t)dt}}$ is about 0.084, which is lower than the pre-defined disturbance attenuation level $\gamma = 0.5$.

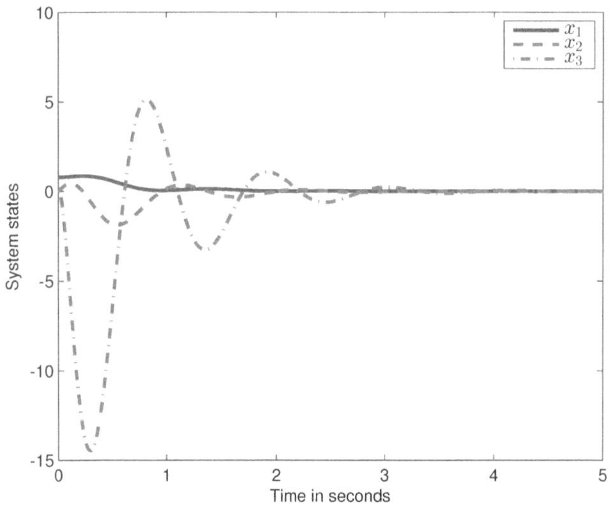

Figure 6.7 Responses of the closed-loop system states in Example 6.2.

All these simulation examples clearly show the effectiveness of the proposed dynamic SMC design method for a class of T-S fuzzy singularly perturbed systems.

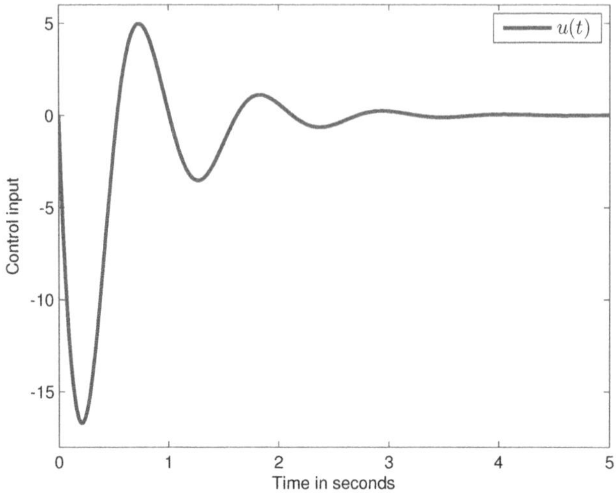

Figure 6.8 The control input in Example 6.2.

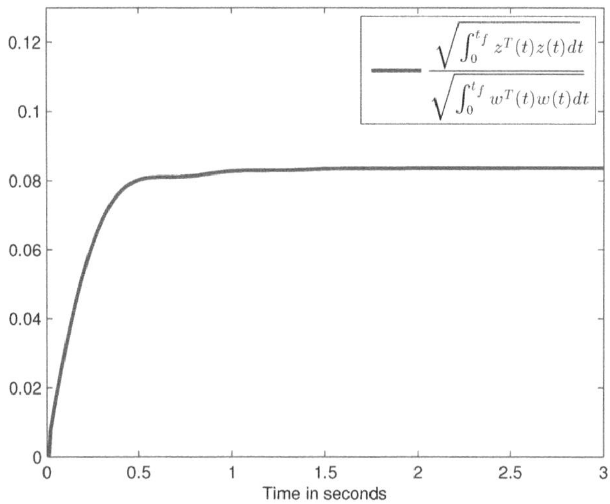

Figure 6.9 Response of the ratio $\frac{\sqrt{\int_0^{t_f} z^T(t)z(t)dt}}{\sqrt{\int_0^{t_f} w^T(t)w(t)dt}}$ in Example 6.2.

6.5 CONCLUSION

This chapter studied the integral sliding mode control issue for T-S fuzzy singular perturbed systems. Via some elegant convexification procedures, a new existence criterion for the integral sliding surface is given in a convex optimization framework. A dynamic sliding mode controller design method is developed to ensure the finite-time convergence of the sliding surface. Simulation studies are presented to illustrate the validity of the proposed approach. In view of the full system states of T-S fuzzy SPSs being not available, the output feedback integral SMC design problem for T-S fuzzy SPSs can be remained as a future study topic for the readers.

7 Fuzzy SMC of Continuous-Time Nonlinear Two-Dimensional Systems

7.1 INTRODUCTION

Chapter 2 - Chapter 6 have fully investigated the T-S fuzzy model-based state feedback and output feedback sliding mode control problems for some nonlinear systems, which include general nonlinear systems and multiple time-scale nonlinear systems. It should be also noted that the system information of the aforementioned nonlinear systems evolves along only one direction, and these systems are actually known as one-dimensional (1-D) systems. Recent decades have witnessed an increasing development on multi-dimensional (m-D) dynamical systems as many practical processes or systems intrinsically contain m-D properties, such as image deblurring, long transmission line, and so on [115, 116, 118, 119, 120, 121]. Differing from 1-D systems, m-D systems tend to be more complex in system structure, modelling and design. Among various m-D systems, two-dimensional (2-D) systems are recognized as a typical class of m-D systems, and also contain highly complex nonlinearities generally. Consequently, many researchers have devoted efforts to 2-D systems and a substantial body of valuable results have been reported [122, 123, 124, 125, 126, 127]. For instance, [126] addressed the robust $\mathscr{H}_\infty$ control problem for the gas absorption, water stream heating and air drying processes based on Roesser model.

In view of the theoretical advancement and engineering application of SMC, great efforts have been also concentrated on SMC for 2-D systems [122, 132, 133, 134, 135, 136, 137]. The authors in [132] addressed the SMC problem for the gas absorption, water stream heating and air drying processes via discrete-time 2-D Roesser model *without considering external disturbance*. In [134], a direct *state feedback* SMC law was developed for 2-D systems through the Fornasini-Marchesini model. When *matched exogenous disturbance exists*, the authors in [135] investigated the SMC issue for discrete-time linear 2-D systems. For the unavailable full system states situation, in [122], an observer-based SMC method was proposed for differential linear repetitive processes, *where the control input matrix was not allowed to have parameter uncertainties*.

Notice that the existing SMC results [132, 133, 134, 135] for 2-D systems were basically obtained through the full state feedback method. Unfortunately, the full system state variables are not accessible in most complex engineering applications, especially for some nonlinear 2-D systems. Furthermore, all those results in

DOI: 10.1201/9781032695716-7

[132, 122, 133, 134, 135] were mainly derived through a common quadratic Lyapunov function, which also tends to remain conservative. In addition, note that T-S fuzzy affine models considered in this monograph inherently have much improved function approximation capability, while there have been few fuzzy SMC results for nonlinear 2-D systems. To our best knowledge, few efforts have been devoted to the SMC problems for both continuous-time and discrete-time nonlinear 2-D processes through T-S fuzzy affine models and piecewise quadratic Lyapunov functions (PQLFs), which inspires the study in Chapter 7 and Chapter 8.

In this chapter, the dynamic SMC design problem for continuous-time nonlinear 2-D systems with exogenous disturbance will be firstly investigated by T-S fuzzy affine models. Associated with sliding surface, a singular 2-D system is first constructed to describe the sliding mode dynamical features. Based on PQLFs, some novel stability and robust performance analysis results on the sliding mode dynamical system are proposed, and the analysis conservatism can be further reduced. The gain of the sliding surface can be attained by solving a set of LMIs. In sequel, a robust state feedback dynamic sliding mode controller and an output feedback dynamic dynamic sliding mode controller are designed, respectively, such that the finite-time reachability of the sliding surface can be guaranteed.

7.2 OUTPUT FEEDBACK DYNAMIC SMC DESIGN VIA 2-D T-S FUZZY MODELS

7.2.1 MODEL DESCRIPTION

Differing from 1-D systems, the system information evolves along two independent directions for 2-D systems. In view of these structural properties, consider a general nonlinear 2-D system as follows,

$$\begin{cases} \mathscr{D}x(t_1,t_2) = f_1(x(t_1,t_2)) + g_1(x(t_1,t_2))u(t_1,t_2) + h(x(t_1,t_2))w(t_1,t_2) \\ y(t_1,t_2) = f_2(x(t_1,t_2)) \\ z(t_1,t_2) = f_3(x(t_1,t_2)) + g_2(x(t_1,t_2))u(t_1,t_2), \end{cases} \tag{7.1}$$

where $x^h(t_1,t_2) = \left[x_1^h(t_1,t_2)\cdots x_{n_x^h}^h(t_1,t_2)\right] \in \mathbb{X}^h = \prod_{l=1}^{n_x^h} \mathbb{X}_l^h \subset \Re^{n_x^h}$ and $x^v(t_1,t_2) = \left[x_1^v(t_1,t_2)\cdots x_{n_x^v}^v(t_1,t_2)\right] \in \mathbb{X}^v = \prod_{l=1}^{n_x^v} \mathbb{X}_l^v \subset \Re^{n_x^v}$ are the horizontal states and vertical states, respectively, and $n_x = n_x^h + n_x^v$, $x(t_1,t_2) = \left[\begin{array}{cc} x^{h\mathrm{T}}(t_1,t_2) & x^{v\mathrm{T}}(t_1,t_2) \end{array}\right]^{\mathrm{T}}$; $y(t_1,t_2) \in \Re^{n_y}$ is the measurement output; $w(t_1,t_2) \in L_2\{[0,\infty),[0,\infty)\}$ is the exogenous disturbance with $w(t_1,t_2) \in \Re^{n_w}$; $z(t_1,t_2) \in \Re^{n_z}$ is the regulated output; $u(t_1,t_2) \in \Re^{n_u}$ is the control input. Nonlinear functions $f_1(x(t_1,t_2))$, $f_2(x(t_1,t_2))$, $f_3(x(t_1,t_2))$, $g_1(x(t_1,t_2))$ $g_2(x(t_1,t_2))$, and $h(x(t_1,t_2))$ are with respect to variables $x^h(t_1,t_2)$ and $x^v(t_1,t_2)$; $f_1(x(t_1,t_2))$, $g_1(x(t_1,t_2))$, and $h(x(t_1,t_2))$ are continuously differentiable on $\mathbb{X}^h \times \mathbb{X}^v$. It is assumed that $\mathbb{X}^h \times \mathbb{X}^v$ is a compact set on $\Re^{n_x^h} \times \Re^{n_x^v}$.

Approximate the system (7.1) via the following 2-D T-S fuzzy affine model with r fuzzy **IF-THEN** rules as,

Plant Rule $\mathscr{R}^l$: IF $\theta_1^h(t_1,t_2)$ is $\mathscr{F}_{l1}^h$ and $\cdots$ and $\theta_{\varphi^h}^h(t_1,t_2)$ is $\mathscr{F}_{l\varphi^h}^h$; $\theta_1^v(t_1,t_2)$ is $\mathscr{F}_{l1}^v$ and $\cdots$ and $\theta_{\varphi^v}^v(t_1,t_2)$ is $\mathscr{F}_{l\varphi^v}^v$, THEN

$$\begin{cases} \mathscr{D}x(t_1,t_2) = (A_l+\Delta A_l)x(t_1,t_2)+a_l+\Delta a_l+(B_l+\Delta B_l)u(t_1,t_2) \\ \qquad\qquad +D_l w(t_1,t_2) \\ y(t_1,t_2) = Cx(t_1,t_2) \\ z(t_1,t_2) = L_l x(t_1,t_2)+N_l u(t_1,t_2),\ l\in\mathscr{L}=\{1,\ldots,r\}, \end{cases} \tag{7.2}$$

where $\mathscr{R}^l$ denotes the l-th fuzzy inference rule; $\mathscr{F}_{l\phi^h}^h(\phi^h=1,\ldots,\varphi^h)$ and $\mathscr{F}_{l\phi^v}^v(\phi^v=1,\ldots,\varphi^v)$ are fuzzy sets; $\theta^h(t_1,t_2) := \left[\theta_1^h(t_1,t_2),\ldots,\theta_{\varphi^h}^h(t_1,t_2)\right]$ and $\theta^v(t_1,t_2) := \left[\theta_1^v(t_1,t_2),\ldots,\theta_{\varphi^v}^v(t_1,t_2)\right]$ are the measurable premise variables; r is the number of inference rules; $(A_l,a_l,B_l,D_l,C,L_l,N_l)$ is the model of the l-th 2-D subsystem; ΔA_l, Δa_l and ΔB_l are the parameter uncertainties subject to

$$\left[\begin{array}{ccc} \Delta A_l & \Delta a_l & \Delta B_l \end{array}\right] = H_{l1}\Delta_l(t_1,t_2)\left[\begin{array}{ccc} H_{l2} & H_{l3} & H_{l4} \end{array}\right],\ l\in\mathscr{L} \tag{7.3}$$

with H_{l1},H_{l2},H_{l3} and H_{l4} being known matrices. $\Delta_l(t_1,t_2)\in\Re^{s_1\times s_2}$ refer to unknown matrices subject to $\Delta_l^{\mathrm{T}}(t_1,t_2)\Delta_l(t_1,t_2)\le\mathbf{I}$, $l\in\{1,\ldots,r\}$.

Different from the traditional model rule for 1-D systems, the premise variables of fuzzy 2-D systems are vector-valued functions rather than scalar functions. Denote $\mu_l\left[\theta^h(t_1,t_2),\theta^v(t_1,t_2)\right]$ as the normalized membership function (MF),

$$\begin{aligned} &\mu_l\left[\theta^h(t_1,t_2),\theta^v(t_1,t_2)\right] \\ &\quad= \frac{\Pi_{\phi^h=1}^{\varphi^h}\mu_{l\phi^h}\left[\theta_{\phi^h}^h(t_1,t_2)\right]\Pi_{\phi^v=1}^{\varphi^v}\mu_{l\phi^v}\left[\theta_{\phi^v}^v(t_1,t_2)\right]}{\Sigma_{v=1}^r\Pi_{\phi^h=1}^{\varphi^h}\mu_{v\phi^h}\left[\theta_{\phi^h}^h(t_1,t_2)\right]\Pi_{\phi^v=1}^{\varphi^v}\mu_{v\phi^v}\left[\theta_{\phi^v}^v(t_1,t_2)\right]}\ge 0, \\ &\sum_{l=1}^r\mu_l\left[\theta^h(t_1,t_2),\theta^v(t_1,t_2)\right]=1, \end{aligned} \tag{7.4}$$

where $\mu_{l\phi^h}\left[\theta_{\phi^h}^h(t_1,t_2)\right]$ and $\mu_{l\phi^v}\left[\theta_{\phi^v}^v(t_1,t_2)\right]$ represent the grade of membership of $\theta_{\phi^h}^h(t_1,t_2)$ in $\mathscr{F}_{l\phi^h}^h$ and $\theta_{\phi^v}^v(t_1,t_2)$ in $\mathscr{F}_{l\phi^v}^v$, respectively. Denote $\mu_l := \mu_l\left[\theta^h(t_1,t_2),\theta^v(t_1,t_2)\right]$ for brevity.

Based on (7.4), the fuzzy 2-D system (7.2) can be inferred as

$$\begin{cases} \mathscr{D}x(t_1,t_2) = (A(\mu)+\Delta A(\mu))x(t_1,t_2)+a(\mu)+\Delta a(\mu) \\ \qquad\qquad +(B(\mu)+\Delta B(\mu))u(t_1,t_2)+D(\mu)w(t_1,t_2) \\ y(t_1,t_2) = Cx(t_1,t_2), \\ z(t_1,t_2) = L(\mu)x(t_1,t_2)+N(\mu)u(t_1,t_2) \end{cases} \tag{7.5}$$

where

$$\begin{cases} A(\mu) = \sum_{l=1}^{r} \mu_l A_l, \Delta A(\mu) = \sum_{l=1}^{r} \mu_l \Delta A_l, \\ a(\mu) = \sum_{l=1}^{r} \mu_l a_l, \Delta a(\mu) = \sum_{l=1}^{r} \mu_l \Delta a_l, \\ B(\mu) = \sum_{l=1}^{r} \mu_l B_l, \Delta B(\mu) = \sum_{l=1}^{r} \mu_l \Delta B_l, \\ D(\mu) = \sum_{l=1}^{r} \mu_l D_l, L(\mu) = \sum_{l=1}^{r} \mu_l L_l, N(\mu) = \sum_{l=1}^{r} \mu_l N_l. \end{cases} \tag{7.6}$$

As fuzzy rules and membership functions induce a polyhedral partition of the system state-space, the global fuzzy 2-D model in (7.5) can be viewed as a convex combination of local models in individual regions. Similar to Chapter 2, the premise variable space can be also decomposed into two kinds of regions: crisp regions and fuzzy regions. The crisp region is the region that contains only one rule, that is, for some l, $\mu_l[\theta^h(t_1,t_2),\theta^v(t_1,t_2)] = 1$ and all other membership functions are zero. The system dynamics in the crisp region are governed by the l-th local model of (7.2). The fuzzy regions denote the regions where $0 < \mu_l[\theta^h(t_1,t_2),\theta^v(t_1,t_2)] < 1$ and the system dynamics are characterized by a convex combination of some local models.

Specify the indices of regions $\mathscr{I} = \mathscr{I}_0 \bigcup \mathscr{I}_1$, and $\mathscr{I}_0$ includes the index set of regions covering the origin, while $\mathscr{I}_1$ involves the index set of regions without the origin. Introduce the set

$$\mathscr{N}(i) := \left\{ m | \mu_m > 0, m \in \mathscr{L}, (\theta^h(t_1,t_2),\theta^v(t_1,t_2)) \in S_i,\ i \in \mathscr{I} \right\} \tag{7.7}$$

to characterize the indices adopted for every region S_i.

Based on (7.7), the fuzzy 2-D system (7.5) is formulated as

$$\begin{cases} \mathscr{D}x(t_1,t_2) = (\mathscr{A}_i + \Delta\mathscr{A}_i)x(t_1,t_2) + \mathrm{a}_i + \Delta\mathrm{a}_i + (\mathscr{B}_i + \Delta\mathscr{B}_i)u(t_1,t_2) \\ \qquad + \mathscr{D}_i w(t_1,t_2) \\ y(t_1,t_2) = Cx(t_1,t_2), \\ z(t_1,t_2) = \mathscr{L}_i x(t_1,t_2) + \mathscr{N}_i u(t_1,t_2), (\theta^h(t_1,t_2),\theta^v(t_1,t_2)) \in S_i, i \in \mathscr{I}, \end{cases} \tag{7.8}$$

where

$$\begin{cases} \mathscr{A}_i := \sum_{m\in\mathscr{N}(i)} \mu_m A_m, \Delta\mathscr{A}_i := \sum_{m\in\mathscr{N}(i)} \mu_m \Delta A_m, \\ \mathrm{a}_i := \sum_{m\in\mathscr{N}(i)} \mu_m a_m, \Delta\mathrm{a}_i := \sum_{m\in\mathscr{N}(i)} \mu_m \Delta a_m, \\ \mathscr{B}_i := \sum_{m\in\mathscr{N}(i)} \mu_m B_m, \Delta\mathscr{B}_i := \sum_{m\in\mathscr{N}(i)} \mu_m \Delta B_m, \\ \mathscr{D}_i := \sum_{m\in\mathscr{N}(i)} \mu_m D_m, \mathscr{L}_i := \sum_{m\in\mathscr{N}(i)} \mu_m L_m, \mathscr{N}_i := \sum_{m\in\mathscr{N}(i)} \mu_m N_m \end{cases} \tag{7.9}$$

where $\sum_{m\in\mathscr{N}(i)} \mu_m = 1, 0 < \mu_m \leq 1$.

This chapter will firstly propose a novel output feedback dynamic SMC synthesis scheme for fuzzy 2-D system (7.2), and it will be shown that the system states can be forced onto the sliding surface in finite time. Through PQLFs, new existence conditions for sliding surface will be obtained. For a prescribed robust $\mathscr{H}_\infty$ performance $\gamma > 0$, the asymptotic stability of the sliding mode dynamical system can be guaranteed when $w(t_1,t_2) \equiv 0$. It is also satisfied that

$$\mathscr{H}_\infty^{zw} = \sup \frac{\|z\|_2}{\|w\|_2} < \gamma \tag{7.10}$$

under zero boundary conditions for $0 \neq w(t_1,t_2) \in L_2\{[0,\ \infty),[0,\ \infty)\}$.

7.2.2 SLIDING SURFACE DESIGN

Relying on the control input and measurement output, design the sliding surface for the fuzzy 2-D system (7.2) as

$$s(t_1,t_2) = Ky(t_1,t_2) + u(t_1,t_2), \tag{7.11}$$

where $K \in \Re^{n_u \times n_y}$ is the sliding surface gain to be determined. After the sliding surface is reached, one has $s(t_1,t_2) = 0$, and then it can be seen that

$$u(t_1,t_2) = -KCx(t_1,t_2). \tag{7.12}$$

Based on system (7.8) and (7.12), one can derive the sliding mode dynamical system as

$$\begin{cases} \mathscr{D}x(t_1,t_2) = (\mathscr{A}_i + \Delta\mathscr{A}_i)x(t_1,t_2) + \mathrm{a}_i + \Delta\mathrm{a}_i \\ \qquad\qquad\qquad +(\mathscr{B}_i + \Delta\mathscr{B}_i)u(t_1,t_2) + \mathscr{D}_i w(t_1,t_2) \\ 0 \times \frac{\partial u(t_1,t_2)}{\partial t_1} = -KCx(t_1,t_2) - u(t_1,t_2) \\ z(t_1,t_2) = \mathscr{L}_i x(t_1,t_2) + \mathscr{N}_i u(t_1,t_2), i \in \mathscr{I}. \end{cases} \tag{7.13}$$

Reformulate system (7.13) as

$$\begin{cases} \bar{E}\mathscr{D}\bar{x}(t_1,t_2) = (\bar{\mathscr{A}_i} + \Delta\bar{\mathscr{A}_i})\bar{x}(t_1,t_2) + \bar{\mathscr{D}}_i w(t_1,t_2) \\ z(t_1,t_2) = \bar{\mathscr{L}}_i \bar{x}(t_1,t_2), i \in \mathscr{I}, \end{cases} \tag{7.14}$$

where

$$\begin{cases} \left.\begin{array}{l} \bar{E} = \mathrm{diag}\{\mathbf{I}_{n_x}, \mathbf{0}_{n_u}\}, \\ \mathscr{D}\bar{x}(t_1,t_2) = \begin{bmatrix} \frac{\partial}{\partial t_1} x_h^{\mathrm{T}}(t_1,t_2) & \frac{\partial}{\partial t_2} x_v^{\mathrm{T}}(t_1,t_2) & \frac{\partial}{\partial t_1} u^{\mathrm{T}}(t_1,t_2) \end{bmatrix}^{\mathrm{T}}, \\ \bar{\mathscr{A}_i} + \Delta\bar{\mathscr{A}_i} = \begin{bmatrix} \mathscr{A}_i + \Delta\mathscr{A}_i & \mathscr{B}_i + \Delta\mathscr{B}_i \\ -KC & -\mathbf{I} \end{bmatrix}, \\ \bar{\mathscr{D}}_i = \begin{bmatrix} \mathscr{D}_i^{\mathrm{T}} & \mathbf{0} \end{bmatrix}^{\mathrm{T}}, \bar{\mathscr{L}}_i = \begin{bmatrix} \mathscr{L}_i & \mathscr{N}_i \end{bmatrix}, \end{array}\right\} i \in \mathscr{I}_0, \\ \left.\begin{array}{l} \bar{E} = \mathrm{diag}\{\mathbf{I}_{(n_x+1)}, \mathbf{0}_{n_u}\}, \\ \mathscr{D}\bar{x}(t_1,t_2) = \begin{bmatrix} \mathrm{i} & \frac{\partial}{\partial t_1} x_h^{\mathrm{T}}(t_1,t_2) & \frac{\partial}{\partial t_2} x_v^{\mathrm{T}}(t_1,t_2) & \frac{\partial}{\partial t_1} u^{\mathrm{T}}(t_1,t_2) \end{bmatrix}^{\mathrm{T}}, \\ \bar{\mathscr{A}_i} + \Delta\bar{\mathscr{A}_i} = \begin{bmatrix} 0 & \mathbf{0} & \mathbf{0} \\ \mathrm{a}_i + \Delta\mathrm{a}_i & \mathscr{A}_i + \Delta\mathscr{A}_i & \mathscr{B}_i + \Delta\mathscr{B}_i \\ \mathbf{0} & -KC & -\mathbf{I} \end{bmatrix}, \\ \bar{\mathscr{D}}_i = \begin{bmatrix} \mathbf{0} & \mathscr{D}_i^{\mathrm{T}} & \mathbf{0} \end{bmatrix}^{\mathrm{T}}, \bar{\mathscr{L}}_i = \begin{bmatrix} \mathbf{0} & \mathscr{L}_i & \mathscr{N}_i \end{bmatrix}, \end{array}\right\} i \in \mathscr{I}_1. \end{cases} \tag{7.15}$$

Via augmenting the system (7.8) with the sliding surface (7.11), the sliding motion dynamics are embedded into the augmented singular 2-D system (7.14). In the sequel, we will derive the existence criterion for the sliding surface (7.11), and then design a new output feedback dynamic sliding mode controller to guarantee the local finite-time reachability of the sliding surface.

7.2.3 STABILITY AND ROBUST PERFORMANCE ANALYSIS OF SLIDING MOTION THROUGH PQLFS

In this subsection, some new results for the asymptotic stability and robust $\mathscr{H}_\infty$ performance analysis of the sliding motion will be proposed based on PQLFs. Along the procedures in [40], to guarantee the PQLFs to be continuous across the subspace boundaries, matrices $\bar{F}_i = \begin{bmatrix} f_i^h & F_i^h & \mathbf{0} \\ f_i^v & \mathbf{0} & F_i^v \end{bmatrix}$, $i \in \mathscr{I}$ with $f_i^h = 0$ and $f_i^v = 0$ for $i \in \mathscr{I}_0$ are constructed to characterize the boundary across the subspaces,

$$\bar{F}_i \begin{bmatrix} 1 \\ x^h(t_1,t_2) \\ x^v(t_1,t_2) \end{bmatrix} = \bar{F}_j \begin{bmatrix} 1 \\ x^h(t_1,t_2) \\ x^v(t_1,t_2) \end{bmatrix}, x(t_1,t_2) \in S_i \cap S_j, i,j \in \mathscr{I}. \tag{7.16}$$

For further conservatism reduction, we will also adopt the S-procedure by establishing matrices $\bar{G}_i = \begin{bmatrix} g_i^h & G_i^h & \mathbf{0} \\ g_i^v & \mathbf{0} & G_i^v \end{bmatrix}$, $i \in \mathscr{I}$, with $g_i^h = 0$ and $g_i^v = 0$ for $i \in \mathscr{I}_0$ and

$$\bar{G}_i \begin{bmatrix} 1 \\ x^h(t_1,t_2) \\ x^v(t_1,t_2) \end{bmatrix} \succeq 0,\ x(t_1,t_2) \in S_i, i \in \mathscr{I} \tag{7.17}$$

with $\succeq$ indicating each entry being nonnegative.

Reference [40] gives some systematic methods to calculate the matrices $\bar{F}_i$ and $\bar{G}_i$ for T-S fuzzy affine systems associated with its space partition. The readers can see the references therein for more details.

Theorem 7.1. Given a disturbance attenuation level γ, the sliding mode dynamical system (7.14) is asymptotically stable, if matrices $X_h = X_h^{\mathrm{T}}$, $X_v = X_v^{\mathrm{T}}$, $U_{hi} = U_{hi}^{\mathrm{T}} \succeq 0$, $U_{vi} = U_{vi}^{\mathrm{T}} \succeq 0$, $W_{hi} = W_{hi}^{\mathrm{T}} \succeq 0$, $W_{vi} = W_{vi}^{\mathrm{T}} \succeq 0$, $K \in \Re^{n_u \times n_y}$, $P_2 \in \Re^{n_u \times n_u}$, and scalars $\varepsilon_i > 0$, $i \in \mathscr{I}$, exist such that conditions (7.18)–(7.19) hold,

$$\bar{F}_i^{\mathrm{T}} X \bar{F}_i - \bar{G}_i^{\mathrm{T}} W_i \bar{G}_i > 0, i \in \mathscr{I}, \tag{7.18}$$

$$\begin{bmatrix} \bar{\Gamma}_1 + \bar{G}_i^{\mathrm{T}} U_i \bar{G}_i & \star & \star & \star \\ \bar{\Gamma}_2 & \bar{\Gamma}_3 & \star & \star \\ D_m^{\mathrm{T}} \bar{Q} \bar{F}_i^{\mathrm{T}} X \bar{F}_i & \mathbf{0} & -\gamma^2 \mathbf{I} & \star \\ H_{m1}^{\mathrm{T}} \bar{Q} \bar{F}_i^{\mathrm{T}} X \bar{F}_i & \mathbf{0} & \mathbf{0} & -\varepsilon_i \mathbf{I} \end{bmatrix} < 0, m \in \mathscr{N}(i), i \in \mathscr{I}, \tag{7.19}$$

where

$$\left\{\begin{array}{l} \bar{\Gamma}_1 = \bar{Q}^{\mathrm{T}} L_m^{\mathrm{T}} L_m \bar{Q} + \varepsilon_i \bar{\Gamma}_{1(1)} \bar{\Gamma}_{1(1)}^{\mathrm{T}} + \mathrm{Sym}\{\bar{F}_i^{\mathrm{T}} X \bar{F}_i \bar{A}_{m1} + \bar{J}^{\mathrm{T}} \bar{K} C \bar{Q}\}, \\ \bar{\Gamma}_2 = -\bar{K} C \bar{Q} + B_m^{\mathrm{T}} \bar{Q} \bar{F}_i^{\mathrm{T}} X \bar{F}_i + P_2 \bar{J} + N_m^{\mathrm{T}} L_m \bar{Q} + \varepsilon_i H_{m4}^{\mathrm{T}} \bar{\Gamma}_{1(1)}^{\mathrm{T}}, \\ \bar{\Gamma}_3 = -P_2 - P_2^{\mathrm{T}} + N_m^{\mathrm{T}} N_m + \varepsilon_i H_{m4}^{\mathrm{T}} H_{m4}, \\ X = \mathrm{diag}\{X_h, X_v\}, W_i = \mathrm{diag}\{W_{hi}, W_{vi}\},\ U_i = \mathrm{diag}\{U_{hi}, U_{vi}\}, \\ \left.\bar{\Gamma}_{1(1)} = H_{m2}^{\mathrm{T}},\ \bar{A}_{m1} = A_m,\ \bar{Q} = \mathbf{I}_{n_x}, \bar{J} \in \mathfrak{R}^{n_u \times n_x},\ \right\} i \in \mathscr{I}_0, \\ \left.\begin{array}{l} \bar{\Gamma}_{1(1)} = \begin{bmatrix} H_{m3}^{\mathrm{T}} \\ H_{m2}^{\mathrm{T}} \end{bmatrix}, \bar{A}_{m1} = \begin{bmatrix} 0 & \mathbf{0} \\ a_m & A_m \end{bmatrix}, \\ \bar{Q} = \begin{bmatrix} \mathbf{0} & \mathbf{I}_{n_x} \end{bmatrix},\ \bar{J} \in \mathfrak{R}^{n_u \times (n_x+1)} \end{array}\right\} i \in \mathscr{I}_1 \end{array}\right. \tag{7.20}$$

and $\bar{J} \in \mathfrak{R}^{n_u \times n_x}$, for $i \in \mathscr{I}_0$, $\bar{J} \in \mathfrak{R}^{n_u \times (n_x+1)}$, for $i \in \mathscr{I}_1$, are arbitrary matrices.

The gain of the sliding surface (7.11) can be computed by

$$K = P_2^{-\mathrm{T}} \bar{K}. \tag{7.21}$$

Proof. Consider the following PQLFs being continuous across the region boundaries

$$V(\bar{x}(t_1,t_2),t_1,t_2) = V_1(\bar{x}_h(t_1,t_2),t_1,t_2) + V_2(\bar{x}_v(t_1,t_2),t_1,t_2), \tag{7.22}$$

where

$$\left\{\begin{array}{l} V_1(\bar{x}_h(t_1,t_2),t_1,t_2) = \bar{x}_h^{\mathrm{T}}(t_1,t_2) \bar{E}_h^{\mathrm{T}} \bar{\mathscr{P}}_{hi} \bar{x}_h(t_1,t_2), \\ V_2(\bar{x}_v(t_1,t_2),t_1,t_2) = \bar{x}_v^{\mathrm{T}}(t_1,t_2) \bar{E}_v^{\mathrm{T}} \bar{\mathscr{P}}_{vi} \bar{x}_v(t_1,t_2), \\ \bar{\mathscr{P}}_{hi} = \begin{bmatrix} \bar{F}_{hi}^{\mathrm{T}} X_h \bar{F}_{hi} & \mathbf{0} \\ \bar{P}_{h3} & P_{h2} \end{bmatrix}, \bar{\mathscr{P}}_{vi} = \begin{bmatrix} \bar{F}_{vi}^{\mathrm{T}} X_v \bar{F}_{vi} & \mathbf{0} \\ \bar{P}_{v3} & P_{v2} \end{bmatrix}, \\ \left.\begin{array}{l} \bar{x}_h(t_1,t_2) = \begin{bmatrix} \bar{x}_h^{\mathrm{T}}(t_1,t_2) & u^{\mathrm{T}}(t_1,t_2) \end{bmatrix}^{\mathrm{T}}, \\ \bar{x}_v(t_1,t_2) = \begin{bmatrix} \bar{x}_v^{\mathrm{T}}(t_1,t_2) & u^{\mathrm{T}}(t_1,t_2) \end{bmatrix}^{\mathrm{T}}, \\ \bar{E}_h = \mathrm{diag}\{\mathbf{I}_{n_x^h}, \mathbf{0}\}, \bar{E}_v = \mathrm{diag}\{\mathbf{I}_{n_x^v}, \mathbf{0}\}, \\ \bar{F}_{hi} = \begin{bmatrix} f_i^h & F_i^h \end{bmatrix}, \bar{F}_{vi} = \begin{bmatrix} f_i^v & F_i^v \end{bmatrix}, \end{array}\right\} i \in \mathscr{I}_0, \\ \left.\begin{array}{l} \bar{x}_h(t_1,t_2) = \begin{bmatrix} 1 & \bar{x}_h^{\mathrm{T}}(t_1,t_2) & u^{\mathrm{T}}(t_1,t_2) \end{bmatrix}^{\mathrm{T}}, \\ \bar{x}_v(t_1,t_2) = \begin{bmatrix} 1 & \bar{x}_v^{\mathrm{T}}(t_1,t_2) & u^{\mathrm{T}}(t_1,t_2) \end{bmatrix}^{\mathrm{T}}, \\ \bar{E}_h = \mathrm{diag}\{\mathbf{I}_{(1+n_x^h)}, \mathbf{0}\}, \bar{E}_v = \mathrm{diag}\{\mathbf{I}_{(1+n_x^v)}, \mathbf{0}\}, \\ \bar{F}_{hi} = \begin{bmatrix} 1 & f_i^h & F_i^h \end{bmatrix}, \bar{F}_{vi} = \begin{bmatrix} 1 & f_i^v & F_i^v \end{bmatrix}, \end{array}\right\} i \in \mathscr{I}_1 \end{array}\right. \tag{7.23}$$

and $X_h = X_h^{\mathrm{T}}$, $X_v = X_v^{\mathrm{T}}$, $\{P_{h2}, P_{v2}\} \in \mathfrak{R}^{n_u \times n_u}$ are arbitrary nonsingular matrices, and $\bar{P}_{h3} \in \mathfrak{R}^{n_u \times n_x^h}$, $\bar{P}_{v3} \in \mathfrak{R}^{n_u \times n_x^v}$, for $i \in \mathscr{I}_0$, and $\bar{P}_{h3} \in \mathfrak{R}^{n_u \times (n_x^h+1)}$, $\bar{P}_{v3} \in \mathfrak{R}^{n_u \times (n_x^v+1)}$, for $i \in \mathscr{I}_1$.

The inequality in (7.18) ensures the PQLFs to be positive definite in each subspace. With the PQLFs given in (7.22), if it is satisfied

$$\mathscr{D}V(\bar{x}(t_1,t_2),t_1,t_2) + \|z(t_1,t_2)\|^2 - \gamma^2 \|w(t_1,t_2)\|^2 < 0 \tag{7.24}$$

with $\mathscr{D}V(\bar{x}(t_1,t_2),t_1,t_2) = \frac{\partial}{\partial t_1} V_1(\bar{x}_h(t_1,t_2),t_1,t_2) + \frac{\partial}{\partial t_2} V_2(\bar{x}_v(t_1,t_2),t_1,t_2)$, the system (7.14) is asymptotically stable in an $\mathscr{H}_\infty = \gamma$ setup.

Noticing $\bar{\mathscr{P}}_{hi}$ and $\bar{\mathscr{P}}_{vi}$ shown in (7.23), one has $\bar{E}_h^{\mathrm{T}}\bar{\mathscr{P}}_{hi} = \bar{\mathscr{P}}_{hi}^{\mathrm{T}}\bar{E}_h$ and $\bar{E}_v^{\mathrm{T}}\bar{\mathscr{P}}_{vi} = \bar{\mathscr{P}}_{vi}^{\mathrm{T}}\bar{E}_v$. Along the two directional trajectories of the system (7.14), one has

$$\begin{aligned}\mathscr{D}V(\bar{x}(t_1,t_2),t_1,t_2) = {} & 2\bar{x}_h^{\mathrm{T}}(t_1,t_2)\bar{P}_{hi}^{\mathrm{T}}\bar{E}_h\frac{\partial}{\partial t_1}\bar{x}_h(t_1,t_2)\\ & +2\bar{x}_v^{\mathrm{T}}(t_1,t_2)\bar{P}_{vi}^{\mathrm{T}}\bar{E}_v\frac{\partial}{\partial t_2}\bar{x}_v(t_1,t_2).\end{aligned} \tag{7.25}$$

Based on (7.25), the inequality (7.24) is derived as

$$\begin{bmatrix} \bar{x}(t_1,t_2) \\ w(t_1,t_2) \end{bmatrix}^{\mathrm{T}} \begin{bmatrix} \bar{\Sigma}_1 & \star \\ \mathscr{D}_i^{\mathrm{T}}\bar{\mathscr{P}}_i & -\gamma^2\mathbf{I} \end{bmatrix} \begin{bmatrix} \bar{x}(t_1,t_2) \\ w(t_1,t_2) \end{bmatrix} < 0, \tag{7.26}$$

where

$$\begin{cases} \bar{\Sigma}_1 = \mathtt{Sym}\{\bar{\mathscr{P}}_i^{\mathrm{T}}(\bar{\mathscr{A}}_i + \Delta\bar{\mathscr{A}}_i)\} + \bar{\mathscr{L}}_i^{\mathrm{T}}\bar{\mathscr{L}}_i, \\ \bar{\mathscr{P}}_i = \begin{bmatrix} \bar{F}_i^{\mathrm{T}}X\bar{F}_i & \mathbf{0} \\ \bar{P}_3 & P_2 \end{bmatrix}, X = \mathrm{diag}\{X_h, X_v\}, \\ \bar{P}_3 = \begin{bmatrix} \bar{P}_{h3} & \bar{P}_{v3} \end{bmatrix}, P_2 = P_{h2} + P_{v2}. \end{cases} \tag{7.27}$$

Denote $\bar{K} = P_2^{\mathrm{T}}K$. Employing the S-procedure based on the structure of $\bar{\mathscr{P}}_i$, the following inequality indicates (7.26),

$$\begin{bmatrix} \bar{\Lambda}_1 + \bar{G}_i^{\mathrm{T}}U_i\bar{G}_i & \star & \star \\ \bar{\Lambda}_2 & -P_2 - P_2^{\mathrm{T}} + \mathscr{N}_i^{\mathrm{T}}\mathscr{N}_i & \star \\ \mathscr{D}_i^{\mathrm{T}}\bar{Q}\bar{F}_i^{\mathrm{T}}X\bar{F}_i & \mathbf{0} & -\gamma^2\mathbf{I} \end{bmatrix} < 0, \tag{7.28}$$

where $U_i \succeq 0$ and

$$\begin{cases} \bar{\Lambda}_1 = \mathtt{Sym}\{\bar{F}_i^{\mathrm{T}}X\bar{F}_i(\bar{\mathscr{A}}_{i1} + \Delta\bar{\mathscr{A}}_{i1}) - \bar{P}_3^{\mathrm{T}}KC\bar{Q}\} + \bar{Q}^{\mathrm{T}}\mathscr{L}_i^{\mathrm{T}}\mathscr{L}_i\bar{Q} \\ \bar{\Lambda}_2 = -\bar{K}C\bar{Q} + (\mathscr{B}_i^{\mathrm{T}} + \Delta\mathscr{B}_i^{\mathrm{T}})\bar{Q}\bar{F}_i^{\mathrm{T}}X\bar{F}_i - \bar{P}_3 + \mathscr{N}_i^{\mathrm{T}}\mathscr{L}_i\bar{Q}, \\ \bar{\mathscr{A}}_{i1} + \Delta\bar{\mathscr{A}}_{i1} = \mathscr{A}_i + \Delta\mathscr{A}_i,\ \bar{Q} = \mathbf{I}_{n_x}, \ \}\, i \in \mathscr{I}_0, \\ \bar{\mathscr{A}}_{i1} + \Delta\bar{\mathscr{A}}_{i1} = \begin{bmatrix} 0 & \mathbf{0} \\ \mathrm{a}_i + \Delta\mathrm{a}_i & \mathscr{A}_i + \Delta\mathscr{A}_i \end{bmatrix}, \bar{Q} = \begin{bmatrix} \mathbf{0} & \mathbf{I}_{n_x} \end{bmatrix}, \Big\}\, i \in \mathscr{I}_1. \end{cases} \tag{7.29}$$

Expanding the MFs and employing Lemma A1 to handle the uncertainty terms, then for scalars $\varepsilon_i > 0$, the following inequality implies (7.28),

$$\begin{bmatrix} \bar{\Psi}_1 + \bar{G}_i^{\mathrm{T}}U_i\bar{G}_i & \star & \star & \star \\ \bar{\Psi}_2 & \bar{\Psi}_3 & \star & \star \\ D_m^{\mathrm{T}}\bar{Q}\bar{F}_i^{\mathrm{T}}X\bar{F}_i & \mathbf{0} & -\gamma^2\mathbf{I} & \star \\ H_{m1}^{\mathrm{T}}\bar{Q}\bar{F}_i^{\mathrm{T}}X\bar{F}_i & \mathbf{0} & \mathbf{0} & -\varepsilon_i\mathbf{I} \end{bmatrix} < 0, m \in \mathscr{N}(i), i \in \mathscr{I}, \tag{7.30}$$

where

$$\begin{cases} \bar{\Psi}_1 = \bar{Q}^{\mathrm{T}}L_m^{\mathrm{T}}L_m\bar{Q} + \varepsilon_i\bar{\Psi}_{1(1)}\bar{\Psi}_{1(1)}^{\mathrm{T}} + \mathtt{Sym}\{\bar{F}_i^{\mathrm{T}}X\bar{F}_i\bar{A}_{m1} - \bar{P}_3^{\mathrm{T}}KC\bar{Q}\}, \\ \bar{\Psi}_2 = -\bar{K}C\bar{Q} + B_m^{\mathrm{T}}\bar{Q}\bar{F}_i^{\mathrm{T}}X\bar{F}_i - \bar{P}_3 + N_m^{\mathrm{T}}L_m\bar{Q} + \varepsilon_iH_{m4}^{\mathrm{T}}\bar{\Psi}_{1(1)}^{\mathrm{T}}, \\ \bar{\Psi}_3 = -P_2 - P_2^{\mathrm{T}} + N_m^{\mathrm{T}}N_m + \varepsilon_iH_{m4}^{\mathrm{T}}H_{m4}, \\ \bar{\Psi}_{1(1)} = H_{m2}^{\mathrm{T}}, \bar{A}_{m1} = A_m, \ \}\, i \in \mathscr{I}_0, \\ \bar{\Psi}_{1(1)} = \begin{bmatrix} H_{m3}^{\mathrm{T}} \\ H_{m2}^{\mathrm{T}} \end{bmatrix}, \bar{A}_{m1} = \begin{bmatrix} 0 & \mathbf{0} \\ a_m & A_m \end{bmatrix}, \Big\}\, i \in \mathscr{I}_1. \end{cases} \tag{7.31}$$

For numerical convexification, specify $\bar{P}_3$ to be

$$\bar{P}_3 = -P_2\bar{J}, \tag{7.32}$$

where $\bar{J} \in \Re^{n_u \times n_x}$, for $i \in \mathscr{I}_0$, and $\bar{J} \in \Re^{n_u \times (n_x+1)}$, for $i \in \mathscr{I}_1$, are arbitrary matrices.

Then it is inferred that (7.19) implies (7.30) based on Schur complement. The proof is accomplished. ■

Remark 7.2.1. In Theorem 7.1, a singular 2-D system approach is adopted to tackle the coupling issue in the control input matrices with the sliding surface gain. By a singular-system-based method, we can derive the existence conditions for the sliding surface (7.11) with some convexification techniques.

Remark 7.2.2. Through a 2-D PQLFs-based method, some new stability and robust performance analysis results on the sliding mode dynamical system are proposed in Theorem 7.1. Note that when calculating the partial differential $V_2(\bar{x}_v(t_1,t_2),t_1,t_2)$ in (7.25) for $i \in \mathscr{I}_1$, one has that

$$\begin{aligned}\frac{\partial}{t_2}V_2(\bar{x}_v(t_1,t_2),t_1,t_2) &= 2\bar{x}_v^{\mathrm{T}}(t_1,t_2)\bar{P}_{vi}^{\mathrm{T}}\bar{E}_v\frac{\partial}{\partial t_2}\bar{x}_v(t_1,t_2)\\ &= 2\bar{x}_v^{\mathrm{T}}(t_1,t_2)\bar{P}_{vi}^{\mathrm{T}}\bar{E}_v\begin{bmatrix} 1 & \frac{\partial}{t_2}x_v^{\mathrm{T}}(t_1,t_2) & \frac{\partial}{t_2}u^{\mathrm{T}}(t_1,t_2)\end{bmatrix}^{\mathrm{T}}.\end{aligned} \tag{7.33}$$

It is observed from system (7.13) that only the partial differential of the control input $u(t_1,t_2)$ with respect to directional variable t_1 is introduced, i.e., $0 \cdot \frac{\partial}{t_1}u(t_1,t_2)$. Nevertheless, the partial differential of the control input $u(t_1,t_2)$ about the other directional variable t_2 exists in (7.33). In fact, the algebraic computation of $\bar{E}_v\begin{bmatrix} 1 & \frac{\partial}{t_2}x_v^{\mathrm{T}}(t_1,t_2) & \frac{\partial}{t_2}u^{\mathrm{T}}(t_1,t_2)\end{bmatrix}^{\mathrm{T}}$ results in the term $0 \cdot \frac{\partial}{t_2}u(t_1,t_2)$, which has no influence on stability and robust performance analysis for the sliding motion. From this perspective, $0 \cdot \frac{\partial}{t_2}u(t_1,t_2)$ can be replaced by $0 \cdot \frac{\partial}{t_1}u(t_1,t_2)$ without loss of generality in (7.33).

Remark 7.2.3. In Theorem 7.1, it is noted that the Lyapunov functions in (7.22) are piecewise Lyapunov functions in fact, as $V_1(\bar{x}_h(t_1,t_2),t_1,t_2) = \bar{x}_h^{\mathrm{T}}(t_1,t_2)\bar{E}_h^{\mathrm{T}}\bar{\mathscr{P}}_{hi}\bar{x}_h(t_1,t_2)$ and $V_2(\bar{x}_v(t_1,t_2),t_1,t_2) = \bar{x}_v^{\mathrm{T}}(t_1,t_2)\bar{E}_v^{\mathrm{T}}\bar{\mathscr{P}}_{vi}\bar{x}_v(t_1,t_2)$ contain the piecewise Lyapunov matrices $\bar{\mathscr{P}}_{hi}$ and $\bar{\mathscr{P}}_{vi}$. If setting the matrices $\begin{bmatrix} f_i^h & F_i^h & \mathbf{0} \\ f_i^v & \mathbf{0} & F_i^v \end{bmatrix} \equiv \begin{bmatrix} \mathbf{0} & \mathbf{I} \end{bmatrix}$ in Theorem 7.1, corresponding stability and robust $\mathscr{H}_\infty$ performance analysis results can be easily obtained based on a CQLF as $V_1(\bar{x}_h(t_1,t_2),t_1,t_2) = \bar{x}_h^{\mathrm{T}}(t_1,t_2)\bar{E}_h^{\mathrm{T}}\bar{\mathscr{P}}_h\bar{x}_h(t_1,t_2)$, $V_2(\bar{x}_v(t_1,t_2),t_1,t_2) = \bar{x}_v^{\mathrm{T}}(t_1,t_2)\bar{E}_v^{\mathrm{T}}\bar{\mathscr{P}}_{vi}\bar{x}_v(t_1,t_2)$, $\bar{\mathscr{P}}_h = \begin{bmatrix} X_h & \mathbf{0} \\ \bar{P}_{h3} & P_{h2}\end{bmatrix}$, $\bar{\mathscr{P}}_v = \begin{bmatrix} X_v & \mathbf{0} \\ \bar{P}_{v3} & P_{v2}\end{bmatrix}$, $\{X_h, X_v\} > 0$. The readers can derive the CQLF-based analysis results by themselves, which will not be presented here. On the other hand, the CQLF can be regard as a special case of the PQLF, and the results obtained in Theorem 7.1 tend to be more general and less conservative.

In the following section, a new output feedback dynamic SMC approach will be proposed to drive the 2-D system states onto the sliding surface in finite time locally.

7.2.4 OUTPUT FEEDBACK DYNAMIC SLIDING MODE CONTROLLER DESIGN

In this subsection, a new output feedback dynamic SMC design approach will be proposed to drive the 2-D system states onto the sliding surface in finite time. Along a similar line in [72], introduce a sliding patch Ξ as

$$\Xi := \{x(t_1,t_2) \in \Re^{n_x} : \|x(t_1,t_2)\| \leq \rho\}. \tag{7.34}$$

Synthesize the following output feedback dynamic sliding mode controller as

$$\mathscr{D}u(t_1,t_2) = -\sum_{l=1}^{r} \mu_l KC\{B_l u(t_1,t_2) + a_l\} - (\beta + \eta(t_1,t_2))\mathrm{sgn}(s(t_1,t_2)), \tag{7.35}$$

where $s(t_1,t_2) = Ky(t_1,t_2) + u(t_1,t_2)$ given in (7.11), and $\beta > 0$ is a given constant, and

$$\begin{aligned}\eta(t_1,t_2) \geq \sum_{l=1}^{r} \mu_l \{&(\|KCA_l\| + \|KCH_{l1}\| \cdot \|H_{l2}\|)\rho \\ &+\|KCH_{l1}\| \cdot \|H_{l4}\| \cdot \|u(t_1,t_2)\| \\ &+\|KCH_{l1}\| \cdot \|H_{l3}\| + \|KCD_l\|\sigma(t_1,t_2)\}\end{aligned} \tag{7.36}$$

with $\sigma(t_1,t_2)$ representing the uniform upper-bound of $w(t_1,t_2)$.

Since the designed output feedback dynamic sliding mode controller (7.35) is in a dynamic form, one does not need to conduct any singular valuable decomposition to the control input matrices for design purpose as in [132]. From this perspective, the control input channels are also permitted to have parameter uncertainties. Notice that the proposed dynamic SMC synthesis method can also handle the mismatched disturbances.

Theorem 7.2. For the sliding surface (7.11) and the patch Ξ (7.34), the output feedback dynamic sliding mode controller presented in (7.35) can force the state trajectories of the fuzzy 2-D system (7.2) onto the sliding surface in finite time locally.

Proof. Construct the Lyapunov function as

$$V(t_1,t_2) = \frac{1}{2}s^{\mathrm{T}}(t_1,t_2)s(t_1,t_2). \tag{7.37}$$

Based on the sliding surface (7.11), one has,

$$\mathscr{D}s(t_1,t_2) = KC\mathscr{D}x(t_1,t_2) + \mathscr{D}u(t_1,t_2). \tag{7.38}$$

Based on the dynamic sliding mode controller (7.35), one has

$$\begin{aligned}&\mathscr{D}V(t_1,t_2) = s^{\mathrm{T}}(t_1,t_2)\mathscr{D}s(t_1,t_2) \\ &= s^{\mathrm{T}}(t_1,t_2)\sum_{l=1}^{r} \mu_l KC\{(A_l+\Delta A_l)x(t_1,t_2) + a_l + \Delta a_l + (B_l+\Delta B_l)u(t_1,t_2) \\ &\quad +D_l w(t_1,t_2)\} - s^{\mathrm{T}}(t_1,t_2)\sum_{l=1}^{r} \mu_l KC\{B_l u(t_1,t_2) + a_l\} \\ &\quad -(\beta+\eta(t_1,t_2))\|s(t_1,t_2)\| \\ &= s^{\mathrm{T}}(t_1,t_2)\sum_{l=1}^{r} \mu_l KC\{(A_l+\Delta A_l)x(t_1,t_2) + \Delta a_l + \Delta B_l u(t_1,t_2) \\ &\quad +D_l w(t_1,t_2)\} - \eta(t_1,t_2)\|s(t_1,t_2)\| - \beta\|s(t_1,t_2)\|.\end{aligned} \tag{7.39}$$

Based on the sliding patch Ξ defined as in (7.34) and the parameter uncertainties in the form of (7.3), one has

$$\begin{aligned}\mathscr{D}V(t_1,t_2) \leq \sum_{l=1}^{r} \mu_l \|s(t_1,t_2)\| \{(\|KCA_l\| + \|KCH_{l1}\| \cdot \|H_{l2}\|)\rho \\ +\|KCH_{l1}\| \cdot \|H_{l3}\| + \|KCH_{l1}\| \cdot \|H_{l4}\| \cdot \|u(t_1,t_2)\| \\ +\|KCD_l\|\sigma(t_1,t_2)\} - \eta(t_1,t_2)\|s(t_1,t_2)\| - \beta\|s(t_1,t_2)\|.\end{aligned} \tag{7.40}$$

Notice that

$$\begin{aligned}\eta(t_1,t_2) \geq \sum_{l=1}^{r} \mu_l \{(\|KCA_l\| + \|KCH_{l1}\| \cdot \|H_{l2}\|)\rho \\ +\|KCH_{l1}\| \cdot \|H_{l4}\| \cdot \|u(t_1,t_2)\| \\ +\|KCH_{l1}\| \cdot \|H_{l3}\| + \|KCD_l\|\sigma(t_1,t_2)\},\end{aligned} \tag{7.41}$$

and then we have

$$\mathscr{D}V(t_1,t_2) \leq -\beta\|s(t_1,t_2)\|, \tag{7.42}$$

which ensures the finite-time reachability of the sliding surface (7.11) in the sliding patch Ξ. The proof is completed. ■

7.2.5 ILLUSTRATIVE EXAMPLES

In this subsection, two illustrative simulation examples will be given to show the effectiveness of the proposed stability and robust $\mathscr{H}_\infty$ performance analysis method and the output feedback dynamic SMC design approach.

Example 7.1. Consider the nonlinear gas absorption, water stream heating, and air drying process which is a thermal process and has been extensively used in chemical reactors. The dynamics of this thermal process are characterized by Darboux equation as

$$\frac{\partial^2 T(x,t)}{\partial x \partial t} = a_1 \frac{\partial T(x,t)}{\partial t} + a_2 \frac{\partial T(x,t)}{\partial x} + a_0 T(x,t) + b_0 u(x,t), \tag{7.43}$$

where $T(x,t)$ represents the temperature in the reactant concentration; $a_1 = -3$ and $a_2 = -1$ are constant parameters; $a_0 = b_0 = \cos T(x,t)$ are nonlinear functions at time $t \in [0,\infty)$ and the state space $x \in [0,x_f]$; $u(x,t)$ is the spatially uniform temperature in the jacket.

Similar to [116, 132], define $R(x,t) = -a_2 T(x,t) + \frac{\partial T(x,t)}{\partial t}$. Assume $T(x,t)$ subject to exogenous disturbance $w(x,t)$, and then the system (7.43) is formulated as

$$\begin{bmatrix} \frac{\partial T(x,t)}{\partial t} \\ \frac{\partial R(x,t)}{\partial x} \end{bmatrix} = \begin{bmatrix} a_2 & 1 \\ a_1 a_2 + a_0 & a_1 \end{bmatrix} \begin{bmatrix} T(x,t) \\ R(x,t) \end{bmatrix} + \begin{bmatrix} 0 \\ b_0 \end{bmatrix} u(x,t) + \begin{bmatrix} 1 \\ 0 \end{bmatrix} w(x,t), \tag{7.44}$$

where $w(x,t)$ represents the exogenous disturbance.

Denote the state variables as $W_1(x,t) = T(x,t)$, $W_2(x,t) = R(x,t)$, and $W(x,t) = \begin{bmatrix} W_1^{\mathrm{T}}(x,t) & W_2^{\mathrm{T}}(x,t) \end{bmatrix}^{\mathrm{T}}$. Linearize system (7.44) at three operating points

$\begin{bmatrix} \pm\frac{\pi}{3} & 0 \end{bmatrix}^{\mathrm{T}}$ and $\begin{bmatrix} 0 & 0 \end{bmatrix}^{\mathrm{T}}$, and the system (7.44) can be characterized by a 2-D T-S fuzzy affine model as,

Plant Rule $\mathscr{R}^l$: IF $W_1(x,t)$ is $\mathscr{F}_{l1}$, THEN

$$\begin{cases} \mathscr{D}W(x,t) = (A_l + \Delta A_l)W(x,t) + a_l + \Delta a_l \\ \qquad\qquad + (B_l + \Delta B_l)u(x,t) + D_l w(x,t) \\ y(x,t) = CW(x,t), \\ z(x,t) = L_l W(x,t) + N_l u(x,t),\ l \in \{1,2,3\}, \end{cases} \tag{7.45}$$

where the system matrices are

$$\begin{cases} A_1 = \begin{bmatrix} -1 & 1 \\ 4 & -3 \end{bmatrix}, A_2 = A_3 = \begin{bmatrix} -1 & 1 \\ 2.5931 & -3 \end{bmatrix}, \\ a_1 = \begin{bmatrix} 0 \\ 0 \end{bmatrix}, a_2 = \begin{bmatrix} 0 \\ -0.9497 \end{bmatrix}, a_3 = \begin{bmatrix} 0 \\ 0.9497 \end{bmatrix}, \\ B_1 = \begin{bmatrix} 0 \\ 1 \end{bmatrix}, B_2 = B_3 = \begin{bmatrix} 0 \\ 0.5 \end{bmatrix}, D_l = \begin{bmatrix} 1 \\ 0 \end{bmatrix}, \\ C = \begin{bmatrix} 1 & 0 \end{bmatrix}, L_l = \begin{bmatrix} 1 & 1 \end{bmatrix}, N_l = 0.5,\ l \in \{1,2,3\} \end{cases} \tag{7.46}$$

and the parameter uncertainties ΔA_l, Δa_l and ΔB_l are given in the form of (7.3) with

$$\begin{cases} H_{l1} = \begin{bmatrix} 0 \\ 0.2 \end{bmatrix}, H_{l2} = \begin{bmatrix} 0.1 & 0 \end{bmatrix}, \\ H_{l3} = H_{l4} = 0.1,\ l \in \{1,2,3\}. \end{cases} \tag{7.47}$$

Figure 7.1 shows the MFs with $d_1 = \frac{\pi}{3}$ and $d_2 = \frac{\pi}{2}$. By observing Figure 7.1, one can see that the premise variable space is decomposed as,

$$\begin{aligned} S_1 &= \{W(x,t) \in \Re^2 | -d_2 \le W_1(x,t) \le -d_1\}, \\ S_2 &= \{W(x,t) \in \Re^2 | -d_1 \le W_1(x,t) \le d_1\}, \\ S_3 &= \{W(x,t) \in \Re^2 | d_1 \le W_1(x,t) \le d_2\}. \end{aligned} \tag{7.48}$$

We attempt to synthesize a sliding surface (7.11) and a new output feedback dynamic sliding mode controller (7.35) to drive the state trajectories of the nonlinear 2-D process in (7.44) to the sliding surface in finite time and guarantee the sliding mode dynamical system to be asymptotically stable with robust performance γ. *The full process states are assumed to be not available in this example. Additionally, the control input matrices presented in (7.46) are different in each local models and possess parameter uncertainties. Considering these issues, the SMC synthesis methods in [94, 95, 122, 132, 133, 134, 135, 72] are not applicable to this example.*

By using Theorem 7.1 with $\bar{J} = \begin{bmatrix} 0 & -5 & 0 \end{bmatrix}$, for $i \in \mathscr{I}_1$, $\bar{J} = \begin{bmatrix} -5 & 0 \end{bmatrix}$, for $i \in \mathscr{I}_0$, the robust $\mathscr{H}_\infty$ performance is obtained as $\gamma_{\min} = 1.7893$ with sliding surface gain $K = 4.8676$. Note that when applying Theorem 7.1 with PQLFs in (7.22), the

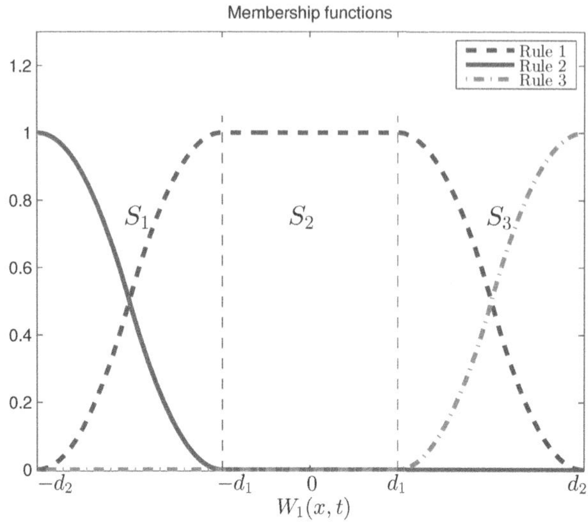

Figure 7.1 Membership functions of Example 7.1-Example 7.4.

matrices $\bar{G}_i$ and $\bar{F}_i$ are

$$\begin{array}{l}
\left[\begin{array}{c|c|c|c|c|c} F_1^h & F_2^h & F_3^h & f_1^h & f_2^h & f_3^h \end{array}\right] = \left[\begin{array}{c|c|c|c|c|c} -1 & 0 & 0 & -d_1 & 0 & 0 \\ 0 & 0 & 1 & 0 & 0 & -d_1 \\ 1 & 1 & 1 & 0 & 0 & 0 \end{array}\right], \\
\left[\begin{array}{c|c|c|c|c|c} G_1^h & G_2^h & G_3^h & g_1^h & g_2^h & g_3^h \end{array}\right] = \left[\begin{array}{c|c|c|c|c|c} 1 & 0 & 1 & d_2 & 0 & -d_1 \\ -1 & 0 & -1 & -d_1 & 0 & d_2 \end{array}\right], \\
\left[\begin{array}{cc} F_i^v & f_i^v \end{array}\right] = \left[\begin{array}{cc} G_i^v & g_i^v \end{array}\right] = \left[\begin{array}{cc} 1 & 0 \end{array}\right], i \in \mathscr{I}.
\end{array} \tag{7.49}$$

To demonstrate the validity of the developed method, simulations are carried out under various initial conditions. In particular, with boundary condition $W(0,t) = \left[\begin{array}{cc} 0 & 0.2 \end{array}\right]^{\mathrm{T}}$, $W(x,0) = \left[\begin{array}{cc} 0.5 & 0 \end{array}\right]^{\mathrm{T}}$, $0 \leq t \leq 0.5$s, $0 \leq x \leq 0.5$m, $W(0,t) = W(x,0) = \left[\begin{array}{cc} 0 & 0 \end{array}\right]^{\mathrm{T}}$, $t > 0.5$s, $x > 0.5$m, and nonlinear disturbance $w(x,t) = 12\mathrm{e}^{-0.1x}\sin(0.3x)$, by applying the output feedback dynamic sliding mode controller (7.35) with $\beta = 1.2$ and $\rho = 0.8$, the closed-loop system states are, respectively, presented in Figure 7.2 and Figure 7.3. Figure 7.4 shows the control input.

Example 7.2. Consider a fuzzy 2-D system in the form of (7.2) with three rules, and the system matrices are

$$\left\{\begin{array}{l}
A_1 = \left[\begin{array}{cc} 0.2 & 0.5 \\ 1 & -0.8 \end{array}\right], A_2 = \left[\begin{array}{cc} 0.5 & 0.4 \\ 1.5 & -1 \end{array}\right], A_3 = \left[\begin{array}{cc} 1.2 & 1 \\ -0.7 & -0.2 \end{array}\right], \\
a_1 = \left[\begin{array}{c} 0 \\ 0 \end{array}\right], a_2 = \left[\begin{array}{c} 0 \\ -0.5 \end{array}\right], a_3 = \left[\begin{array}{c} 0 \\ 0.5 \end{array}\right], \\
B_1 = \left[\begin{array}{c} 1 \\ 1 \end{array}\right], B_2 = \left[\begin{array}{c} 1 \\ 1.5 \end{array}\right], B_3 = \left[\begin{array}{c} 1.5 \\ 1 \end{array}\right], D_l = \left[\begin{array}{c} 1 \\ 0 \end{array}\right], \\
C = \left[\begin{array}{cc} 1 & 0 \end{array}\right], L_l = \left[\begin{array}{cc} 1 & 0.5 \end{array}\right], N_l = 1, \ l \in \{1,2,3\},
\end{array}\right. \tag{7.50}$$

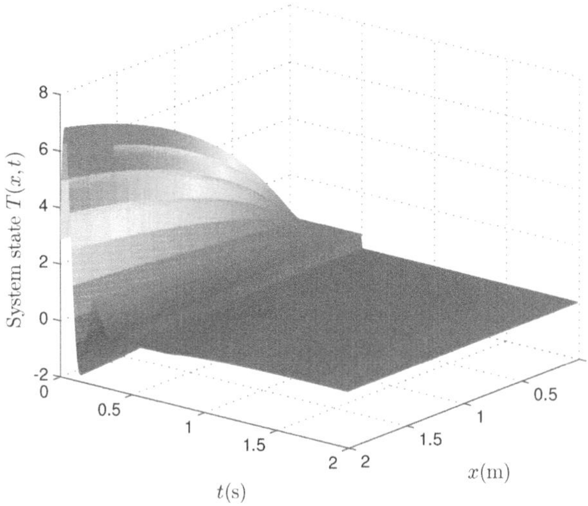

Figure 7.2 Response of the state $T(x,t)$ with dynamic sliding mode controller (7.35) in Example 7.1.

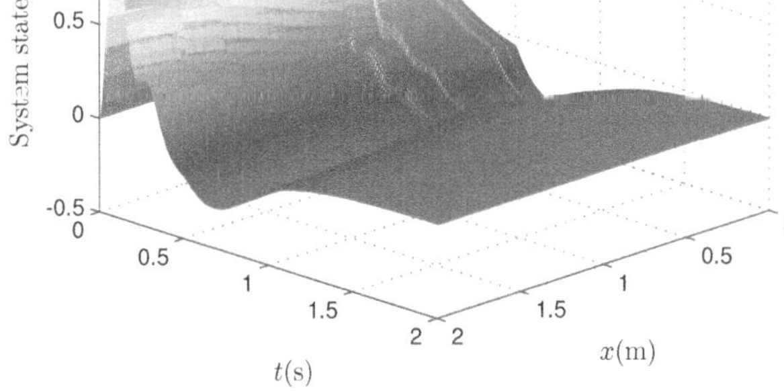

Figure 7.3 Response of the state $R(x,t)$ with dynamic sliding mode controller (7.35) in Example 7.1.

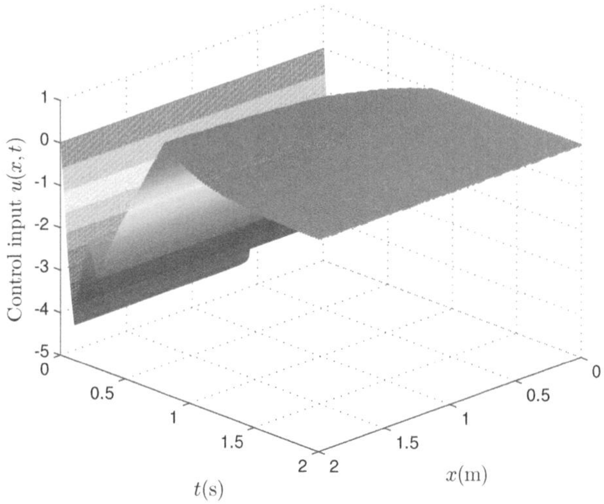

Figure 7.4 The control input with dynamic sliding mode controller (7.35) in Example 7.1.

and the parameter uncertainties ΔA_l, Δa_l, and ΔB_l are given in the form of (7.3) with

$$\begin{cases} H_{l1} = \begin{bmatrix} 0.1 \\ 0.1 \end{bmatrix}, H_{l2} = \begin{bmatrix} 0.1 & 0.2 \end{bmatrix}, \\ H_{l3} = 0.3, H_{l4} = 0.1,\ l \in \{1,2,3\}. \end{cases} \tag{7.51}$$

The MFs are also given as in Figure 7.1 with $d_1 = 30$ and $d_2 = 100$. Choosing $\bar{J} = \begin{bmatrix} 1 & -2 & 0 \end{bmatrix}$, for $i \in \mathscr{I}_1$, $\bar{J} = \begin{bmatrix} -2 & 0 \end{bmatrix}$, for $i \in \mathscr{I}_0$, and applying Theorem 7.1, the gain of the sliding surface (7.11) is computed as $K = 1.0162$ with robust $\mathscr{H}_\infty$ performance $\gamma_{\min} = 1.4293$. Simulations are conducted to show the effectiveness of the proposed dynamic SMC design method. For boundary condition $x^h(t_1,0) = \begin{bmatrix} 0.5\sin(0.2t_1) & 0 \end{bmatrix}^{\mathrm{T}}, 0 \le t_1 \le 0.5, x^v(0,t_2) = \begin{bmatrix} 0 & 0.5\cos(t_2) \end{bmatrix}^{\mathrm{T}}$, $0 \le t_2 \le 0.5, x^h(t_1,0) = x^v(0,t_2) = \begin{bmatrix} 0 & 0 \end{bmatrix}^{\mathrm{T}}, t_1 > 0.5, t_2 > 0.5$, and nonlinear disturbance $w(x,t) = 15\mathrm{e}^{-0.5t_1}\sin(0.1t_1)$, using the output feedback dynamic sliding mode controller (7.35) with $\beta = 0.5$ and $\rho = 0.5$, the closed-loop system states are shown in Figure 7.5 and Figure 7.6, respectively. The sliding surface is presented in Figure 7.7.

These simulation results clearly show the validity of the proposed stability and robust performance analysis method for the sliding motion and the dynamic SMC design approach.

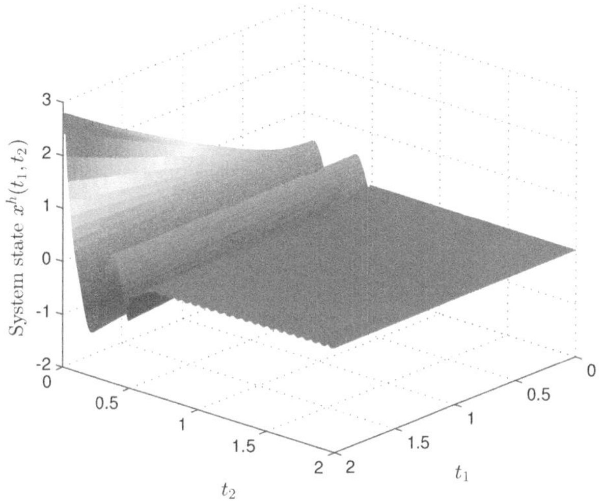

Figure 7.5 Response of the state $x^h(t_1,t_2)$ with dynamic sliding mode controller (7.35) in Example 7.2.

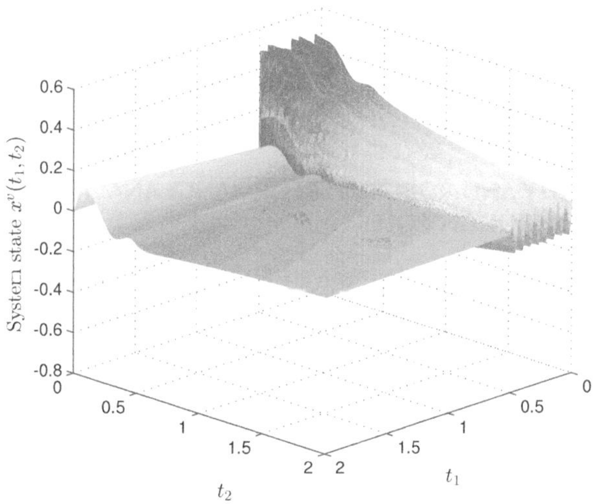

Figure 7.6 Response of the state $x^v(t_1,t_2)$ with dynamic sliding mode controller (7.35) in Example 7.2.

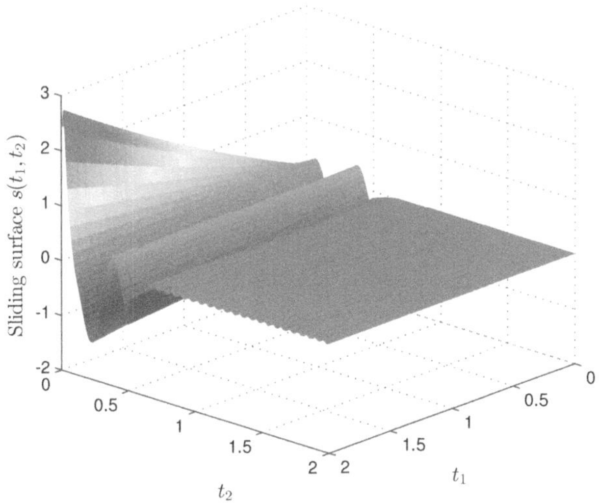

Figure 7.7 The sliding surface in with dynamic sliding mode controller (7.35) Example 7.2.

7.3 STATE FEEDBACK SMC DESIGN VIA 2-D T-S FUZZY MODELS

In the case of the full fuzzy 2-D system states being available, a state feedback dynamic sliding mode controller design method and the corresponding stability and robust $\mathscr{H}_\infty$ performance analysis results on the sliding motion will be proposed in this section.

7.3.1 STABILITY AND ROBUST PERFORMANCE ANALYSIS OF SLIDING MOTION THROUGH PQLFS

Based on the control input and system state, design the sliding surface for the fuzzy 2-D system (7.2) as

$$s(t_1,t_2) = Kx(t_1,t_2) + u(t_1,t_2), \tag{7.52}$$

where $K \in \Re^{n_u \times n_x}$ is the sliding surface gain to be determined. After the sliding surface is reached, one has $s(t_1,t_2) = 0$, and then one can see that $u(t_1,t_2) = -Kx(t_1,t_2)$.

Based on the sliding surface (7.52), the sliding motion can be described as follows,

$$\begin{cases} \bar{E}\mathscr{D}\bar{x}(t_1,t_2) = (\bar{\mathscr{A}_i} + \Delta\bar{\mathscr{A}_i})\bar{x}(t_1,t_2) + \bar{\mathscr{D}}_i w(t_1,t_2) \\ z(t_1,t_2) = \bar{\mathscr{L}}_i \bar{x}(t_1,t_2), i \in \mathscr{I}, \end{cases} \tag{7.53}$$

where

$$\left\{\begin{array}{l}\left.\begin{array}{l}\bar{E}=\mathrm{diag}\{\mathbf{I}_{n_x},\mathbf{0}_{n_u}\},\\ \mathscr{D}\bar{x}(t_1,t_2)=\left[\begin{array}{ccc}\frac{\partial}{\partial t_1}x_h^{\mathrm{T}}(t_1,t_2) & \frac{\partial}{\partial t_2}x_v^{\mathrm{T}}(t_1,t_2) & \frac{\partial}{\partial t_1}u^{\mathrm{T}}(t_1,t_2)\end{array}\right]^{\mathrm{T}},\\ \bar{\mathscr{A}_i}+\Delta\bar{\mathscr{A}_i}=\left[\begin{array}{cc}\mathscr{A}_i+\Delta\mathscr{A}_i & \mathscr{B}_i+\Delta\mathscr{B}_i\\ -K & -\mathbf{I}\end{array}\right],\\ \bar{\mathscr{D}}_i=\left[\begin{array}{cc}\mathscr{D}_i^{\mathrm{T}} & \mathbf{0}\end{array}\right]^{\mathrm{T}},\bar{\mathscr{L}}_i=\left[\begin{array}{cc}\mathscr{L}_i & \mathscr{N}_i\end{array}\right],\end{array}\right\} i\in\mathscr{I}_0,\\ \left.\begin{array}{l}\bar{E}=\mathrm{diag}\{\mathbf{I}_{(n_x+1)},\mathbf{0}_{n_u}\},\\ \mathscr{D}\bar{x}(t_1,t_2)=\left[\begin{array}{cccc}\dot{\mathrm{1}} & \frac{\partial}{\partial t_1}x_h^{\mathrm{T}}(t_1,t_2) & \frac{\partial}{\partial t_2}x_v^{\mathrm{T}}(t_1,t_2) & \frac{\partial}{\partial t_1}u^{\mathrm{T}}(t_1,t_2)\end{array}\right]^{\mathrm{T}},\\ \bar{\mathscr{A}_i}+\Delta\bar{\mathscr{A}_i}=\left[\begin{array}{ccc}0 & \mathbf{0} & \mathbf{0}\\ \mathrm{a}_i+\Delta\mathrm{a}_i & \mathscr{A}_i+\Delta\mathscr{A}_i & \mathscr{B}_i+\Delta\mathscr{B}_i\\ \mathbf{0} & -K & -\mathbf{I}\end{array}\right],\\ \bar{\mathscr{D}}_i=\left[\begin{array}{ccc}\mathbf{0} & \mathscr{D}_i^{\mathrm{T}} & \mathbf{0}\end{array}\right]^{\mathrm{T}},\bar{\mathscr{L}}_i=\left[\begin{array}{ccc}\mathbf{0} & \mathscr{L}_i & \mathscr{N}_i\end{array}\right],\end{array}\right\} i\in\mathscr{I}_1.\end{array}\right. \tag{7.54}$$

The stability and robust $\mathscr{H}_\infty$ performance analysis results for sliding motion are proposed in Theorem 7.3.

Theorem 7.3. The sliding mode dynamical system (7.53) is asymptotically stable with robust $\mathscr{H}_\infty$ performance γ, if matrices $X_h = X_h^{\mathrm{T}}$, $X_v = X_v^{\mathrm{T}}$, $U_{hi} = U_{hi}^{\mathrm{T}} \succeq 0$, $U_{vi} = U_{vi}^{\mathrm{T}} \succeq 0$, $W_{hi} = W_{hi}^{\mathrm{T}} \succeq 0$, $W_{vi} = W_{vi}^{\mathrm{T}} \succeq 0$, $\bar{K} \in \Re^{n_u\times n_x}$, $P_2 \in \Re^{n_u\times n_u}$, and scalars $\varepsilon_i > 0$, $i \in \mathscr{I}$, exist such that the following conditions hold,

$$\bar{F}_i^{\mathrm{T}} X \bar{F}_i - \bar{G}_i^{\mathrm{T}} W_i \bar{G}_i > 0, i \in \mathscr{I}, \tag{7.55}$$

$$\left[\begin{array}{cccc}\bar{\Gamma}_1+\bar{G}_i^{\mathrm{T}}U_i\bar{G}_i & \star & \star & \star\\ \bar{\Gamma}_2 & \bar{\Gamma}_3 & \star & \star\\ D_m^{\mathrm{T}}\bar{Q}\bar{F}_i^{\mathrm{T}}X\bar{F}_i & \mathbf{0} & -\gamma^2\mathbf{I} & \star\\ H_{m1}^{\mathrm{T}}\bar{Q}\bar{F}_i^{\mathrm{T}}X\bar{F}_i & \mathbf{0} & \mathbf{0} & -\varepsilon_i\mathbf{I}\end{array}\right] < 0, m\in\mathscr{N}(i), i\in\mathscr{I}, \tag{7.56}$$

where

$$\left\{\begin{array}{l}\bar{\Gamma}_1=\bar{Q}^{\mathrm{T}}L_m^{\mathrm{T}}L_m\bar{Q}+\varepsilon_i\bar{\Gamma}_{1(1)}\bar{\Gamma}_{1(1)}^{\mathrm{T}}+\mathrm{Sym}\{\bar{F}_i^{\mathrm{T}}X\bar{F}_i\bar{A}_{m1}+\bar{J}^{\mathrm{T}}\bar{K}\bar{Q}\},\\ \bar{\Gamma}_2=-\bar{K}\bar{Q}+B_m^{\mathrm{T}}\bar{Q}\bar{F}_i^{\mathrm{T}}X\bar{F}_i+P_2\bar{J}+N_m^{\mathrm{T}}L_m\bar{Q}+\varepsilon_iH_{m4}^{\mathrm{T}}\bar{\Gamma}_{1(1)}^{\mathrm{T}},\\ \bar{\Gamma}_3=-P_2-P_2^{\mathrm{T}}+N_m^{\mathrm{T}}N_m+\varepsilon_iH_{m4}^{\mathrm{T}}H_{m4},\\ X=\mathrm{diag}\{X_h,X_v\},W_i=\mathrm{diag}\{W_{hi},W_{vi}\},\\ U_i=\mathrm{diag}\{U_{hi},U_{vi}\},\\ \bar{\Gamma}_{1(1)}=H_{m2}^{\mathrm{T}},\ \bar{A}_{m1}=A_m,\ \bar{Q}=\mathbf{I}_{n_x},\bar{J}\in\Re^{n_u\times n_x},\ \}i\in\mathscr{I}_0,\\ \left.\begin{array}{l}\bar{\Gamma}_{1(1)}=\left[\begin{array}{c}H_{m3}^{\mathrm{T}}\\ H_{m2}^{\mathrm{T}}\end{array}\right],\bar{A}_{m1}=\left[\begin{array}{cc}0 & \mathbf{0}\\ a_m & A_m\end{array}\right],\\ \bar{Q}=\left[\begin{array}{cc}\mathbf{0} & \mathbf{I}_{n_x}\end{array}\right],\bar{J}\in\Re^{n_u\times(n_x+1)}\end{array}\right\}i\in\mathscr{I}_1\end{array}\right. \tag{7.57}$$

and $\bar{J} \in \Re^{n_u\times n_x}$, for $i \in \mathscr{I}_0$, $\bar{J} \in \Re^{n_u\times(n_x+1)}$, for $i \in \mathscr{I}_1$, are arbitrary matrices.

The gain of the sliding surface (7.52) can be computed by

$$K = P_2^{-\mathrm{T}}\bar{K}. \tag{7.58}$$

Remark 7.3.1. It is noted that the PQLFs-based stability and robust $\mathscr{H}_\infty$ performance analysis of sliding motion for the full state feedback situation can be deduced by a similar way to the output feedback case discussed as in subsection 7.2.3, and the PQLFs can be also constructed as in (7.22). Thus, the proof of Theorem 7.3 is omitted and readers can also derive the CQLF-based sliding motion stability analysis results for the state feedback case by themselves.

7.3.2 STATE FEEDBACK DYNAMIC SLIDING MODE CONTROLLER DESIGN

Theorem 7.4. Consider the first and third equation of system (7.2) with the sliding surface designed as in (7.52). The following state feedback dynamic sliding mode controller can guarantee the finite-time reachability of the sliding surface.

$$\begin{aligned}\mathscr{D}u(t_1,t_2) = &-\sum_{l=1}^{r}\mu_l K\{A_l x(t_1,t_2)+a_l+B_l u(t_1,t_2)\}\\ &-(\beta+\zeta(t_1,t_2))\mathrm{sgn}(s(t_1,t_2)),\end{aligned} \tag{7.59}$$

where $s(t_1,t_2) = Kx(t_1,t_2)+u(t_1,t_2)$ given in (7.52), and $\beta > 0$ is a given constant, and

$$\begin{aligned}\zeta(t_1,t_2) \geq \sum_{l=1}^{r}\mu_l\{&\|KH_{l1}\|\cdot\|H_{l2}\|\cdot\|x(t_1,t_2)\|\\ &+\|KH_{l1}\|\cdot\|H_{l4}\|\cdot\|u(t_1,t_2)\|\\ &+\|KH_{l1}\|\cdot\|H_{l3}\|+\|KD_l\|\sigma(t_1,t_2)\}\end{aligned} \tag{7.60}$$

with $\sigma(t_1,t_2)$ being the uniform upper-bound of $w(t_1,t_2)$.

Proof. Construct the Lyapunov function as

$$V(t_1,t_2) = \frac{1}{2}s^{\mathrm{T}}(t_1,t_2)s(t_1,t_2). \tag{7.61}$$

Based on the sliding surface (7.52), one has,

$$\mathscr{D}s(t_1,t_2) = K\mathscr{D}x(t_1,t_2)+\mathscr{D}u(t_1,t_2). \tag{7.62}$$

Substitute the state feedback dynamic sliding mode controller (7.59) into (7.62), and then yields

$$\begin{aligned}&\mathscr{D}V(t_1,t_2) = s^{\mathrm{T}}(t_1,t_2)\mathscr{D}s(t_1,t_2)\\ &= s^{\mathrm{T}}(t_1,t_2)\sum_{l=1}^{r}\mu_l K\{(A_l+\Delta A_l)x(t_1,t_2)+a_l+\Delta a_l+(B_l+\Delta B_l)u(t_1,t_2)\\ &\quad +D_l w(t_1,t_2)\}-s^{\mathrm{T}}(t_1,t_2)\sum_{l=1}^{r}\mu_l K\{A_l x(t_1,t_2)+a_l+B_l u(t_1,t_2)\}\\ &\quad -(\beta+\zeta(t_1,t_2))\|s(t_1,t_2)\|\\ &= s^{\mathrm{T}}(t_1,t_2)\sum_{l=1}^{r}\mu_l K\{\Delta A_l x(t_1,t_2)+\Delta a_l+\Delta B_l u(t_1,t_2)+D_l w(t_1,t_2)\}\\ &\quad -\zeta(t_1,t_2)\|s(t_1,t_2)\|-\beta\|s(t_1,t_2)\|.\end{aligned} \tag{7.63}$$

Considering the condition in (7.3), one has

$$\begin{aligned}\mathscr{D}V(t_1,t_2) \leq \sum_{l=1}^{r} \mu_l \|s(t_1,t_2)\| \{ &\|KH_{l1}\| \cdot \|H_{l2}\| \cdot \|x(t_1,t_2)\| \\ &+\|KH_{l1}\| \cdot \|H_{l3}\| + \|KH_{l1}\| \cdot \|H_{l4}\| \cdot \|u(t_1,t_2)\| \\ &+\|KD_l\|\sigma(t_1,t_2)\} - \zeta(t_1,t_2)\|s(t_1,t_2)\| - \beta\|s(t_1,t_2)\|.\end{aligned} \tag{7.64}$$

Notice the parameter $\zeta(t_1,t_2)$ designed as in (7.60) and then we have $\mathscr{D}V(t_1,t_2) \leq -\beta\|s(t_1,t_2)\|$, which indicates that the sliding surface (7.52) can be reached in finite time. The proof is ended. ■

7.3.3 AN ILLUSTRATIVE EXAMPLE

In this subsection, simulation on the nonlinear gas absorption, water stream heating, and air drying process will be conducted to show the effectiveness of the proposed state feedback dynamic SMC design method.

Example 7.3. Consider the nonlinear gas absorption, water stream heating, and air drying process described as in (7.43). By conducting a similar fuzzy modelling process in Example 7.1, the system (7.43) can be described by a 2-D T-S fuzzy affine model as,

Plant Rule $\mathscr{R}^l$: IF $W_1(x,t)$ is $\mathscr{F}_{l1}$, THEN

$$\begin{cases} \mathscr{D}W(x,t) = (A_l + \Delta A_l)W(x,t) + a_l + \Delta a_l \\ \qquad\qquad + (B_l + \Delta B_l)u(x,t) + D_l w(x,t) \\ z(x,t) = L_l W(x,t) + N_l u(x,t), \ l \in \{1,2,3\}, \end{cases} \tag{7.65}$$

where the system matrices are given as in (7.46), and the parameter uncertainties ΔA_l, Δa_l and ΔB_l are given in the form of (7.3) with matrices H_{l1}, H_{l2}, and H_{l3} given as in (7.47).

The normalized MFs of this example are also shown as in Figure 7.1 with $d_1 = \frac{\pi}{3}$ and $d_2 = \frac{\pi}{2}$. The purpose of this example is to design a sliding surface (7.52) and a state feedback dynamic sliding mode controller (7.59) to guarantee the finite-time reachability of the sliding surface and the asymptotic stability of the sliding mode dynamical system with robust $\mathscr{H}_\infty$ performance γ. By using Theorem 7.3 with $\bar{J} = \begin{bmatrix} 0 & -3 & 0 \end{bmatrix}$, for $i \in \mathscr{I}_1$, $\bar{J} = \begin{bmatrix} -3 & 0 \end{bmatrix}$, for $i \in \mathscr{I}_0$, the robust $\mathscr{H}_\infty$ performance is obtained as $\gamma_{\min} = 0.6728$ with sliding surface gain $K = \begin{bmatrix} 2.3748 & 1.4293 \end{bmatrix}$. It is also worth mentioning that when using Theorem 7.3 with PQLFs, the matrices $\bar{G}_i$ and $\bar{F}_i$ can be also computed via (7.49).

To show the effectiveness of the developed state feedback dynamic SMC design method, simulations are carried out. With boundary condition $W(0,t) = \begin{bmatrix} 0 & 0.2 \end{bmatrix}^{\mathrm{T}}$, $W(x,0) = \begin{bmatrix} 0.5 & 0 \end{bmatrix}^{\mathrm{T}}$, $0 \leq t \leq 0.5$s, $0 \leq x \leq 0.5$m, $W(0,t) = W(x,0) = \begin{bmatrix} 0 & 0 \end{bmatrix}^{\mathrm{T}}$, $t > 0.5$s, $x > 0.5$m, and nonlinear disturbance $w(x,t) = 12\mathrm{e}^{-0.1x}\sin(0.3x)$, using the state feedback dynamic sliding mode controller (7.59) with $\beta = 1.2$ and $\rho = 0.8$, the closed-loop system states are, respectively, presented in Figure 7.8 and Figure 7.9. The control input is presented in Figure 7.10, and the sliding surface is presented in Figure 7.11.

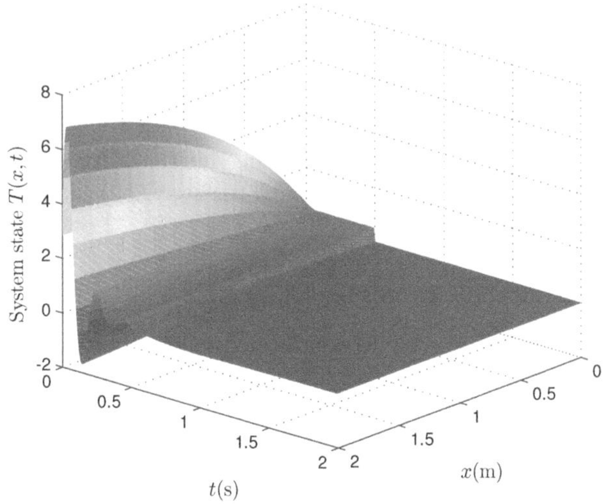

Figure 7.8 Response of the state $T(x,t)$ with dynamic sliding mode controller (7.59) in Example 7.3.

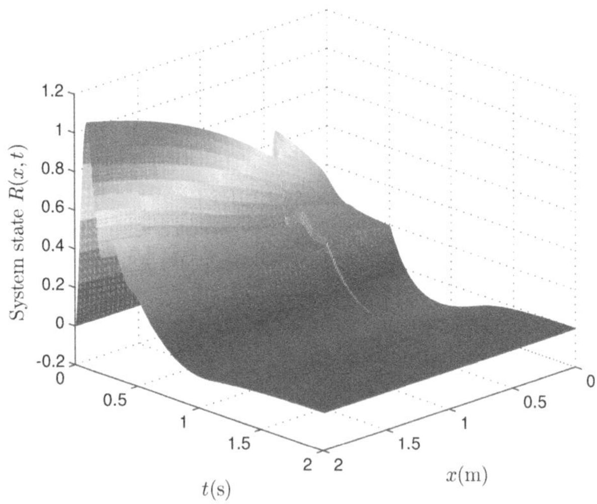

Figure 7.9 Response of the state $R(x,t)$ with dynamic sliding mode controller (7.59) in Example 7.3.

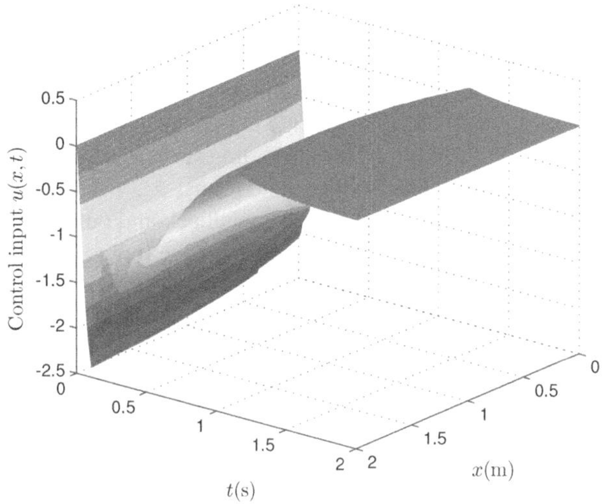

Figure 7.10 The control input with dynamic sliding mode controller (7.59) in Example 7.3.

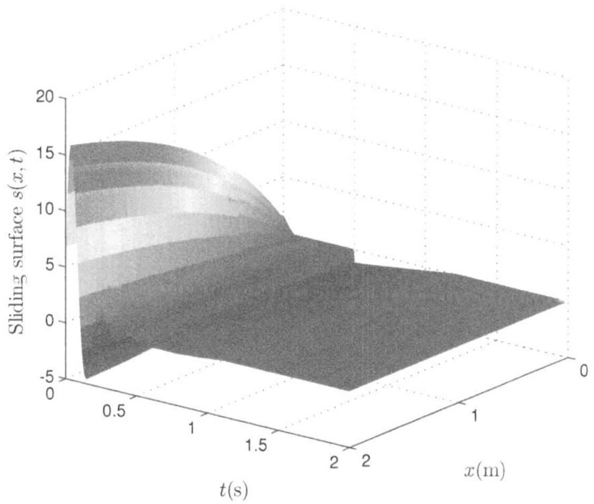

Figure 7.11 The sliding surface with dynamic sliding mode controller (7.59) in Example 7.3.

7.4 CONCLUSION

In this chapter, the state feedback and output feedback dynamic SMC design problem for continuous-time nonlinear 2-D systems has been studied through T-S fuzzy affine models. By combining with the sliding surface, the sliding motion dynamics are embedded into an augmented singular 2-D system. Through a 2-D PQLFs-based approach, new existence conditions for the sliding surface are derived under a convex optimization framework, and the stability and robust $\mathscr{H}_\infty$ performance analysis conservatism can be further reduced. Both state feedback and output feedback dynamic SMC synthesis schemes are proposed to guarantee the finite-time reachability of the sliding surface. As the designed sliding mode controller exists in a dynamic form, the control input matrices are permitted to have parameter uncertainties. Additionally, the mismatched disturbances can be also handled simultaneously. Simulation studies are finally presented to verify the effectiveness of the developed approaches.

8 Fuzzy SMC of Discrete-Time Nonlinear Two-Dimensional Systems

8.1 INTRODUCTION

This chapter investigates the fuzzy SMC problem for discrete-time nonlinear 2-D systems by T-S fuzzy affine models. By using a singular system-based method and combining with the sliding surface, the sliding mode dynamical properties are described by a singular piecewise affine 2-D system. Stability and robust $\mathscr{H}_\infty$ performance analysis of the sliding motion are carried out via the piecewise quadratic Lyapunov functions (PQLFs), and the sliding surface gains can be calculated through a convex optimization process. Both state feedback and output feedback dynamic sliding mode controllers are designed to guarantee that the 2-D system states can converge into a neighbourhood of the sliding surface. Simulation studies are given to verify the validity of the proposed schemes. Main contributions of this chapter are summarized as follows. 1) A singular piecewise affine 2-D system scheme is proposed to tackle the couplings between the control input matrices and the sliding surface gain, and the stability and robust $\mathscr{H}_\infty$ performance analysis results on the sliding motion are proposed under a unified convex optimization setup; 2) Based on the system measurement output, the designed output feedback sliding mode controller is suggested to be more applicable; 3) The designed state feedback and output feedback sliding mode controllers are both in a dynamic form, and matched and mismatched external disturbances can be handled successfully; 4) The conservatism of the stability and robust performance analysis of sliding motion can be further relaxed via adopting the PQLFs.

8.2 OUTPUT FEEDBACK DYNAMIC SMC DESIGN VIA 2-D T-S FUZZY MODELS

8.2.1 MODEL DESCRIPTION

Consider a general nonlinear 2-D system described as in (7.1). Through a fuzzy modelling and discretizing process, the system (7.1) can be described by the following discrete-time 2-D T-S fuzzy affine model with r fuzzy **IF-THEN** rules as,

Plant Rule $\mathscr{R}^l$: **IF** $\zeta_1(x(i,j))$ **is** $\mathscr{F}_1^l$ **and** $\cdots$ **and** $\zeta_\varphi(x(i,j))$ **is** $\mathscr{F}_\varphi^l$, **THEN**

$$\begin{cases} x^+(i,j) = (A_l + \Delta A_l)x(i,j) + a_l + \Delta a_l + B_l u(i,j) + D_l w(i,j) \\ y(i,j) = Cx(i,j) \\ z(i,j) = L_l x(i,j) + N_l u(i,j), \ l \in \mathscr{L} = \{1,\ldots,r\}, \end{cases} \tag{8.1}$$

DOI: 10.1201/9781032695716-8

where $\mathscr{F}_\phi^l(\phi = 1,\ldots,\varphi)$ are fuzzy sets; $\mathscr{R}^l$ is the l-th fuzzy inference rule; $\zeta(x(i,j)) := \left[\zeta_1(x(i,j)),\ldots,\zeta_\varphi(x(i,j))\right]$ denote the measurable premise variables; r is the number of inference rules; $x(i,j) = \begin{bmatrix} x^{h\mathrm{T}}(i,j) & x^{v\mathrm{T}}(i,j) \end{bmatrix}^\mathrm{T}$, $x^+(i,j) = \begin{bmatrix} x^{h\mathrm{T}}(i+1,j) & x^{v\mathrm{T}}(i,j+1) \end{bmatrix}^\mathrm{T}$, $x^h(i,j) \in \Re^{n_x^h}$ and $x^v(i,j) \in \Re^{n_x^v}$ are the horizontal states and vertical states, respectively, and $n_x = n_x^h + n_x^v$; $y(i,j) \in \Re^{n_y}$ is the measurement output; $w(i,j) \in l_2\{[0,\infty),[0,\infty)\}$ is the exogenous disturbance with $w(i,j) \in \Re^{n_w}$; $z(i,j) \in \Re^{n_z}$ is the regulated output; $u(i,j) \in \Re^{n_u}$ is the control input; ΔA_l and Δa_l are the uncertainty terms satisfying

$$\begin{bmatrix} \Delta A_l & \Delta a_l \end{bmatrix} = H_{l1}\Delta_l(i,j)\begin{bmatrix} H_{l2} & H_{l3} \end{bmatrix}, l \in \mathscr{L}, \tag{8.2}$$

where H_{l1}, H_{l2}, and H_{l3} are known matrices. $\Delta_l(i,j) \in \Re^{s_1 \times s_2}$ represent unknown matrices satisfying

$$\|\Delta_l(i,j)\| \leq \mathbf{I},\ l \in \mathscr{L}. \tag{8.3}$$

Denote $\mu_l\left[\zeta(x(i,j))\right] = \mu_l\left[x^h(i,j), x^v(i,j)\right]$ as the normalized membership function,

$$\mu_l\left[\zeta(x(i,j))\right] := \frac{\prod_{\phi=1}^{\varphi}\mu_{l\phi}\left[\zeta_\phi(x(i,j))\right]}{\sum_{v=1}^{r}\prod_{\phi=1}^{\varphi}\mu_{v\phi}\left[\zeta_\phi(x(i,j))\right]} \geq 0,\ \sum_{l=1}^{r}\mu_l\left[\zeta(x(i,j))\right] = 1 \tag{8.4}$$

with $\mu_{l\phi}\left[\zeta_\phi(x(i,j))\right]$ being the grade of membership of $\zeta_\phi(x(i,j))$ in $\mathscr{F}_\phi^l$. For brevity, denote $\mu_l := \mu_l\left[\zeta(x(i,j))\right]$.

By a product inference, center-average defuzzifier, and singleton fuzzifier, one can formulate the system (8.1) as

$$\begin{cases} x^+(i,j) = (A(\mu) + \Delta A(\mu))x(i,j) + a(\mu) + \Delta a(\mu) + B(\mu)u(i,j) \\ \qquad\qquad +D(\mu)w(i,j) \\ y(i,j) = Cx(i,j) \\ z(i,j) = L(\mu)x(i,j) + N(\mu)u(i,j) \end{cases} \tag{8.5}$$

where

$$\begin{cases} A(\mu) = \sum_{l=1}^{r}\mu_l A_l, \Delta A(\mu) = \sum_{l=1}^{r}\mu_l \Delta A_l, a(\mu) = \sum_{l=1}^{r}\mu_l a_l, \Delta a(\mu) = \sum_{l=1}^{r}\mu_l \Delta a_l, \\ B(\mu) = \sum_{l=1}^{r}\mu_l B_l, D(\mu) = \sum_{l=1}^{r}\mu_l D_l, L(\mu) = \sum_{l=1}^{r}\mu_l L_l, N(\mu) = \sum_{l=1}^{r}\mu_l N_l. \end{cases} \tag{8.6}$$

The premise variable space can be decomposed by a similar line in Chapter 7. To describe the indices in each region S_i, introduce the following set

$$\mathscr{N}(n) := \{m | \mu_m > 0, m \in \mathscr{L}, x(i,j) \in S_n, n \in \mathscr{I}\}. \tag{8.7}$$

Based on (8.7), reformulate the system (8.5) as

$$\begin{cases} x^+(i,j) = (\mathscr{A}_n + \Delta\mathscr{A}_n)x(i,j) + \mathrm{a}_n + \Delta\mathrm{a}_n + \mathscr{B}_n u(i,j) + \mathscr{D}_n w(i,j) \\ y(i,j) = Cx(i,j) \\ z(i,j) = \mathscr{L}_n x(i,j) + \mathscr{N}_n u(i,j),\ x(i,j) \in S_n, n \in \mathscr{I}, \end{cases} \tag{8.8}$$

where

$$\begin{cases} \mathscr{A}_n := \sum\limits_{m\in\mathscr{N}(n)} \mu_m A_m, \Delta\mathscr{A}_n := \sum\limits_{m\in\mathscr{N}(n)} \mu_m \Delta A_m, \\ \mathrm{a}_n := \sum\limits_{m\in\mathscr{N}(n)} \mu_m a_m, \Delta\mathrm{a}_n := \sum\limits_{m\in\mathscr{N}(n)} \mu_m \Delta a_m, \\ \mathscr{B}_n := \sum\limits_{m\in\mathscr{N}(n)} \mu_m B_m, \mathscr{D}_n := \sum\limits_{m\in\mathscr{N}(n)} \mu_m D_m, \\ \mathscr{L}_n := \sum\limits_{m\in\mathscr{N}(n)} \mu_m L_m, \mathscr{N}_n := \sum\limits_{m\in\mathscr{N}(n)} \mu_m N_m \end{cases} \tag{8.9}$$

with $0 < \mu_m \leq 1$, $\sum_{m\in\mathscr{N}(n)} \mu_m = 1$.

This section attempts to propose a novel output feedback dynamic SMC design approach for discrete-time 2-D fuzzy-affine system (8.1), and drive the system states into a neighbourhood of the sliding surface. A new existence criterion for the sliding surface will be derived through PQLFs. For a prescribed disturbance attenuation level γ, the sliding motion can be ensured to be asymptotically stable when $w(i,j) \equiv 0$, and it is satisfied that

$$\sum_{i=0}^{\infty}\sum_{j=0}^{\infty} z^{\mathrm{T}}(i,j)z(i,j) < \gamma^2 \sum_{i=0}^{\infty}\sum_{j=0}^{\infty} w^{\mathrm{T}}(i,j)w(i,j) \tag{8.10}$$

for $0 \neq w(i,j) \in l_2\{[0,\ \infty),[0,\ \infty)\}$ with zero boundary conditions $x^h(0,j) = 0$ and $x^v(i,0) = 0$.

8.2.2 SLIDING SURFACE DESIGN

For the discrete-time fuzzy 2-D system (8.1), design a sliding surface based on the measurement output and control input as

$$s(i,j) = Ky(i,j) + u(i,j), \tag{8.11}$$

where $K \in \Re^{n_u \times n_y}$ is the sliding surface gain to be determined.

Similar to [132, 135], it is satisfied that $s(i,j) = 0$ for the ideal quasi sliding motion. Then one has

$$u(i,j) = -KCx(i,j). \tag{8.12}$$

Based on system (8.8) and (8.12), one can derive the sliding motion as

$$\begin{cases} x^+(i,j) = (\mathscr{A}_n + \Delta\mathscr{A}_n)x(i,j) + \mathrm{a}_n + \Delta\mathrm{a}_n + \mathscr{B}_n u(i,j) + \mathscr{D}_n w(i,j) \\ 0 \times u^+(i,j) = -KCx(i,j) - u(i,j) \\ z(i,j) = \mathscr{L}_n x(i,j) + \mathscr{N}_n u(i,j), n \in \mathscr{I}. \end{cases} \tag{8.13}$$

Reformulate system (8.13) as

$$\begin{cases} E\bar{x}^+(i,j) = (\bar{\mathscr{A}}_n + \Delta\bar{\mathscr{A}}_n)\bar{x}(i,j) + \bar{\mathrm{a}}_n + \Delta\bar{\mathrm{a}}_n + \bar{\mathscr{D}}_n w(i,j) \\ z(i,j) = \bar{\mathscr{L}}_n \bar{x}(i,j), n \in \mathscr{I}, \end{cases} \tag{8.14}$$

where

$$\begin{cases} E = \mathrm{diag}\{\mathbf{I}_{n_x}, \mathbf{0}_{n_u}\}, \\ \bar{x}^+(i,j) = \begin{bmatrix} x^+(i,j) \\ u^+(i,j) \end{bmatrix}, \bar{x}(i,j) = \begin{bmatrix} x(i,j) \\ u(i,j) \end{bmatrix}, \\ \bar{\mathscr{A}}_n + \Delta\bar{\mathscr{A}}_n = \begin{bmatrix} \mathscr{A}_n + \Delta\mathscr{A}_n & \mathscr{B}_n \\ -KC & -\mathbf{I} \end{bmatrix}, \\ \bar{\mathrm{a}}_n + \Delta\bar{\mathrm{a}}_n = \begin{bmatrix} \mathrm{a}_n + \Delta\mathrm{a}_n \\ \mathbf{0} \end{bmatrix}, \bar{\mathscr{D}}_n = \begin{bmatrix} \mathscr{D}_n^{\mathrm{T}} & \mathbf{0} \end{bmatrix}^{\mathrm{T}}, \bar{\mathscr{L}}_n = \begin{bmatrix} \mathscr{L}_n & \mathscr{N}_n \end{bmatrix}. \end{cases} \tag{8.15}$$

By augmenting the sliding surface function (8.11) with the system (8.8), the singular piecewise affine 2-D system (8.14) can fully characterize the dynamical features of the sliding motion. Then the existence criterion for the sliding surface will be derived. For the fuzzy 2-D system (8.1), an output feedback dynamic SMC synthesis approach will be also proposed.

8.2.3 STABILITY AND ROBUST PERFORMANCE ANALYSIS OF SLIDING MOTION THROUGH PQLFS

In this subsection, through PQLFs, the robust stability of the sliding motion will be analyzed under a convex optimization setup. Then a novel output feedback dynamic SMC design method will be proposed to ensure the local convergence of the sliding surface. We first introduce the subsequent set Ω to describe the subspace transitions,

$$\Omega := \{(n,k) | x(i,j) \in S_n,\ x^+(i,j) \in S_k, n,k \in \mathscr{I}\}. \tag{8.16}$$

In this chapter, according to [40], assume that an ellipsoid $\mathscr{E}_i$ can be adopted to outer approximate each polyhedral subspace S_n in (8.16) as

$$S_n \subseteq \mathscr{E}_n,\ \mathscr{E}_n := \{x(i,j) | \|Q_n x(i,j) + q_n\| \leq 1\}. \tag{8.17}$$

The aforementioned outer approximation is useful especially when S_n are slab subspaces. For this situation, parameters Q_n and q_n in (8.17) are ensured to exist, and the covering is precise, namely, $\mathscr{E}_n \subseteq S_n$ and $S_n \subseteq \mathscr{E}_n$. In particular, when S_n are slabs with

$$S_n = \{x(i,j) | \alpha_n \leq \upsilon_n^{\mathrm{T}} x(i,j) \leq \beta_n\},\ n \in \mathscr{I}, \tag{8.18}$$

where $\upsilon_n \in \Re^{n_x}$, $\{\alpha_n, \beta_n\} \in \Re$, one can adopt a degenerate ellipsoid shown as in (8.17) to exactly characterize each subspace S_n with

$$Q_n = \frac{2\upsilon_n^{\mathrm{T}}}{\beta_n - \alpha_n},\ q_n = -\frac{\beta_n + \alpha_n}{\beta_n - \alpha_n}. \tag{8.19}$$

Relying on (8.19), for each subspace, the following relationship holds,

$$\begin{bmatrix} 1 \\ x(i,j) \end{bmatrix}^{\mathrm{T}} \begin{bmatrix} q_n^{\mathrm{T}} q_n - 1 & \star \\ Q_n^{\mathrm{T}} q_n & Q_n^{\mathrm{T}} Q_n \end{bmatrix} \begin{bmatrix} 1 \\ x(i,j) \end{bmatrix} \leq 0,\ n \in \mathscr{I}. \tag{8.20}$$

Theorem 8.1. For a robust $\mathscr{H}_\infty$ performance γ, the sliding mode dynamical system (8.14) is asymptotically stable, if matrices $0 < P_{n1}^h = P_{n1}^{h\mathrm{T}} \in \Re^{n_x^h \times n_x^h}$, $0 < P_{n1}^v = P_{n1}^{v\mathrm{T}} \in \Re^{n_x^v \times n_x^v}$, $P_{n2} \in \Re^{n_u \times n_x}$, $P_{n3} \in \Re^{n_u \times n_u}$, $n \in \mathscr{I}$, $\bar{K} \in \Re^{n_u \times n_y}$, and $\{V_{n1}, W_{n1}\} \in \Re^{n_x \times n_x}$, $\{V_{n2}, W_{n2}\} \in \Re^{n_u \times n_x}$, $V_0 \in \Re^{n_u \times n_u}$, and scalars $\{\varepsilon_n > 0, n \in \mathscr{I}\}$, $\{\rho_n < 0, n \in \mathscr{I}_1\}$ exist, such that the following conditions hold,

$$\begin{bmatrix} \Psi_{11} & \star & \star & \star \\ \Psi_{21} - W_n^{\mathrm{T}} & P_k - V_n - V_n^{\mathrm{T}} & \star & \star \\ \Psi_{31} & \Psi_{32} & -\gamma^2 \mathbf{I} & \star \\ \Psi_{41} & \Psi_{42} & \mathbf{0} & -\varepsilon_n \mathbf{I} \end{bmatrix} < 0, n \in \mathscr{I}_0, m \in \mathscr{N}(n), \quad (8.21)$$

$$\begin{bmatrix} \rho_n(q_n^{\mathrm{T}} q_n - 1) + \varepsilon_n H_{m3}^{\mathrm{T}} H_{m3} & \star & \star & \star & \star \\ \rho_n \bar{Q}_n^{\mathrm{T}} q_n + \Psi_{10} & \rho_n \bar{Q}_n^{\mathrm{T}} \bar{Q}_n + \Psi_{11} & \star & \star & \star \\ \Psi_{20} & \Psi_{21} - W_n^{\mathrm{T}} & P_k - V_n - V_n^{\mathrm{T}} & \star & \star \\ \mathbf{0} & \Psi_{31} & \Psi_{32} & -\gamma^2 \mathbf{I} & \star \\ \mathbf{0} & \Psi_{41} & \Psi_{42} & \mathbf{0} & -\varepsilon_n \mathbf{I} \end{bmatrix} < 0,$$
$$n \in \mathscr{I}_1, m \in \mathscr{N}(n), \quad (8.22)$$

where

$$\begin{cases} \Psi_{11} = \mathtt{Sym}\left\{ \begin{bmatrix} W_{n1} \\ W_{n2} \end{bmatrix} \begin{bmatrix} A_m & B_m \end{bmatrix} - \begin{bmatrix} J_2 \\ \rho_2 \mathbf{I} \end{bmatrix} \begin{bmatrix} \bar{K}C & V_0 \end{bmatrix} \right\} + \varepsilon_n \begin{bmatrix} H_{m2}^{\mathrm{T}} \\ \mathbf{0} \end{bmatrix} (\star) \\ \quad - E^{\mathrm{T}} P_n E + \bar{L}_m^{\mathrm{T}} \bar{L}_m, \\ \Psi_{21} = \begin{bmatrix} V_{n1} \\ V_{n2} \end{bmatrix} \begin{bmatrix} A_m & B_m \end{bmatrix} - \begin{bmatrix} J_1 \\ \rho_1 \mathbf{I} \end{bmatrix} \begin{bmatrix} \bar{K}C & V_0 \end{bmatrix}, \\ \Psi_{31} = D_m^{\mathrm{T}} \begin{bmatrix} W_{n1}^{\mathrm{T}} & W_{n2}^{\mathrm{T}} \end{bmatrix}, \Psi_{32} = D_m^{\mathrm{T}} \begin{bmatrix} V_{n1}^{\mathrm{T}} & V_{n2}^{\mathrm{T}} \end{bmatrix}, \\ \Psi_{41} = H_{m1}^{\mathrm{T}} \begin{bmatrix} W_{n1}^{\mathrm{T}} & W_{n2}^{\mathrm{T}} \end{bmatrix}, \Psi_{42} = H_{m1}^{\mathrm{T}} \begin{bmatrix} V_{n1}^{\mathrm{T}} & V_{n2}^{\mathrm{T}} \end{bmatrix}, \\ \Psi_{10} = \begin{bmatrix} W_{n1} \\ W_{n2} \end{bmatrix} a_m + \varepsilon_n \begin{bmatrix} H_{m2}^{\mathrm{T}} \\ \mathbf{0} \end{bmatrix} H_{m3}, \\ \Psi_{20} = \begin{bmatrix} V_{n1} \\ V_{n2} \end{bmatrix} a_m, \ P_n = \begin{bmatrix} \mathrm{diag}\{P_{n1}^h, P_{n1}^v\} & \star \\ P_{n2} & P_{n3} \end{bmatrix}, \\ V_n = \begin{bmatrix} V_{n1} & J_1 V_0 \\ V_{n2} & \rho_1 V_0 \end{bmatrix}, W_n = \begin{bmatrix} W_{n1} & J_2 V_0 \\ W_{n2} & \rho_2 V_0 \end{bmatrix}, \\ \bar{L}_m = \begin{bmatrix} L_m & N_m \end{bmatrix}, \bar{Q}_n = \begin{bmatrix} Q_n & \mathbf{0}_{1 \times n_u} \end{bmatrix} \end{cases} \quad (8.23)$$

and $\{J_1, J_2\} \in \Re^{n_x \times n_u}$ are arbitrary matrices, and $\rho_1 \neq 0$, ρ_2 are constant scalars.

Additionally, the gain of the sliding surface (8.11) can be calculated by

$$K = V_0^{-1} \bar{K}. \quad (8.24)$$

Proof. Notice that the condition (8.21) for $n \in \mathscr{I}_0$ denotes a special case of the condition (8.22) for $n \in \mathscr{I}_1$. In sequence, only the more complicated case for $n \in \mathscr{I}_1$ is proved without loss of generality. Construct a PQLF as

$$V(\bar{x}(i,j), i, j) = \bar{x}^{\mathrm{T}}(i,j) E^{\mathrm{T}} P_n E \bar{x}(i,j), \quad (8.25)$$

where

$$P_n = \begin{bmatrix} \mathrm{diag}\{P_{n1}^h, P_{n1}^v\} & \star \\ P_{n2} & P_{n3} \end{bmatrix}, \ n \in \mathscr{I} \tag{8.26}$$

and $0 < P_{n1}^h = P_{n1}^{h\mathrm{T}} \in \Re^{n_x^h \times n_x^h}$, $0 < P_{n1}^v = P_{n1}^{v\mathrm{T}} \in \Re^{n_x^v \times n_x^v}$, $P_{n2} \in \Re^{n_u \times n_x}$, $P_{n3} \in \Re^{n_u \times n_u}$.

In view of (8.25), if the following condition holds,

$$\begin{aligned}(\bar{x}^+(i,j))^{\mathrm{T}} E^{\mathrm{T}} P_k E \bar{x}^+(i,j) - \bar{x}^{\mathrm{T}}(i,j) E^{\mathrm{T}} P_n E \bar{x}(i,j) + z^{\mathrm{T}}(i,j) z(i,j) \\ - \gamma^2 w^{\mathrm{T}}(i,j) w(i,j) < 0, \ (n,k) \in \Omega,\end{aligned} \tag{8.27}$$

the system (8.14) is asymptotically stable with a satisfied $\mathscr{H}_\infty$ performance γ.

Denote

$$\xi(i,j) = \begin{bmatrix} 1 & \bar{x}^{\mathrm{T}}(i,j) & (E\bar{x}^+(i,j))^{\mathrm{T}} & w^{\mathrm{T}}(i,j) \end{bmatrix}^{\mathrm{T}} \tag{8.28}$$

and then rewrite the inequality (8.27) as,

$$\xi^{\mathrm{T}}(i,j) \begin{bmatrix} 0 & \star & \star & \star \\ \mathbf{0} & \bar{\mathscr{L}}_n^{\mathrm{T}} \bar{\mathscr{L}}_n - E^{\mathrm{T}} P_n E & \star & \star \\ \mathbf{0} & \mathbf{0} & P_k & \star \\ \mathbf{0} & \mathbf{0} & \mathbf{0} & -\gamma^2 \mathbf{I} \end{bmatrix} \xi(i,j) < 0, (n,k) \in \Omega. \tag{8.29}$$

Adopting the S-procedure based on the partition information in (8.20), the subsequent inequality implies (8.29) for $\rho_n < 0$, $n \in \mathscr{I}_1$,

$$\xi^{\mathrm{T}}(i,j) \Sigma \xi(i,j) < 0, n \in \mathscr{I}_1, (n,k) \in \Omega, \tag{8.30}$$

where

$$\begin{cases} \Sigma = \begin{bmatrix} \rho_n (q_n^{\mathrm{T}} q_n - 1) & \star & \star & \star \\ \rho_n \bar{Q}_n^{\mathrm{T}} q_n & \Sigma_{11} & \star & \star \\ \mathbf{0} & \mathbf{0} & P_k & \star \\ \mathbf{0} & \mathbf{0} & \mathbf{0} & -\gamma^2 \mathbf{I} \end{bmatrix}, \\ \Sigma_{11} = \rho_n \bar{Q}_n^{\mathrm{T}} \bar{Q}_n + \bar{\mathscr{L}}_n^{\mathrm{T}} \bar{\mathscr{L}}_n - E^{\mathrm{T}} P_n E, \bar{Q}_n = \begin{bmatrix} Q_n & \mathbf{0}_{1 \times n_u} \end{bmatrix}. \end{cases} \tag{8.31}$$

Based on sliding mode dynamical system (8.14), it is satisfied for any matrices $\mathscr{G}_n$, $n \in \mathscr{I}$ that

$$\begin{aligned} 0 &= 2\mathscr{G}_n \left\{ \bar{\mathrm{a}}_n + \Delta \bar{\mathrm{a}}_n + (\bar{\mathscr{A}}_n + \Delta \bar{\mathscr{A}}_n) \bar{x}(i,j) + \bar{\mathscr{D}}_n w(i,j) - E\bar{x}^+(i,j) \right\} \\ &= 2\mathscr{G}_n \begin{bmatrix} \bar{\mathrm{a}}_n + \Delta \bar{\mathrm{a}}_n & \bar{\mathscr{A}}_n + \Delta \bar{\mathscr{A}}_n & -\mathbf{I} & \bar{\mathscr{D}}_n \end{bmatrix} \xi(i,j). \end{aligned} \tag{8.32}$$

Combining (8.30) with (8.32), one has,

$$\xi^{\mathrm{T}}(i,j) \left\{ \Sigma + \mathtt{Sym}\{\mathscr{G}_n \mathbb{A}_n\} \right\} \xi(i,j) < 0, n \in \mathscr{I}_1, (n,k) \in \Omega, \tag{8.33}$$

where $\mathbb{A}_n = \begin{bmatrix} \bar{\mathrm{a}}_n + \Delta \bar{\mathrm{a}}_n & \bar{\mathscr{A}}_n + \Delta \bar{\mathscr{A}}_n & -\mathbf{I} & \bar{\mathscr{D}}_n \end{bmatrix}$.

It follows from (8.33) that the following inequality implies (8.27),

$$\Sigma + \mathtt{Sym}\{\mathscr{G}_n \mathbb{A}_n\} < 0, n \in \mathscr{I}_1, (n,k) \in \Omega. \tag{8.34}$$

Expanding the MFs in (8.34), yields

$$\begin{bmatrix} \rho_n(q_n^{\mathrm{T}}q_n-1) & \star & \star & \star \\ \rho_n\bar{Q}_n^{\mathrm{T}}q_n & \Xi_{11} & \star & \star \\ \mathbf{0} & \mathbf{0} & P_k & \star \\ \mathbf{0} & \mathbf{0} & \mathbf{0} & -\gamma^2\mathbf{I} \end{bmatrix} + \mathtt{Sym}\{\mathscr{G}_n\mathscr{A}_m\} < 0, \quad m\in\mathscr{N}(n), n\in\mathscr{I}_1, (n,k)\in\Omega, \tag{8.35}$$

where

$$\begin{cases} \mathscr{A}_m = \begin{bmatrix} \bar{a}_m+\Delta\bar{a}_m & \bar{A}_m+\Delta\bar{A}_m & -\mathbf{I} & \bar{D}_m \end{bmatrix}, \\ \bar{A}_m+\Delta\bar{A}_m = \begin{bmatrix} A_m+\Delta A_m & B_m \\ -KC & -\mathbf{I} \end{bmatrix}, \bar{a}_m+\Delta\bar{a}_m = \begin{bmatrix} a_m+\Delta a_m \\ \mathbf{0} \end{bmatrix}, \\ \Xi_{11} = \rho_n\bar{Q}_n^{\mathrm{T}}\bar{Q}_n + \bar{L}_m^{\mathrm{T}}\bar{L}_m - E^{\mathrm{T}}P_nE, \bar{L}_m = \begin{bmatrix} L_m & N_m \end{bmatrix}. \end{cases} \tag{8.36}$$

Notice that the first row of $\bar{a}_m+\Delta\bar{a}_m$ and $\bar{A}_m+\Delta\bar{A}_m$ does not contain the sliding surface gain K. For numerical convexification, specify $\mathscr{G}_n = \begin{bmatrix} \mathbf{0} & W_n^{\mathrm{T}} & V_n^{\mathrm{T}} & \mathbf{0} \end{bmatrix}^{\mathrm{T}}$, wherein the slack variable matrices V_n and W_n are given as

$$V_n = \begin{bmatrix} V_{n1} & J_1V_0 \\ V_{n2} & \rho_1V_0 \end{bmatrix}, W_n = \begin{bmatrix} W_{n1} & J_2V_0 \\ W_{n2} & \rho_2V_0 \end{bmatrix} \tag{8.37}$$

with $\rho_1 \neq 0$ and $\rho_2 \neq 0$ being scalar parameters, and $\{J_1, J_2\} \in \Re^{n_x\times n_u}$ are arbitrary matrices.

Define

$$\bar{K} = V_0K. \tag{8.38}$$

Considering (8.37)–(8.38), the inequality (8.35) can be derived as

$$\begin{bmatrix} \rho_n(q_n^{\mathrm{T}}q_n-1) & \star & \star & \star \\ \rho_n\bar{Q}_n^{\mathrm{T}}q_n+\Lambda_{10} & \Lambda_{11} & \star & \star \\ \Lambda_{20} & \Lambda_{21}-W_n^{\mathrm{T}} & \Lambda_{22} & \star \\ \mathbf{0} & \Lambda_{31} & \Lambda_{32} & -\gamma^2\mathbf{I} \end{bmatrix} < 0, m\in\mathscr{N}(n), n\in\mathscr{I}_1, (n,k)\in\Omega, \tag{8.39}$$

where

$$\begin{cases} \Lambda_{10} = \begin{bmatrix} W_{n1} \\ W_{n2} \end{bmatrix}(a_m+\Delta a_m), \\ \Lambda_{11} = \rho_n\bar{Q}_n^{\mathrm{T}}\bar{Q}_n - E^{\mathrm{T}}P_nE + \bar{L}_m^{\mathrm{T}}\bar{L}_m \\ \qquad + \mathtt{Sym}\left\{ \begin{bmatrix} W_{n1} \\ W_{n2} \end{bmatrix} \begin{bmatrix} A_m+\Delta A_m & B_m \end{bmatrix} - \begin{bmatrix} J_2 \\ \rho_2\mathbf{I} \end{bmatrix} \begin{bmatrix} \bar{K}C & V_0 \end{bmatrix} \right\}, \\ \Lambda_{20} = \begin{bmatrix} V_{n1} \\ V_{n2} \end{bmatrix}(a_m+\Delta a_m), \\ \Lambda_{21} = \begin{bmatrix} V_{n1} \\ V_{n2} \end{bmatrix} \begin{bmatrix} A_m+\Delta A_m & B_m \end{bmatrix} - \begin{bmatrix} J_1 \\ \rho_1\mathbf{I} \end{bmatrix} \begin{bmatrix} \bar{K}C & V_0 \end{bmatrix}, \\ \Lambda_{22} = P_k - V_n - V_n^{\mathrm{T}}, \Lambda_{31} = D_m^{\mathrm{T}}\begin{bmatrix} W_{n1}^{\mathrm{T}} & W_{n2}^{\mathrm{T}} \end{bmatrix}, \Lambda_{32} = D_m^{\mathrm{T}}\begin{bmatrix} V_{n1}^{\mathrm{T}} & V_{n2}^{\mathrm{T}} \end{bmatrix}. \end{cases} \tag{8.40}$$

Utilizing Lemma A1 to tackle the parameter uncertainties in (8.39), then for scalars $\varepsilon_n > 0$, the inequality (8.22) implies (8.39) by Schur complement.

Additionally, the conditions in (8.21)–(8.22) imply $\rho_1 V_0 + \rho_1 V_0^{\mathrm{T}} - P_{k3} > 0$, which indicates that V_0 is nonsingular. Consequently, one can compute the sliding surface gain through (8.24). This ends the proof. ■

Via augmenting the control input with system states, a singular piecewise affine 2-D system method is proposed to handle the coupling problem between the system matrices and the sliding surface gain. By means of introducing auxiliary matrix multipliers, stability and robust $\mathscr{H}_\infty$ performance analysis for the sliding motion can be carried out by using some convexification techniques.

Remark 8.2.1. Note that due to the introduction of the auxiliary matrix multipliers W_n and V_n, the sliding surface gain is also not coupled with the Lyapunov matrices in Theorem 8.1. The conservatism has been also further reduced in light of the increased freedoms introduced by the auxiliary matrix multipliers. The aforementioned points have distinguished our results from the prevalent ones in [132, 133, 134, 135, 122]. It is also noted that when $P_{n1}^h = P_{k1}^h \equiv P_1^h \in \Re^{n_x^h \times n_x^h}$, $P_{n1}^v = P_{k1}^v \equiv P_1^v \in \Re^{n_x^v \times n_x^v}$, $P_{n2} = P_{k2} \equiv P_2 \in \Re^{n_u \times n_x}$, and $P_{n3} = P_{v3} \equiv P_3 \in \Re^{n_u \times n_u}$, the corresponding stability analysis results in Theorem 8.1 can be easily derived through a CQLF.

Remark 8.2.2. Due to the application of the singular piecewise affine 2-D system method and the introduction of auxiliary matrix multipliers, the stability and robust performance analysis results proposed in Theorem 8.1 are derived in terms of linear matrix inequalities under a convex optimization framework, which can be solved efficiently by commercially available software. It is noted that analysis and design conservatism can be also further reduced due to the increased freedom from the slack variables.

8.2.4 OUTPUT FEEDBACK DYNAMIC SLIDING MODE CONTROLLER DESIGN

Similar to [72], we introduce a sliding patch Υ,

$$\Upsilon := \{x(i,j) \in \Re^{n_x} : \|x(i,j)\| \leq \lambda\} \tag{8.41}$$

with $\lambda > 0$.

Then an output feedback dynamic sliding mode controller is designed as

$$u(i+1,j+1) = u(i,j) - \sum_{l=1}^{r} \mu_l KCB_l u(i,j) - (\sigma + \vartheta(i,j))\mathrm{sgn}(s(i,j)), \tag{8.42}$$

where $s(i,j) = Ky(i,j) + u(i,j)$ is given in (8.11), and $\sigma > 0$ is a given scalar, and

$$\begin{aligned}\vartheta(i,j) = \sum_{l=1}^{r} \mu_l \{&(\|KCA_l\| + \|KCH_{l1}\| \cdot \|H_{l2}\| + \|KC\|)\lambda \\ &+ \|KCa_l\| + \|KCH_{l1}\| \cdot \|H_{l3}\| + \|KCD_l\| \eta(i,j)\}\end{aligned} \tag{8.43}$$

with $\eta(i,j)$ representing the uniform upper-bound of $w(i,j)$.

Theorem 8.2. In view of the sliding patch Υ (8.41) with the sliding surface (8.11), the output feedback dynamic sliding mode controller (8.42) can drive the system states of the 2-D system (8.1) into a neighbourhood of the sliding surface.

Proof. Consider the Lyapunov function,

$$V(s(i,j),i,j) = \frac{1}{2}s^{\mathrm{T}}(i,j)s(i,j). \tag{8.44}$$

The incremental $\Delta V(s(i,j),i,j)$ is

$$\begin{aligned}\Delta V(s(i,j),i,j) &= V(s^{+}(i,j),i+1,j+1) - V(s(i,j),i,j)\\ &= s^{\mathrm{T}}(i,j)\Delta s(i,j) + \Gamma\end{aligned} \tag{8.45}$$

with $\Delta s(i,j) = s^{+}(i,j) - s(i,j)$, $s^{+}(i,j) = KCx^{+}(i,j) + u(i+1,j+1)$, $\Gamma = \frac{1}{2}\Delta s^{\mathrm{T}}(i,j)\Delta s(i,j)$.

Considering the sliding mode controller (8.42) and noticing the sliding patch Υ given as in (8.41), one has

$$\begin{aligned}&\Delta V(s(i,j),i,j)\\ &= s^{\mathrm{T}}(i,j)\left\{KCx^{+}(i,j) - KCx(i,j) + u(i+1,j+1) - u(i,j)\right\} + \Gamma\\ &= s^{\mathrm{T}}(i,j)\left\{\sum_{l=1}^{r}\mu_l KC\left[(A_l + \Delta A_l - \mathbf{I})x(i,j) + a_l + \Delta a_l + B_l u(i,j)\right.\right.\\ &\qquad\left.\left. + D_l w(i,j)\right] + u(i+1,j+1) - u(i,j)\right\} + \Gamma\\ &= s^{\mathrm{T}}(i,j)\sum_{l=1}^{r}\mu_l KC\left\{(A_l + \Delta A_l - \mathbf{I})x(i,j) + a_l + \Delta a_l + D_l w(i,j)\right\}\\ &\quad - \vartheta(i,j)\|s(i,j)\| - \sigma\|s(i,j)\| + \Gamma\\ &\le \|s(i,j)\|\sum_{l=1}^{r}\mu_l\left\{(\|KCA_l\| + \|KC\Delta A_l\| + \|KC\|)\lambda + \|KCa_l\|\right.\\ &\quad\left. + \|KC\Delta a_l\| + \|KCD_l\|\eta(i,j)\right\} - \vartheta(i,j)\|s(i,j)\| - \sigma\|s(i,j)\| + \Gamma\\ &\le \|s(i,j)\|\sum_{l=1}^{r}\mu_l\left\{(\|KCA_l\| + \|KCH_{l1}\|\cdot\|H_{l2}\| + \|KC\|)\lambda + \|KCa_l\|\right.\\ &\quad\left. + \|KCH_{l1}\|\cdot\|H_{l3}\| + \|KCD_l\|\eta(i,j)\right\} - \vartheta(i,j)\|s(i,j)\| - \sigma\|s(i,j)\| + \Gamma.\end{aligned} \tag{8.46}$$

Notice that

$$\begin{aligned}\vartheta(i,j) = \sum_{l=1}^{r}\mu_l\{&(\|KCA_l\| + \|KCH_{l1}\|\cdot\|H_{l2}\| + \|KC\|)\lambda\\ &+ \|KCa_l\| + \|KCH_{l1}\|\cdot\|H_{l3}\| + \|KCD_l\|\eta(i,j)\},\end{aligned} \tag{8.47}$$

and then we have

$$\Delta V(s(i,j),i,j) \le -\sigma\|s(i,j)\| + \Gamma. \tag{8.48}$$

Notice that

$$\begin{aligned}\Delta s(i,j) &= KCx^{+}(i,j) - KCx(i,j) + \Delta u(i,j) \\ &\leq \sum_{l=1}^{r} \mu_l \{(\|KCA_l\| + \|KCH_{l1}\| \cdot \|H_{l2}\| + \|KC\|)\lambda \\ &\quad + \|KCa_l\| + \|KCH_{l1}\| \cdot \|H_{l3}\| + \|KCD_l\| \eta(i,j)\} \\ &\quad - \vartheta(i,j)\mathrm{sgn}(s(i,j)) - \sigma \mathrm{sgn}(s(i,j)).\end{aligned} \tag{8.49}$$

Then one can see that

$$\|\Delta s(i,j)\| \leq \sigma + 2\vartheta(i,j), \tag{8.50}$$

which means that $\|\Delta s(i,j)\|$ is bounded.

By tuning the parameter σ, $\Delta V(s(i,j),i,j) < 0$ can be ensured. Thus, the states of the fuzzy 2-D system (8.1) can be forced into a neighbourhood of the sliding surface and maintained hereafter by the output feedback sliding mode controller (8.42) in the patch Υ. The proof is completed. ■

As discussed in Theorem 8.2, the convergence of the sliding surface (8.11) can be guaranteed by the developed output feedback dynamic SMC design approach, which also proposes a more general method to investigate the output feedback SMC issue for both matched and mismatched exogenous disturbance cases. The above-mentioned properties also distinguish the obtained results in this work from the prevalent ones in [122, 132, 133, 134, 135].

8.2.5 ILLUSTRATIVE EXAMPLES

Example 8.1. Recall a nonlinear water stream heating, gas absorption, and air drying process, whose dynamical features are described by the Darboux equation as

$$\frac{\partial^2 T(x,t)}{\partial x \partial t} = b_1 \frac{\partial T(x,t)}{\partial t} + b_2 \frac{\partial T(x,t)}{\partial x} + b_0 T(x,t) + c_0 u(x,t), \tag{8.51}$$

where $b_1 = -3$, $b_2 = -1$, $c_0 = 1$ are constant parameters; $T(x,t)$ represents the temperature within the reactant concentration; $b_0 = \cos T(x,t)$ represents a function at state space $x \in [0, x_f]$ and time $t \in [0, \infty)$; $u(x,t)$ represents the spatially uniform temperature in the jacket.

Along the lines in [116, 132], denote $R(x,t) = \frac{\partial T(x,t)}{\partial t} - b_2 T(x,t)$. Considering the external disturbance $w(x,t)$, the system (8.51) can be rewritten as

$$\begin{bmatrix} \frac{\partial R(x,t)}{\partial x} \\ \frac{\partial T(x,t)}{\partial t} \end{bmatrix} = \begin{bmatrix} b_1 & b_1 b_2 + b_0 \\ 1 & b_2 \end{bmatrix} \begin{bmatrix} R(x,t) \\ T(x,t) \end{bmatrix} + \begin{bmatrix} c_0 \\ 0 \end{bmatrix} u(x,t) + \begin{bmatrix} 0.5 \\ 0 \end{bmatrix} w(x,t). \tag{8.52}$$

Define $W(i,j) = \begin{bmatrix} W_1^{\mathrm{T}}(i,j) & W_2^{\mathrm{T}}(i,j) \end{bmatrix}^{\mathrm{T}}$, where $W_1(i,j) = R(i\Delta x, j\Delta t)$, $W_2(i,j) = T(i\Delta x, j\Delta t)$, and $w(i,j) = w(i\Delta x, j\Delta t)$, $u(i,j) = u(i\Delta x, j\Delta t)$. Linearize system (8.52) at $\left(0, \pm\frac{\pi}{3}\right)^{\mathrm{T}}$ and $(0,0)^{\mathrm{T}}$, and discretize the nonlinear process (8.52) with the vertical and horizontal sampling periods $\Delta t = 0.1$ and $\Delta x = 0.1$. Then the system (8.52) can be described by the 2-D T-S fuzzy affine model as follows.

Plant Rule $\mathscr{R}^l$: IF $W_2(i,j)$ is $\mathscr{F}_{l1}$, THEN

$$\begin{cases} W^+(i,j) = (A_l + \Delta A_l)W(i,j) + a_l + \Delta a_l + B_l u(i,j) + D_l w(i,j) \\ y(i,j) = CW(i,j), \\ z(i,j) = L_l W(i,j) + N_l u(i,j),\ l \in \{1,2,3\}, \end{cases} \tag{8.53}$$

where i and j denote the vertical and horizontal direction for the discrete-time fuzzy 2-D process (8.53), and

$$\begin{cases} A_1 = \begin{bmatrix} 0.7 & 0.4 \\ 0.1 & 0.9 \end{bmatrix}, A_2 = A_3 = \begin{bmatrix} 0.7 & 0.2593 \\ 0.1 & 0.9 \end{bmatrix}, \\ a_1 = \begin{bmatrix} 0 \\ 0 \end{bmatrix}, a_2 = \begin{bmatrix} -0.0950 \\ 0 \end{bmatrix}, a_3 = \begin{bmatrix} 0.0950 \\ 0 \end{bmatrix}, B_l = \begin{bmatrix} 0.1 \\ 0 \end{bmatrix}, \\ D_l = \begin{bmatrix} 0.05 \\ 0 \end{bmatrix}, C = \begin{bmatrix} 0 & 1 \end{bmatrix}, L_l = \begin{bmatrix} 1 & 1 \end{bmatrix}, N_l = 0.5,\ l \in \{1,2,3\} \end{cases} \tag{8.54}$$

and the parameter uncertainties terms ΔA_l and Δa_l are given in the form of (8.2) with

$$\begin{cases} \Delta_l(i,j) = \sin(i)\cos(j), \\ H_{l1} = \begin{bmatrix} 0.2 \\ 0 \end{bmatrix}, H_{l2} = \begin{bmatrix} 0 & 0.1 \end{bmatrix}, H_{l3} = 0.1,\ l \in \{1,2,3\}. \end{cases} \tag{8.55}$$

Figure 8.1 shows the fuzzy MFs wherein $d_1 = \dfrac{\pi}{3}$ and $d_2 = \dfrac{\pi}{2}$, and it is found that

$$\begin{cases} S_1 = \{W(i,j) \in \Re^2 \mid -d_2 \le W_2(i,j) \le -d_1\}, \\ S_2 = \{W(i,j) \in \Re^2 \mid -d_1 \le W_2(i,j) \le d_1\}, \\ S_3 = \{W(i,j) \in \Re^2 \mid \ d_1 \le W_2(i,j) \le d_2\}. \end{cases} \tag{8.56}$$

The above-mentioned three subspaces can be precisely described via degenerate ellipsoids in (8.17) with

$$\begin{cases} Q_1 = Q_3 = \begin{bmatrix} 0 & \dfrac{2}{d_2 - d_1} \end{bmatrix}, Q_2 = \begin{bmatrix} 0 & \dfrac{1}{d_1} \end{bmatrix}, \\ q_1 = \dfrac{d_1 + d_2}{d_2 - d_1}, q_2 = 0, q_3 = \dfrac{d_1 + d_2}{d_1 - d_2}. \end{cases} \tag{8.57}$$

The aim is to synthesize an output feedback dynamic sliding mode controller (8.42) to force the system states of the 2-D process (8.53) into a neighbourhood of the sliding surface (8.11) and guarantee the asymptotical stability of the sliding motion with disturbance attenuation level γ. *Assume that the system states are not fully accessible in this example. Additionally, the external disturbance is in a mismatched form. Thus, the prevailing SMC methods in [94, 95, 122, 132, 133, 134, 135] are not applicable for this case.* Nevertheless, by adopting Theorem 8.1 with $\rho_1 = \rho_2 = 1$, $J_1 = J_2 = \begin{bmatrix} 1 & 0 \end{bmatrix}^{\mathrm{T}}$, one can obtain the $\mathscr{H}_\infty$ performance index $\gamma_{\min} = 0.6251$ with

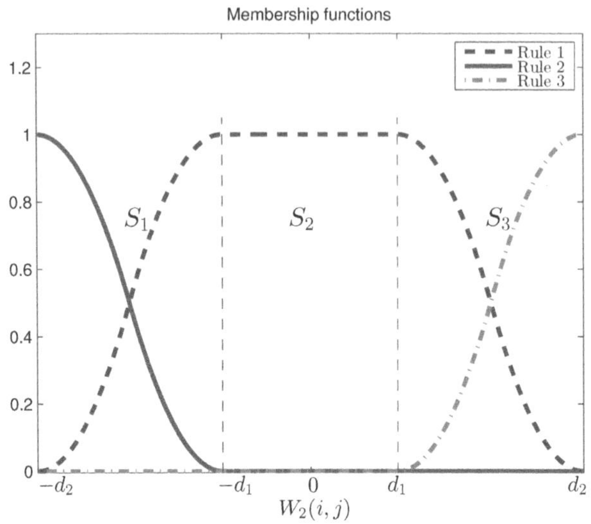

Figure 8.1 Membership functions of Example 8.1-Example 8.4.

sliding surface gain $K = 2.7940$. When it is assumed that $D_l = B_l = \begin{bmatrix} 0.1 & 0 \end{bmatrix}^{\mathrm{T}}$, the external disturbance is in matched form. With the same conditions, one can obtain the $\mathscr{H}_\infty$ performance index $\gamma_{\min} = 1.2505$ with sliding surface gain $K = 2.8135$.

To illustrate the effectiveness of the proposed approach, simulations are conducted. Specifically, with boundary condition $W(0,j) = \begin{bmatrix} 0 & -0.3 \end{bmatrix}^{\mathrm{T}}$, $W(i,0) = \begin{bmatrix} 0.2 & 0 \end{bmatrix}^{\mathrm{T}}$, $0 \le i \le 10$, $0 \le j \le 10$, $W(0,j) = W(i,0) = \begin{bmatrix} 0 & 0 \end{bmatrix}^{\mathrm{T}}$, $i > 10$, $j > 10$, and nonlinear disturbance function $w(i,j) = 50\mathrm{e}^{-0.1i}\sin(0.5i) + 5\mathrm{e}^{-2j}\sin(0.5j)$, adopting the output feedback dynamic sliding mode controller (8.42) with $\sigma = 0.2$ and $\lambda = 0.1$, the system state trajectories and control input are demonstrated in Figure 8.2, Figure 8.3 and Figure 8.4, respectively.

Furthermore, for some practical scenarios, the measured outputs are affected by measurement noises inevitably. Under the same exogenous disturbance and initial system states, the system state trajectories are shown in Figure 8.5 and Figure 8.6.

Example 8.2. Recall a fuzzy 2-D system (8.1) with

$$\begin{cases} A_1 = \begin{bmatrix} 0.5 & 0.3 \\ 0.2 & 1 \end{bmatrix}, A_2 = \begin{bmatrix} 0.9 & 0.5 \\ -0.1 & 0.8 \end{bmatrix}, A_3 = \begin{bmatrix} 0.2 & 0.5 \\ 1 & 1 \end{bmatrix}, \\ a_1 = \begin{bmatrix} 0 \\ 0 \end{bmatrix}, -a_2 = a_3 = \begin{bmatrix} 0.5 \\ 0 \end{bmatrix}, \\ B_1 = \begin{bmatrix} 1.5 \\ 1.2 \end{bmatrix}, B_2 = \begin{bmatrix} 2 \\ 1.5 \end{bmatrix}, B_3 = \begin{bmatrix} 1 \\ 1 \end{bmatrix}, D_l = \begin{bmatrix} 0 \\ 0.1 \end{bmatrix}, \\ C = \begin{bmatrix} 0 & 1 \end{bmatrix}, L_l = \begin{bmatrix} 1 & 0 \end{bmatrix}, N_l = 1,\ l \in \{1,2,3\} \end{cases} \tag{8.58}$$

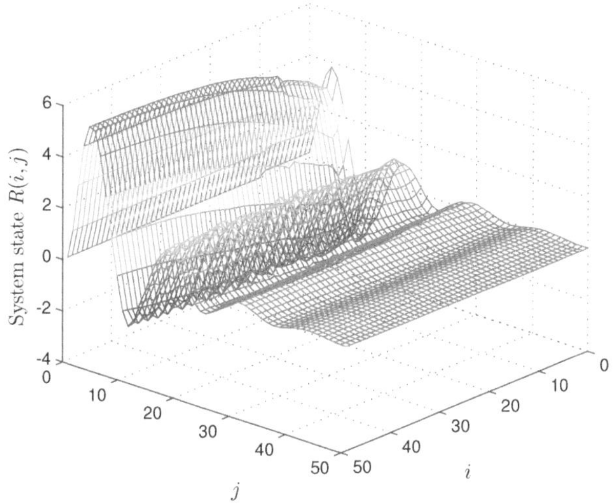

Figure 8.2 State response of $R(i,j)$ without affection of measurement noises in Example 8.1 by dynamic sliding mode controller (8.42).

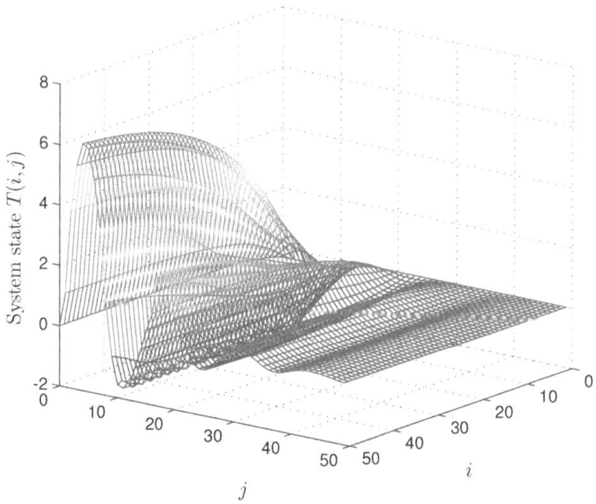

Figure 8.3 State response of $T(i,j)$ without affection of measurement noises in Example 8.1 by dynamic sliding mode controller (8.42).

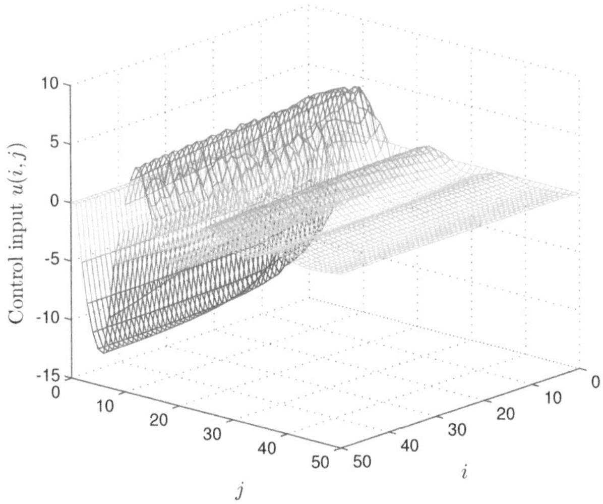

Figure 8.4 Response of the control input without affection of measurement noises in Example 8.1 by dynamic sliding mode controller (8.42).

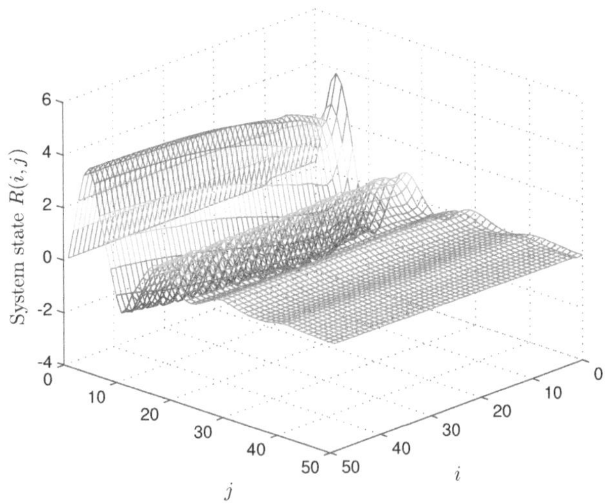

Figure 8.5 State response of $R(i, j)$ affected by measurement noise in Example 8.1 with dynamic sliding mode controller (8.42).

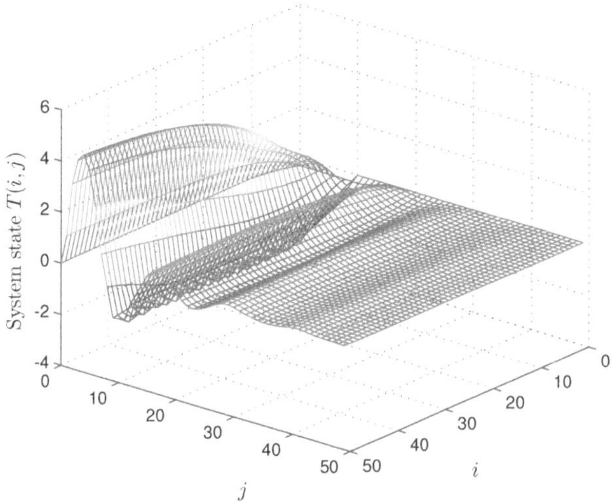

Figure 8.6 State response of $T(i,j)$ affected by measurement noise in Example 8.1 with dynamic sliding mode controller (8.42).

and the uncertainty terms ΔA_l and Δa_l are given in the form of (8.2) with

$$\begin{cases} \Delta_l(i,j) = \sin(i)\cos(j), \\ H_{l1} = \begin{bmatrix} 0 \\ 0.1 \end{bmatrix}, H_{l2} = \begin{bmatrix} 0 & 0.1 \end{bmatrix}, H_{l3} = 0.05, \ l \in \{1,2,3\}. \end{cases} \tag{8.59}$$

The fuzzy MFs of this example are also demonstrated in Figure 8.1 with $d_1 = 70$ and $d_2 = 700$, and the state space is decomposed into three subspaces as in (8.56). *The system states are also assumed to be not fully accessible in this example. By inspecting the system matrices given in (8.58), the control input matrices are different for each local model, and the disturbance is in a mismatched form. Therefore, the prevalent SMC design schemes in [94, 95, 122, 132, 133, 134, 135] are still not applicable for this case.* Fortunately, choosing $\rho_1 = \rho_2 = 1$, $J_1 = J_2 = \begin{bmatrix} 1 & 0 \end{bmatrix}^{\mathrm{T}}$ and using Theorem 8.1, sliding surface gain can be calculated as $K = 0.3571$ with $\mathscr{H}_\infty$ performance index $\gamma_{\min} = 5.7575$. Note that the PQLFs P_n can reduce to a common quadratic Lyapunov function (CQLF) as $P_n \equiv P = \begin{bmatrix} \mathrm{diag}\{P_1^h, P_1^v\} & \star \\ P_2 & P_3 \end{bmatrix}$. Using Theorem 8.1 based on a CQLF (also with the same slack variables), the sliding surface gain can be calculated as $K = 0.3556$ with $\mathscr{H}_\infty$ performance index $\gamma_{\min} = 6.0693$, which can be concluded that the PQLFs-based method brings less conservatism than the CQLF-based one.

In light of the simulation results, it is clearly shown that the developed control design approach is valid. The designed output feedback dynamic sliding mode controller can handle both matched and mismatched disturbances.

8.3 STATE FEEDBACK SMC DESIGN VIA 2-D T-S FUZZY MODELS

When the full system states of the discrete-time 2-D system states are available, this section will propose a state feedback dynamic SMC design method, and the corresponding stability and robust $\mathscr{H}_\infty$ performance analysis for the sliding motion will be also discussed.

8.3.1 STABILITY AND ROBUST PERFORMANCE ANALYSIS OF SLIDING MOTION THROUGH PQLFS

The sliding surface for the discrete-time fuzzy 2-D system (8.1) can be designed based on the control input and system state as follows,

$$s(i,j) = Kx(i,j) + u(i,j), \tag{8.60}$$

where $K \in \Re^{n_u \times n_x}$ is the sliding surface gain to be determined. It is also satisfied that $s(i,j) = 0$ for the ideal quasi sliding motion, and then one has $u(i,j) = -Kx(i,j)$.

Based on the sliding surface (8.60), the dynamics of the sliding motion can be described by the following singular piecewise affine 2-D system as,

$$\begin{cases} E\bar{x}^+(i,j) = (\bar{\mathscr{A}}_n + \Delta\bar{\mathscr{A}}_n)\bar{x}(i,j) + \bar{\mathrm{a}}_n + \Delta\bar{\mathrm{a}}_n + \bar{\mathscr{D}}_n w(i,j) \\ z(i,j) = \bar{\mathscr{L}}_n \bar{x}(i,j), n \in \mathscr{I}, \end{cases} \tag{8.61}$$

where

$$\begin{cases} E = \mathrm{diag}\{\mathbf{I}_{n_x}, \mathbf{0}_{n_u}\}, \\ \bar{x}^+(i,j) = \begin{bmatrix} x^+(i,j) \\ u^+(i,j) \end{bmatrix}, \bar{x}(i,j) = \begin{bmatrix} x(i,j) \\ u(i,j) \end{bmatrix}, \\ \bar{\mathscr{A}}_n + \Delta\bar{\mathscr{A}}_n = \begin{bmatrix} \mathscr{A}_n + \Delta\mathscr{A}_n & \mathscr{B}_n \\ -K & -\mathbf{I} \end{bmatrix}, \\ \bar{\mathrm{a}}_n + \Delta\bar{\mathrm{a}}_n = \begin{bmatrix} \mathrm{a}_n + \Delta\mathrm{a}_n \\ \mathbf{0} \end{bmatrix}, \bar{\mathscr{D}}_n = \begin{bmatrix} \mathscr{D}_n^{\mathrm{T}} & \mathbf{0} \end{bmatrix}^{\mathrm{T}}, \bar{\mathscr{L}}_n = \begin{bmatrix} \mathscr{L}_n & \mathscr{N}_n \end{bmatrix}. \end{cases} \tag{8.62}$$

The corresponding stability and robust $\mathscr{H}_\infty$ performance analysis results for sliding motion are given by the following Theorem 8.3.

Theorem 8.3. For a robust $\mathscr{H}_\infty$ performance γ, the sliding mode dynamical system (8.61) is asymptotically stable, if matrices $0 < P_{n1}^h = P_{n1}^{h\mathrm{T}} \in \Re^{n_x^h \times n_x^h}$, $0 < P_{n1}^v = P_{n1}^{v\mathrm{T}} \in \Re^{n_x^v \times n_x^v}$, $P_{n2} \in \Re^{n_u \times n_x}$, $P_{n3} \in \Re^{n_u \times n_u}$, $n \in \mathscr{I}$, $\bar{K} \in \Re^{n_u \times n_x}$, and $\{V_{n1}, W_{n1}\} \in \Re^{n_x \times n_x}$, $\{V_{n2}, W_{n2}\} \in \Re^{n_u \times n_x}$, $V_0 \in \Re^{n_u \times n_u}$, and scalars $\{\varepsilon_n > 0, n \in \mathscr{I}\}$, $\{\rho_n < 0, n \in \mathscr{I}_1\}$ exist, such that the following conditions hold,

$$\begin{bmatrix} \Psi_{11} & \star & \star & \star \\ \Psi_{21} - W_n^{\mathrm{T}} & P_k - V_n - V_n^{\mathrm{T}} & \star & \star \\ \Psi_{31} & \Psi_{32} & -\gamma^2\mathbf{I} & \star \\ \Psi_{41} & \Psi_{42} & \mathbf{0} & -\varepsilon_n\mathbf{I} \end{bmatrix} < 0, n \in \mathscr{I}_0, m \in \mathscr{N}(n), \tag{8.63}$$

$$\begin{bmatrix} \rho_n(q_n^{\mathrm{T}}q_n-1)+\varepsilon_n H_{m3}^{\mathrm{T}}H_{m3} & \star & \star & \star & \star \\ \rho_n\bar{Q}_n^{\mathrm{T}}q_n+\Psi_{10} & \rho_n\bar{Q}_n^{\mathrm{T}}\bar{Q}_n+\Psi_{11} & \star & \star & \star \\ \Psi_{20} & \Psi_{21}-W_n^{\mathrm{T}} & P_k-V_n-V_n^{\mathrm{T}} & \star & \star \\ \mathbf{0} & \Psi_{31} & \Psi_{32} & -\gamma^2\mathbf{I} & \star \\ \mathbf{0} & \Psi_{41} & \Psi_{42} & \mathbf{0} & -\varepsilon_n\mathbf{I} \end{bmatrix} < 0,$$
$$n\in\mathscr{I}_1, m\in\mathscr{N}(n) \tag{8.64}$$

where

$$\begin{cases} \Psi_{11}=\mathtt{Sym}\left\{\begin{bmatrix} W_{n1} \\ W_{n2}\end{bmatrix}\begin{bmatrix} A_m & B_m\end{bmatrix}-\begin{bmatrix} J_2 \\ \rho_2\mathbf{I}\end{bmatrix}\begin{bmatrix}\bar{K} & V_0\end{bmatrix}\right\}+\varepsilon_n\begin{bmatrix} H_{m2}^{\mathrm{T}} \\ \mathbf{0}\end{bmatrix}(\star) \\ \quad -E^{\mathrm{T}}P_nE+\bar{L}_m^{\mathrm{T}}\bar{L}_m, \\ \Psi_{21}=\begin{bmatrix} V_{n1} \\ V_{n2}\end{bmatrix}\begin{bmatrix} A_m & B_m\end{bmatrix}-\begin{bmatrix} J_1 \\ \rho_1\mathbf{I}\end{bmatrix}\begin{bmatrix}\bar{K} & V_0\end{bmatrix}, \\ \Psi_{31}=D_m^{\mathrm{T}}\begin{bmatrix} W_{n1}^{\mathrm{T}} & W_{n2}^{\mathrm{T}}\end{bmatrix}, \Psi_{32}=D_m^{\mathrm{T}}\begin{bmatrix} V_{n1}^{\mathrm{T}} & V_{n2}^{\mathrm{T}}\end{bmatrix}, \\ \Psi_{41}=H_{m1}^{\mathrm{T}}\begin{bmatrix} W_{n1}^{\mathrm{T}} & W_{n2}^{\mathrm{T}}\end{bmatrix}, \Psi_{42}=H_{m1}^{\mathrm{T}}\begin{bmatrix} V_{n1}^{\mathrm{T}} & V_{n2}^{\mathrm{T}}\end{bmatrix}, \\ \Psi_{10}=\begin{bmatrix} W_{n1} \\ W_{n2}\end{bmatrix}a_m+\varepsilon_n\begin{bmatrix} H_{m2}^{\mathrm{T}} \\ \mathbf{0}\end{bmatrix}H_{m3}, \end{cases}$$

$$\begin{cases} \Psi_{20}=\begin{bmatrix} V_{n1} \\ V_{n2}\end{bmatrix}a_m,\ P_n=\begin{bmatrix} \mathrm{diag}\{P_{n1}^h, P_{n1}^v\} & \star \\ P_{n2} & P_{n3}\end{bmatrix}, \\ V_n=\begin{bmatrix} V_{n1} & J_1V_0 \\ V_{n2} & \rho_1V_0\end{bmatrix}, W_n=\begin{bmatrix} W_{n1} & J_2V_0 \\ W_{n2} & \rho_2V_0\end{bmatrix}, \\ \bar{L}_m=\begin{bmatrix} L_m & N_m\end{bmatrix}, \bar{Q}_n=\begin{bmatrix} Q_n & \mathbf{0}_{1\times n_u}\end{bmatrix} \end{cases} \tag{8.65}$$

and $\{J_1, J_2\}\in\mathfrak{R}^{n_x\times n_u}$ are arbitrary matrices, and $\rho_1\neq 0$, ρ_2 are constant scalars.

Additionally, the gain of the sliding surface (8.60) can be calculated by

$$K=V_0^{-1}\bar{K}. \tag{8.66}$$

It is worth mentioning that in view of the full state feedback circumstance, the stability and robust $\mathscr{H}_\infty$ performance analysis results on the sliding motion can be derived by a similar approach in Theorem 8.1 in the framework of PQLFs, and the proof of Theorem 8.3 is then omitted. As discussed in Remark 8.2.2, the readers can also derive the stability and robust $\mathscr{H}_\infty$ performance analysis results for the sliding motion via a CQLF-based method by themselves.

8.3.2 STATE FEEDBACK DYNAMIC SLIDING MODE CONTROLLER DESIGN

Theorem 8.4. In view of the sliding surface designed as in (8.60), the following state feedback dynamic sliding mode controller can drive the system states of the system (8.1) into a vicinity of the sliding surface (8.60),

$$\begin{aligned} u(i+1,j+1)=u(i,j)+Kx(i,j)-\sum_{l=1}^{r}\mu_lK\left(A_lx(i,j)+a_l+B_lu(i,j)\right) \\ -(\sigma+\zeta(i,j))\mathtt{sgn}(s(i,j)), \end{aligned} \tag{8.67}$$

where $s(i,j) = Kx(i,j) + u(i,j)$ is given as in (8.60), and $\sigma > 0$ is a given scalar, and

$$\zeta(i,j) = \sum_{l=1}^{r} \mu_l \{\|KH_{l1}\| \cdot \|H_{l2}\| \cdot \|x(i,j)\| + \|KH_{l1}\| \cdot \|H_{l3}\| + \|KD_l\| \eta(i,j)\} \tag{8.68}$$

with $\eta(i,j)$ being the uniform upper-bound of the external disturbance $w(i,j)$.

Proof. Construct the following Lyapunov function

$$V(s(i,j),i,j) = \frac{1}{2} s^{\mathrm{T}}(i,j)s(i,j). \tag{8.69}$$

The incremental $\Delta V(s(i,j),i,j)$ is

$$\begin{aligned} \Delta V(s(i,j),i,j) &= V(s^+(i,j),i+1,j+1) - V(s(i,j),i,j) \\ &= s^{\mathrm{T}}(i,j)\Delta s(i,j) + \Gamma, \end{aligned} \tag{8.70}$$

where $\Delta s(i,j) = s^+(i,j) - s(i,j)$, $s^+(i,j) = Kx^+(i,j) + u(i+1,j+1)$, $\Gamma = \frac{1}{2}\Delta s^{\mathrm{T}}(i,j)\Delta s(i,j)$.

Substitute the state feedback dynamic sliding mode controller (8.67) into (8.70), and then yields

$$\begin{aligned} &\Delta V(s(i,j),i,j) \\ &= s^{\mathrm{T}}(i,j)\{Kx^+(i,j) - Kx(i,j) + u(i+1,j+1) - u(i,j)\} + \Gamma \\ &= s^{\mathrm{T}}(i,j)\Big\{\sum_{l=1}^{r} \mu_l K[(A_l + \Delta A_l)x(i,j) + a_l + \Delta a_l + B_l u(i,j) \\ &\qquad + D_l w(i,j)] - Kx(i,j) + u(i+1,j+1) - u(i,j)\Big\} + \Gamma \\ &= s^{\mathrm{T}}(i,j)\sum_{l=1}^{r} \mu_l K\Big\{\Delta A_l x(i,j) + \Delta a_l + D_l w(i,j)\Big\} - \zeta(i,j)\|s(i,j)\| \\ &\quad - \sigma\|s(i,j)\| + \Gamma \\ &\le \|s(i,j)\|\sum_{l=1}^{r} \mu_l \Big\{\|K\Delta A_l\| \cdot \|x(i,j)\| + \|K\Delta a_l\| + \|KD_l\|\eta(i,j)\Big\} \\ &\quad - \zeta(i,j)\|s(i,j)\| - \sigma\|s(i,j)\| + \Gamma \\ &\le \|s(i,j)\|\sum_{l=1}^{r} \mu_l \Big\{\|KH_{l1}\| \cdot \|H_{l2}\| \cdot \|x(i,j) + \|KH_{l1}\| \cdot \|H_{l3}\| \\ &\qquad + \|KD_l\|\eta(i,j)\Big\} - \zeta(i,j)\|s(i,j)\| - \sigma\|s(i,j)\| + \Gamma. \end{aligned} \tag{8.71}$$

Noticing the parameter $\zeta(i,j)$ designed as in (8.68), $\Delta V(s(i,j),i,j) \le -\sigma\|s(i,j)\| + \Gamma$ can be obtained. By tuning the parameter σ, $\Delta V(s(i,j),i,j) < 0$ can be guaranteed, which indicates that the states of the fuzzy 2-D system (8.1) can be driven into a neighbourhood of the sliding surface (8.60) and maintained hereafter by the sliding mode controller (8.67). The proof is completed. ■

8.3.3 ILLUSTRATIVE EXAMPLES

Example 8.3. Consider a nonlinear water stream heating, gas absorption, and air drying process described as in (8.51). By conducting a similar fuzzy modelling process in Example 8.1 and considering the external disturbance, the nonlinear 2-D process (8.51) can be described by the following 2-D T-S fuzzy affine model as,

Plant Rule $\mathscr{R}^l$: IF $W_2(i,j)$ is $\mathscr{F}_{l1}$, THEN

$$\begin{cases} W^+(i,j) = (A_l + \Delta A_l)W(i,j) + a_l + \Delta a_l + B_l u(i,j) + D_l w(i,j) \\ z(i,j) = L_l W(i,j) + N_l u(i,j),\ l \in \{1,2,3\}, \end{cases} \tag{8.72}$$

where the system matrices are given as in (8.54), and the parameter uncertainties ΔA_l and Δa_l are given in the form of (8.2) with the matrices H_{l1}, H_{l2}, and H_{l3} given as in (8.55).

The fuzzy MFs are shown as in Figure 8.1 with $d_1 = \frac{\pi}{3}$ and $d_2 = \frac{\pi}{2}$. These three subspaces can be precisely described via degenerate ellipsoids as in (8.17) with the matrices Q_i and q_i given as in (8.57). The purpose of this example is to design a sliding surface (8.60) and a state feedback dynamic sliding mode controller (8.67) to drive the system states into a vicinity of the sliding surface and guarantee the asymptotical stability of the sliding motion with robust $\mathscr{H}_\infty$ performance γ. Using Theorem 8.3 with $\rho_1 = 1$, $\rho_2 = 1$, $J_1 = J_2 = \begin{bmatrix} 1 & 0 \end{bmatrix}^{\mathrm{T}}$, one can obtain the $\mathscr{H}_\infty$ performance index $\gamma_{\min} = 0.6250$ with the sliding surface gain $K = \begin{bmatrix} 1.6541 & 0.3681 \end{bmatrix}$.

To illustrate the effectiveness of the proposed method, simulations are conducted. With boundary condition $x(0,j) = \begin{bmatrix} 0 & -0.3 \end{bmatrix}^{\mathrm{T}}, 0 \le j \le 10, x(i,0) = \begin{bmatrix} 0.2 & 0 \end{bmatrix}^{\mathrm{T}}$, $0 \le i \le 10$, $x(i,0) = x(0,j) = \begin{bmatrix} 0 & 0 \end{bmatrix}^{\mathrm{T}}$, $i > 10$, $j > 10$, external disturbance $w(i,j) = 50\mathrm{e}^{-0.1i}\sin(0.5i) + 5\mathrm{e}^{-2j}\sin(0.5j)$, using the state feedback dynamic sliding mode controller (8.67) with $\sigma = 0.5$, the system state trajectories and control input are shown in Figure 8.7, Figure 8.8 and Figure 8.9, respectively.

Example 8.4. Consider the following doubly-indexed nonlinear process [127], which is utilized to characterize the thermal processes in many practical applications,

$$\frac{\partial T(s,t)}{\partial s} = -\frac{\partial T(s,t)}{\partial t} - f(s,t)T(s,t) + c_0 u(s,t) + c_1 w(s,t), \tag{8.73}$$

where $c_0 = 2$ and $c_1 = 1$ are constant; $f(s,t) = \cos(T(s,t))$ is a nonlinear function along space $s \in [0, s_f]$ and time $t \in [0, \infty)$; $T(s,t)$ is the temperature; $w(s,t)$ is the external disturbance; $u(s,t)$ is the control input.

Denote $T(i,j) = T(i\Delta s, j\Delta t)$, $w(i,j) = w(i\Delta s, j\Delta t)$ and $u(i,j) = u(i\Delta s, j\Delta t)$. Thus, one has

$$\begin{cases} \dfrac{\partial T(i,j)}{\partial s} = \dfrac{T(i,j) - T(i-1,j)}{\Delta s}, \\ \dfrac{\partial T(i,j)}{\partial t} = \dfrac{T(i,j+1) - T(i,j)}{\Delta t}. \end{cases} \tag{8.74}$$

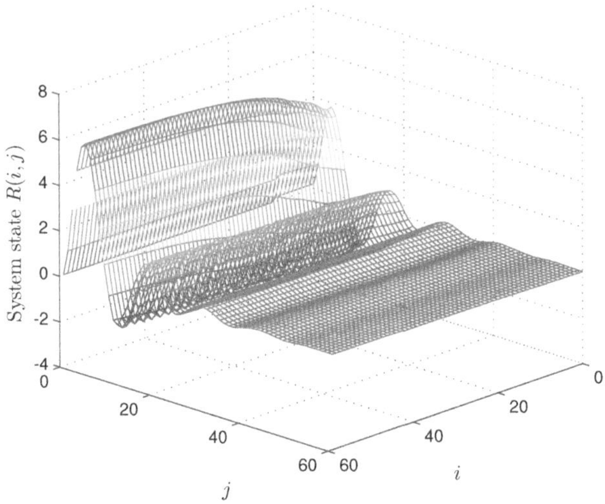

Figure 8.7 State response of $R(i, j)$ with dynamic sliding mode controller (8.67) in Example 8.3.

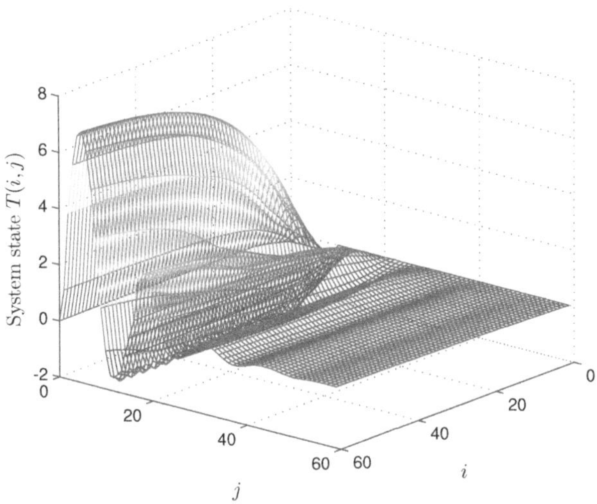

Figure 8.8 State response of $T(i, j)$ with dynamic sliding mode controller (8.67) in Example 8.3.

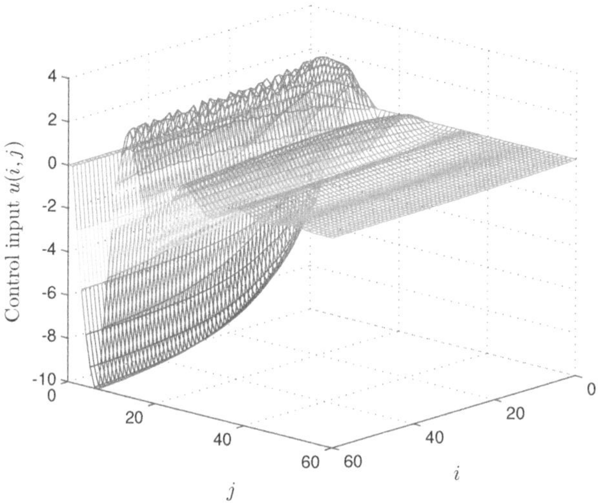

Figure 8.9 Response of the control input with dynamic sliding mode controller (8.67) in Example 8.3.

Define $x(i,j) = \begin{bmatrix} x_1^{\mathrm{T}}(i,j) & x_2^{\mathrm{T}}(i,j) \end{bmatrix}^{\mathrm{T}}$ with $x_1(i,j) = T(i-1,j)$, $x_2(i,j) = T(i,j)$. Then rewrite the nonlinear process (8.73) as

$$\begin{aligned} x^+(i,j) = &\begin{bmatrix} 0 & 1 \\ \frac{\Delta t}{\Delta s} & 1 - \frac{\Delta t}{\Delta s} - \Delta t f(s,t) \end{bmatrix} x(i,j) + \begin{bmatrix} 0 \\ c_0 \Delta t \end{bmatrix} u(i,j) \\ &+ \begin{bmatrix} 0 \\ c_1 \Delta t \end{bmatrix} w(i,j), \end{aligned} \tag{8.75}$$

where $x^+(i,j) = \begin{bmatrix} x_1^{\mathrm{T}}(i+1,j) & x_2^{\mathrm{T}}(i,j+1) \end{bmatrix}^{\mathrm{T}}$.

The regulated output is given as $z(i,j) = \begin{bmatrix} 0 & 0.5 \end{bmatrix} x(i,j) + 0.5u(i,j)$. Linearize the system (8.75) around $(0,0)^{\mathrm{T}}$ and $\left(0, \pm\frac{\pi}{3}\right)^{\mathrm{T}}$. With sampling periods $\Delta t = 0.5$, $\Delta s = 1$, respectively, for the time and space, the system (8.75) can be described via the discrete-time 2-D T-S fuzzy affine model as,

Plant Rule $\mathscr{R}^l$: IF $x_2(i,j)$ is $\mathscr{F}_{l1}$, THEN

$$\begin{cases} x^+(i,j) = (A_l + \Delta A_l)x(i,j) + a_l + \Delta a_l + B_l u(i,j) + D_l w(i,j) \\ z(i,j) = L_l x(i,j) + N_l u(i,j), \end{cases} \tag{8.76}$$

where

$$\begin{cases} A_1 = A_3 = \begin{bmatrix} 0 & 1 \\ 0.5 & 0.7034 \end{bmatrix}, A_2 = \begin{bmatrix} 0 & 1 \\ 0.5 & 0 \end{bmatrix}, \\ a_1 = -a_3 = \begin{bmatrix} 0.4749 \\ 0 \end{bmatrix}, a_2 = \begin{bmatrix} 0 \\ 0 \end{bmatrix}, B_l = \begin{bmatrix} 0 \\ 1 \end{bmatrix}, \\ D_l = \begin{bmatrix} 0 \\ 0.5 \end{bmatrix}, L_l = \begin{bmatrix} 0 & 0.5 \end{bmatrix}, N_l = 0.5,\ l \in \{1,2,3\} \end{cases} \tag{8.77}$$

and the uncertainty terms ΔA_l and Δa_l are given in the form of (8.2) with

$$\begin{cases} \Delta_l(i,j) = \sin(i)\cos(j), \\ H_{l1} = \begin{bmatrix} 0 \\ 0.1 \end{bmatrix}, H_{l2} = \begin{bmatrix} 0 & 0.1 \end{bmatrix}, H_{l3} = 0.1,\ l \in \{1,2,3\}. \end{cases} \tag{8.78}$$

The MFs are also shown as in Figure 8.1 with $d_1 = \frac{\pi}{6}$ and $d_2 = \frac{\pi}{3}$. Using Theorem 8.3 with $\rho_1 = 1$, $\rho_2 = 3$, $J_1 = J_2 = \begin{bmatrix} 0 & 1 \end{bmatrix}^{\mathrm{T}}$, one can obtain the $\mathscr{H}_\infty$ performance index $\gamma_{\min} = 0.9367$ with the sliding surface gain $K = \begin{bmatrix} 0.2810 & 0.1950 \end{bmatrix}$.

To show the effectiveness of the proposed method, simulations are conducted. For boundary condition $x(0,j) = \begin{bmatrix} 0 & 1+\cos(0.4j) \end{bmatrix}^{\mathrm{T}}$, $0 \le j \le 60$, $x(i,0) = \begin{bmatrix} \sin(0.2i) & 0 \end{bmatrix}^{\mathrm{T}}$, $0 \le i \le 60$, and nonlinear disturbance function $w(i,j) = 5\mathrm{e}^{-0.1i}\sin(0.5i) + 5\mathrm{e}^{-2j}\sin(0.5j)$, using the state feedback dynamic sliding mode controller (8.67) with $\sigma = 0.2$, the system states and the control input are shown in Figure 8.10 and Figure 8.11, respectively.

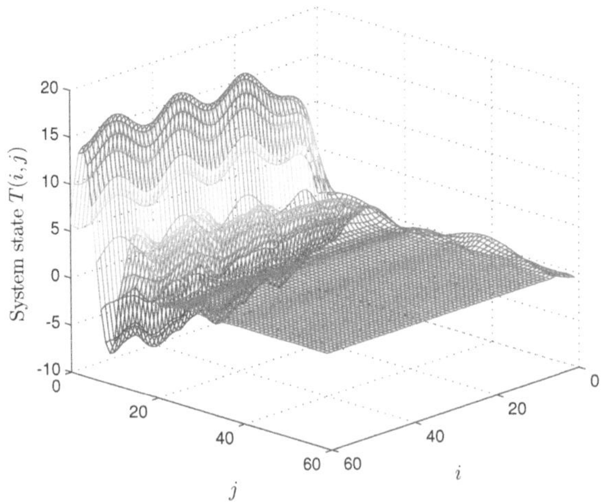

Figure 8.10 State response of $T(i,j)$ with dynamic sliding mode controller (8.67) in Example 8.4.

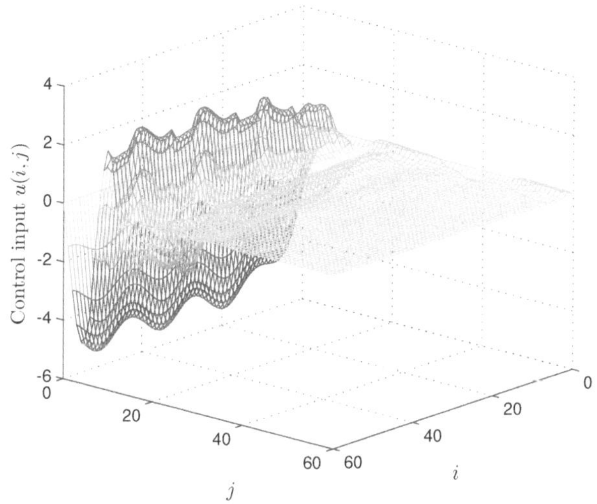

Figure 8.11 Response of the control input with dynamic sliding mode controller (8.67) in Example 8.4.

8.4 CONCLUSION

This chapter investigates the fuzzy SMC of discrete-time nonlinear 2-D through T-S fuzzy affine models. By a 2-D PQLFs-based approach, new existence conditions for the sliding surface are proposed. It is noted that due to the application of the singular piecewise affine 2-D system method and the introduction of auxiliary matrix multipliers, the sliding motion stability and robust performance analysis conservatism has been also further reduced. A state feedback dynamic sliding mode controller and an output feedback dynamic sliding mode controller are designed to drive the 2-D system states into a neighbourhood of the sliding surface. Simulation studies are given to illustrate the effectiveness of the proposed approaches.

References

1. L. A. Zadeh, Outline of a new approach to analysis of complex systems and decision processes, *IEEE Transactions on Systems, Man, and Cybernetics*, vol. SMC-3, no. 1, pp. 28-44, Jan. 1973.
2. M. Sugeno, *Industrial Applications of Fuzzy Control*. New York: Elsevier, 1985.
3. K. Tanaka and M. Sano, A robust stabilization problem of fuzzy control systems and its application to backing up control of a truck-trailer, *IEEE Transactions on Fuzzy Systems*, vol. 2, no. 2, pp. 119-134, May. 1994.
4. T. Tanaka and H. O. Wang, *Fuzzy Control Systems Design and Analysis: A Linear Matrix Inequality Approach*. New York: Wiley, 2001.
5. Y. Bai, H. Zhuang, and Z. S. Roth, Fuzzy logic control to suppress noises and coupling effects in a laser tracking system, *IEEE Transactions on Control System Technology*, vol. 13, no. 1, pp. 113-121, Jan. 2005.
6. S. G. Cao, N. W. Rees, and G. Feng, Stability analysis of fuzzy control systems, *IEEE Transactions on Systems, Man, and Cybernetics, Part B: Cybernetics*, vol. 26, no. 1, pp. 201-204, Feb. 1996.
7. G. Feng and J. Ma, Quadratic stabilization of uncertain discrete-time fuzzy dynamic systems, *IEEE Transactions on Circuits and Systems I: Fundamental Theory and Applications*, vol. 48, no. 11, pp. 1337-1343, Nov. 2001.
8. S.-S. Chen, Y.-C. Chang, S.-F. Su, S.-L. Chung, and T.-T. Lee, Robust static output-feedback stabilization for nonlinear discrete-time systems with time delay via fuzzy control approach, *IEEE Transactions on Fuzzy Systems*, vol. 13, no. 2, pp. 263-272, Apr. 2005.
9. C. L. P. Chen, Y.-J. Liu, and G.-X. Wen, Fuzzy neural network-based adaptive control for a class of uncertain nonlinear stochastic systems, *IEEE Transactions on Cybernetics*, vol. 44, no. 5, pp. 583-593, May 2014.
10. C. L. P. Chen, C.-Y. Zhang, L. Chen, and M. Gan, Fuzzy restricted Boltzmann machine for the enhancement of deep learning, *IEEE Transactions on Fuzzy Systems*, vol. 23, no. 6, pp. 2163-2173, Dec. 2015.
11. G. Feng, Controller design and analysis of uncertain piecewise-linear systems, *IEEE Transactions on Circuits and Systems I: Fundamental Theory and Applications*, vol. 49, no. 2, pp. 224-232, Feb. 2002.
12. W.-J. Chang and C.-C. Sun, Constrained fuzzy controller design of discrete Takagi-Sugeno fuzzy models, *Fuzzy Sets and Systems*, vol. 133, no. 1, pp. 37-55, Jan. 2003.
13. B. Chen and X. Liu, Delay-dependent robust $\mathscr{H}_\infty$ control for T-S fuzzy systems with time delay, *IEEE Transactions on Fuzzy Systems*, vol. 13, no. 4, pp. 544-556, Aug. 2005.
14. C.-L. Chen, G. Feng, D. Sun, and X.-P. Guan, $\mathscr{H}_\infty$ output feedback control of discrete-time fuzzy systems with application to chaos control, *IEEE Transactions on Fuzzy Systems*, vol. 13, no. 4, pp. 531-543, Aug. 2005.
15. G. Feng, Nonsynchronized state estimation of discrete time piecewise linear systems, *IEEE Transactions on Signal Processing*, vol. 54, no. 1, pp. 295-303, Jan. 2006.
16. G. Feng, A survey on analysis and design of model-based fuzzy control systems, *IEEE Transactions on Fuzzy Systems*, vol. 14, no. 5, pp. 676-697, Oct. 2006.

17. C.-H. Fang, Y.-S. Liu, S.-W. Kau, L. Hong, and C.-H. Lee, A new LMI-based approach to relaxed quadratic stabilization of T-S fuzzy control systems, *IEEE Transactions on Fuzzy Systems*, vol. 14, no. 3, pp. 386-397, Jun. 2006.
18. L. Wu and W. Zheng, $\mathscr{L}_2$-$\mathscr{L}_\infty$ control of nonlinear fuzzy Itô stochastic delay systems via dynamic output feedback, *IEEE Transactions on Systems, Man, and Cybernetics, Part B: Cybernetics*, vol. 39, no. 5, pp. 1308-1315, Mar. 2009.
19. J. Qiu, G. Feng, and H. Gao, Asynchronous output-feedback control of networked nonlinear systems with multiple packet dropouts: T-S fuzzy affine model-based approach, *IEEE Transactions on Fuzzy Systems*, vol. 19, no. 6, pp. 1014-1030, Dec. 2011.
20. M. Chadli and T. M. Guerra, LMI solution for robust static output feedback control of discrete Takagi-Sugeno fuzzy models, *IEEE Transactions on Fuzzy Systems*, vol. 20, No. 6, pp. 1160-1165, Dec. 2012.
21. Y.-C. Hsueh and S.-F. Su, Learning error feedback design of direct adaptive fuzzy control systems, *IEEE Transactions on Fuzzy Systems*, vol. 20, no. 3, pp. 536-545, Jun. 2012.
22. M. Chadli and H. R. Karimi, Robust observer design for unknown inputs Takagi-Sugeno models, *IEEE Transactions on Fuzzy Systems*, vol. 21, no. 1, pp. 158-164, Feb. 2013.
23. H. K. Lam, C. Liu, L. Wu, and X. Zhao, Polynomial fuzzy-model-based control systems: Stability analysis via approximated membership functions considering sector nonlinearity of control input, *IEEE Transactions on Fuzzy Systems*, vol. 23, no. 6, pp. 2202-2214, Dec. 2015.
24. J. Qiu, H. Gao, and S. X. Ding, Recent advances on fuzzy-model-based nonlinear networked control systems: A survey, *IEEE Transactions on Industrial Electronics*, vol. 63, no. 2, pp. 1207-1217, Nov. 2016.
25. J. Qiu, G. Feng, and H. Gao, Observer-based piecewise affine output feedback controller synthesis of continuous-time T-S fuzzy affine dynamic systems using quantized measurements, *IEEE Transactions on Fuzzy Systems*, vol. 20, no. 6, pp. 1046-1062, Dec. 2012.
26. F. Cuesta, F. Gordillo, J. Aracil, and A. Ollero, Stability analysis of nonlinear multivariable Takagi-Sugeno fuzzy control systems, *IEEE Transactions on Fuzzy Systems*, vol. 7, no. 5, pp. 508-520, Oct. 1999.
27. G. Feng, Robust $\mathscr{H}_\infty$ filtering of fuzzy dynamic systems, *IEEE Transactions on Aerospace and Electronic Systems*, vol. 41, no. 2, pp. 658-670, Apr. 2005.
28. A. Chatterjee, R. Chatterjee, F. Matsuno, and T. Endo, Augmented stable fuzzy control for flexible robotic arm using LMI approach and neuro-fuzzy state space modeling, *IEEE Transactions on Industrial Electronics*, vol. 55, no. 3, pp. 1256-1270, Mar. 2008.
29. F. Delmotte, T. M. Guerra, and A. Kruszewski, Discrete Takagi-Sugeno's fuzzy models: Reduction of the number of LMI in fuzzy control techniques, *IEEE Transactions on Systems, Man, and Cybernetics, Part B: Cybernetics*, vol. 38, no. 5, pp. 1423-1427, Oct. 2008.
30. H. Gao, X. Liu, and J. Lam, Stability analysis and stabilization for discrete-time fuzzy systems with time-varying delay, *IEEE Transactions on Systems, Man, and Cybernetics, Part B: Cybernetics*, vol. 39, no. 2, pp. 306-317, Apr. 2009.
31. H. Gao, Y. Zhao, and T. Chen, $\mathscr{H}_\infty$ fuzzy control of nonlinear systems under unreliable communication links, *IEEE Transactions on Fuzzy Systems*, vol. 17, no. 2, pp. 265-278, Apr. 2009.

32. H. Gao, Y. Zhao, J. Lam, and K. Chen, $\mathscr{H}_\infty$ fuzzy filtering of nonlinear systems with intermittent measurements, *IEEE Transactions on Fuzzy Systems*, vol. 17, no. 2, pp. 291-300, Apr. 2009.
33. L. Wu and D. W. C. Ho, Fuzzy filter design for Itô stochastic systems with application to sensor fault detection, *IEEE Transactions on Fuzzy Systems*, vol. 17, no. 1, pp. 233-242, Feb. 2009.
34. H. Du and N. Zhang, Fuzzy control for nonlinear uncertain electrohydraulic active suspensions with input constraint, *IEEE Transactions on Fuzzy Systems*, vol. 17, no. 2, pp. 343-356, Apr. 2009.
35. B. Ding, Stabilization of Takagi-Sugeno model via nonparallel distributed compensation law, *IEEE Transactions on Fuzzy Systems*, vol. 18, no. 6, pp. 188-194, Feb. 2010.
36. B. Ding, Homogeneous polynomially nonquadratic stabilization of discrete-time Takagi-Sugeno systems via nonparallel distributed compensation law, *IEEE Transaction on Fuzzy Systems*, vol. 18, no. 5, pp. 994-1000, Oct. 2010.
37. H. Dong, Z. Wang, D. W. C. Ho, and H. Gao, Robust $\mathscr{H}_\infty$ fuzzy output-feedback control with multiple probabilistic delays and multiple missing measurements, *IEEE Transactions on Fuzzy Systems*, vol. 18, no. 4, pp. 712-725, Aug. 2010.
38. R.-E. Precup, R.-C. David, E. M. Petriu, S. Preitl, and M.-B. Radac, Novel adaptive gravitational search algorithm for fuzzy controlled servo systems, *IEEE Transactions on Industrial Informatics*, vol. 8, no. 4, pp. 791-800, Nov. 2012.
39. M. Liu, X. Ca, and P. Shi, Fuzzy model-based fault-tolerant design for nonlinear stochastic systems against simultaneous sensor and actuator faults, *IEEE Transactions on Fuzzy Systems*, vol. 21, no. 5, pp. 789-799, Oct. 2013.
40. J. Qiu, G. Feng, and H. Gao, Static-output-feedback $\mathscr{H}_\infty$ control of continuous-time T-S fuzzy affine systems via piecewise Lyapunov functions, *IEEE Transactions on Fuzzy Systems*, vol. 21, no. 2, pp. 245-261, Apr. 2013.
41. S. F. Derakhshan, A. Fatehi, and M. G. Sharabiany, Nonmonotonic observer-based fuzzy controller designs for discrete time T-S fuzzy systems via LMI, *IEEE Transactions on Cybernetics*, vol. 44, no. 12, pp. 2557-2567, Dec. 2014.
42. C.-E. Ren, L. Chen, and C. L. P. Chen, Adaptive fuzzy leader-following consensus control for stochastic multiagent systems with heterogeneous nonlinear dynamics, *IEEE Transactions on Fuzzy Systems*, vol. 25, no. 1, pp. 181-190, Feb. 2017.
43. H. D. Choi, C. K. Ahn, P. Shi, L. Wu, and M. T. Lim, Dynamic output-feedback dissipative control for T-S fuzzy systems with time-varying input delay and output constraints, *IEEE Transactions on Fuzzy Systems*, vol. 25, no. 3, pp. 511-526, Jun. 2017.
44. A. Chibani, M. Chadli, P. Shi, and N. B. Braiek, Fuzzy fault detection filter design for T-S fuzzy systems in the finite-frequency domain, *IEEE Transactions on Fuzzy Systems*, vol. 25, no. 5, pp. 1051-1061, Oct. 2017.
45. H. K. Lam, A review on stability analysis of continuous-time fuzzy-model-based control systems: From membership-function-independent to membership-function-dependent analysis, *Engineering Applications of Artificial Intelligence*, vol. 67, pp. 390-408, Jan. 2018.
46. N. Wang, S.-F. Su, J. Yin, Z. Zheng, and M. J. Er, Global asymptotic model-free trajectory-independent tracking control of an uncertain marine vehicle: An adaptive universe-based fuzzy control approach, *IEEE Transactions on Fuzzy Systems*, vol. 26, no. 3, pp. 1613-1625, Jun. 2018.
47. J. Xia, G. Chen, J. H. Park, H. Shen, and G. Zhuang, Dissipativity-based sampled-data control for fuzzy switched Markovian jump systems, *IEEE Transactions on Fuzzy Systems*, vol. 29, no. 6, pp. 1325-1339, Jun. 2021.

48. C. Fantuzzi and R. Rovatti, On the approximation capabilities of the homogeneous Takagi-Sugeno model, *in Proceedings of IEEE 5th International Fuzzy Systems*, New Orleans, LA, Sep. 1996, pp. 1067-1072.
49. H. Liu, F. Sun, and Y. Hu, $\mathscr{H}_\infty$ control for fuzzy singularly perturbed systems, *Fuzzy Sets and Systems*, vol. 155, no.2, pp. 272-291, Oct. 2005.
50. C. Lin, Q.-G. Wang, T. H. Lee, and Y. He, Design of observer-based control for fuzzy time-delay systems, *IEEE Transactions on Fuzzy Systems*, vol. 16, no. 2, pp. 534-543, Apr. 2008.
51. S. He and F. Liu, Finite-time $\mathscr{H}_\infty$ fuzzy control of nonlinear jump systems with time delays via dynamic observer-based state feedback, *IEEE Transactions on Fuzzy Systems*, vol. 20, no. 4, pp. 605-614, Aug. 2012.
52. G. Feng, Stability analysis of discrete-time fuzzy dynamic systems based on piecewise Lyapunov functions, *IEEE Transactions on Fuzzy Systems*, vol. 12, no. 1, pp. 22-28, Feb. 2004.
53. G. Feng and D. Sun, Generalized $\mathscr{H}_\infty$ controller synthesis of fuzzy dynamic systems based on piecewise Lyapunov functions, *IEEE Transactions on Circuits and Systems I: Fundamental Theory and Applications*, vol. 49, no. 12, pp. 1843-1850, Dec. 2002.
54. G. Feng, Controller synthesis of fuzzy dynamic systems based on piecewise Lyapunov functions, *IEEE Transactions on Fuzzy Systems*, vol. 11, no. 10, pp. 605-612, Oct. 2003.
55. G. Feng and T. Zhang, Output regulation of discrete-time piecewise-linear systems with application to controlling Chaos, *IEEE Transactions on Sircuits and Systems II: Express Briefs*, vol. 53, no. 4, pp. 249-253, Apr. 2006.
56. H. Zhang, C. Li, and X. Liao, Stability analysis and $\mathscr{H}_\infty$ controller design of fuzzy large-scale systems based on piecewise Lyapunov functions, *IEEE Transactions on Systems, Man, and Cybernetics-Part B: Cybernetics*, vol. 36, no. 3, pp. 685-698, Jun. 2006.
57. H. Zhang and G. Feng, Stability analysis and $\mathscr{H}_\infty$ controller design of discrete-time fuzzy large-scale systems based on piecewise Lyapunov functions, *IEEE Transactions on Systems, Man, and Cybernetics-Part B: Cybernetics*, vol. 38, no. 5, pp. 1390-1401, Oct. 2008.
58. J. Qiu, G. Feng, and H. Gao, Fuzzy-model-based piecewise $\mathscr{H}_\infty$ static-output-feedback controller design for networked nonlinear systems, *IEEE Transactions on Fuzzy Systems*, vol. 18, no. 5, pp. 919-934, Oct. 2010.
59. K. Tanaka, H. Ohtake, and H. O. Wang, A descriptor system approach to fuzzy control system design via fuzzy Lyapunov functions, *IEEE Transactions on Fuzzy Systems*, vol. 15, no. 3, pp. 333-341, Jun. 2007.
60. S. H. Kim, Relaxation technique for a T-S fuzzy control design based on a continuous-time fuzzy weighting-dependent Lyapunov function, *IEEE Transactions on Fuzzy Systems*, vol. 21, no. 4, pp. 761-766, Aug. 2013.
61. Y. Liu, F. Wu, and X. Ban, Dynamic output feedback control for continuous-time T-S fuzzy systems using fuzzy Lyapunov functions, *IEEE Transactions on Fuzzy Systems*, vol. 25, no. 5, pp. 1155-1167, Oct. 2017.
62. V. C. S. Campos, F. O. Souza, L. A. B. Tôrres, and R. M. Palhares, New stability conditions based on piecewise fuzzy Lyapunov functions and tensor product transformations, *IEEE Transactions on Fuzzy Systems*, vol. 21, no. 4, pp. 748-760, Aug. 2013.
63. G. Feng, Stability analysis of piecewise discrete-time linear systems, *IEEE Transactions on Automatic Control*, vol. 47, no. 7, pp. 1008-1112, Jul. 2002.

64. Y.-J. Chen, H. Ohtake, K. Tanaka, W.-J. Wang, and H. O. Wang, Relaxed stabilization criterion for T-S fuzzy systems by minimum-type piecewise-Lyapunov-function-based switching fuzzy controller, *IEEE Transactions on Fuzzy Systems*, vol. 20, no. 6, pp. 1166-1173, Dec. 2012.
65. V. I. Utkin, Variable structure systems with sliding modes, *IEEE Transactions on Automatic Control*, vol. AC-22, no. 2, pp. 212-222, Apr. 1977.
66. V. I. Utkin, *Sliding Modes in Control Optimisation*. Berlin, Germany: Springer-Verlag Press, 1992.
67. W. Gao, Y. Wang, and A. Homaifa, Discrete-time variable structure control systems, *IEEE Transactions on Industrial Electronics*, vol. 42, no. 2, pp. 117-122, Apr. 1995.
68. P. Shi, Y. Xia, G. P. Liu, and D. Rees, On designing of sliding-mode control for stochastic jump systems, *IEEE Transactions on Automatic Control*, vol. 51, no. 1, pp. 97-103, Jan. 2006.
69. X. Yu and O. Kaynak, Sliding-mode control with soft computing: A survey, *IEEE Transactions on Industrial Electronics*, vol. 56, no. 9, pp. 3275-3285, Sep. 2009.
70. J. Hu, Z. Wang, H. Gao, and L. K. Stergioulas, Robust sliding mode control for discrete stochastic systems with mixed time delays, randomly occurring uncertainties, and randomly occurring nonlinearities, *IEEE Transactions on Industrial Electronics*, vol. 59, no. 7, pp. 3008-3015, Jul. 2012.
71. P. Shi, M. Liu, and L. Zhang, Fault-tolerant sliding mode observer synthesis of Markovian jump systems using quantized measurements, *IEEE Transactions on Industrial Electronics*, vol. 62, no. 9, pp. 5910-5918, Sep. 2015.
72. J. Zhang and Y. Xia, Design of static output feedback sliding mode control for uncertain linear systems, *IEEE Transactions on Industrial Electronics*, vol. 57, no. 6, pp. 2161-2170, Jun. 2010.
73. X. Yu, Z. Man, and B. Wu, Design of fuzzy sliding-mode control systems, *Fuzzy Sets and Systems*, vol. 95, no. 3, pp. 295-306, May. 1998.
74. C. Lin, Q.-G. Wang, and T. H. Lee, Stabilization of uncertain fuzzy time-delay systems via variable structure control approach, *IEEE Transactions on Fuzzy Systems*, vol. 13, no. 6, pp. 787-798, Dec. 2005.
75. D. W. C. Ho and Y. Niu, Robust fuzzy design for nonlinear uncertain stochastic systems via sliding-mode control, *IEEE Transactions on Fuzzy Systems*, vol. 15, no. 3, pp. 350-358, Jun. 2007.
76. Z. Xi, G. Feng, and T. Hesketh , Piecewise sliding-mode control for T-S fuzzy systems, *IEEE Transactions on Fuzzy Systems*, vol. 19, no. 4, pp. 707-716, Feb. 2011.
77. Q. Gao, G. Feng, Z. Xi, Y. Wang, and J. Qiu, Robust $\mathscr{H}_\infty$ control of T-S fuzzy time-delay systems via a new sliding-mode control scheme, *IEEE Transactions on Fuzzy Systems*, vol. 22, no. 2, pp. 459-465, Apr. 2014.
78. Q. Gao, G. Feng, Z. Xi, Y. Wang, and J. Qiu, A new design of robust $\mathscr{H}_\infty$ sliding mode control for uncertain stochastic T-S fuzzy time-delay systems, *IEEE Transactions on Cybernetics*, vol. 44, no. 9, pp. 1556-1566, Sep. 2014.
79. J. Zhang, Y. Lin, and G. Feng, Analysis and synthesis of memory-based fuzzy sliding mode controllers, *IEEE Transactions on Cybernetics*, vol. 45, no. 12, pp. 2880-2889, Dec. 2015.
80. B. Pang and Q. Zhang, Sliding mode control for polynomial fuzzy singular systems with time delay, *IET Control Theory and Applications*, vol. 12, no. 10, pp. 1483-1490, Jun. 2018.
81. H. Li, J. Wang, Q. Zhou, and H. Du, Adaptive sliding mode control for interval type-2 fuzzy systems, *IEEE Transactions on Systems, Man, and Cybernetics: Systems*, vol. 46, no. 12, pp. 1654-1663, Dec. 2016.

82. S. Wen, M. Z. Q. Chen, Z. Zeng, X. Yu, and T. Huang, Fuzzy control for uncertain vehicle active suspension systems via dynamic sliding-mode approach, *IEEE Transactions on Systems, Man, and Cybernetics: Systems*, vol. 47, no. 1, pp. 24-32, Jan. 2017.
83. X. Sun and Q. Zhang, Observer-based adaptive sliding mode control for T-S fuzzy singular systems, *IEEE Transactions on Systems, Man, and Cybernetics: Systems*, vol. 50, no. 11, pp. 4438-4446, Nov. 2020.
84. B. Pang and Q. Zhang, Sliding mode control for polynomial fuzzy singular systems with time delay, *IET Control Theorem and Applications*, vol. 12, no. 10, pp. 1483-1490, Jun. 2018.
85. Q. Gao, G. Feng, L. Liu, J. Qiu, and Y. Wang, Robust $\mathscr{H}_\infty$ control for stochastic T-S fuzzy systems via integral sliding-mode approach, *IEEE Transactions on Fuzzy Systems*, vol. 22, no. 4, pp. 870-881, Aug. 2014.
86. S. Janardhanan and B. Bandyopadhyay, Multirate output feedback based robust quasi-sliding mode control of discrete-time systems, *IEEE Transactions on Automatic Control*, vol. 52, no. 3, pp. 499-503, Mar. 2007.
87. H. Choi, Variable structure output feedback control design for a class of uncertain dynamic systems, *Automatica*, vol. 38, no. 2, pp. 335-341, Feb. 2002.
88. J. Zhang, G. Feng, and Y. Xia, Design of estimator-based sliding-mode output-feedback controllers for discrete-time systems, *IEEE Transactions on Industrial Electronics*, vol. 61, no. 5, pp. 2432-2440, May 2014.
89. P. Park, D. J. Choi, and S. G. Kong, Output feedback variable structure control for linear systems with uncertainties and disturbances, *Automatica*, vol. 43, no. 1, pp. 72-79, Jan. 2007.
90. L. Wu, P. Shi, and H. Gao, State estimation and sliding-mode control of Markovian jump singular systems, *IEEE Transactions on Automatic Control*, vol. 55, no. 5, pp. 1213-1219, May 2010.
91. X. Su, X. Liu, P. Shi, and R. Yang, Sliding mode control of discrete-time switched systems with repeated scalar nonlinearities, *IEEE Transactions on Automatic Control*, vol. 62, no. 9, pp. 4604-4610, Sep. 2017.
92. H. Li, P. Shi, D. Yao, and L. Wu, Observer-based adaptive sliding mode control for nonlinear Markovian jump systems, *Automatica*, vol. 64, pp. 133-142, Feb. 2016.
93. W. Qi, G. Zong, and H. R. Karimi, Observer-based adaptive SMC for nonlinear uncertain singular semi-Markov jump systems with applications to DC motor, *IEEE Transactions on Circuits and Systems I: Regular Papers*, vol. 65, no. 9, pp. 2951-2960, Sep. 2018.
94. J. Zhang, P. Shi, and Y. Xia, Robust adaptive sliding-mode control for fuzzy systems with mismatched uncertainties, *IEEE Transactions on Fuzzy Systems*, vol. 18, no. 4, pp. 700-711, Aug. 2010.
95. H. Li, J. Wang, and P. Shi, Output-feedback based sliding mode control for fuzzy systems with actuator saturation, *IEEE Transactions on Fuzzy Systems*, vol. 24, no. 6, pp. 1282-1293, Dec. 2016.
96. W. Ji, J. Qiu, L. Wu, and H.-K. Lam, Fuzzy-affine-model based output feedback dynamic sliding mode controller design of nonlinear systems, *IEEE Transactions on Systems, Man and Cybernetics: Systems*, vol. 51, no. 3, pp. 1652-1661, Mar. 2021.
97. W. Ji, J. Qiu, and H. R. Karimi, Fuzzy-model-based output feedback sliding mode control for discrete-time uncertain nonlinear systems, *IEEE Transactions on Fuzzy Systems*, vol. 28, no. 8, pp. 1519-1530, Aug. 2020.

98. W. Ji, J. Qiu, and H. K. Lam, A new design of output feedback sliding mode controller for T-S fuzzy-affine systems, *Asian Journal of Control*, vol. 23, no. 4, pp. 1658-1670, Jul. 2021.
99. J. Qiu, G. Feng, H. Gao, and J. Yang, "Robust $\mathscr{H}_\infty$ static output feedback control of discrete-time piecewise affine systems," in *2009 IEEE International Conference on Control and Automation*, Christchurch, New Zealand, Dec 9-11, 2009, pp. 206-211.
100. S. Boyd, L. El Ghaoui, E. Feron, and V. Balakrishnan, *Linear Matrix Inequality in Systems and Control Theory*. Philadelphia, PA: SIAM, 1994.
101. B. Samadia and L. Rodrigues, Stability of sampled-data piecewise affine systems: A time-delay approach, *Automatica*, vol. 45, no. 9, pp. 1995-2001, Sep. 2009.
102. J. M. A.-D. Silva, C. Edwards, and S. K. Spurgeon, Sliding-mode output-feedback control based on LMIs for plants with mismatched uncertainties, *IEEE Transactions on Industrial Electronics*, vol. 56, no. 9, pp. 3675-3683, Sep. 2009.
103. W. Ji, J. Qiu, H. R. Karimi, and Y. Fu, New results on fuzzy integral sliding mode control of nonlinear singularly perturbed systems, *IEEE Transactions on Fuzzy Systems*, vol. 29, no. 7, pp. 2062-2067, Jul. 2021.
104. P. V. Kokotovic, J. O'Reilly, and H. K. Khalil, *Singular Perturbation Methods in Control: Analysis and Design*. Orlando, FL: Academic Press, 1986.
105. T.-H. S. Li and K.-J. Lin, Stabilization of singularly perturbed fuzzy systems, *IEEE Transactions on Fuzzy Systems*, vol. 12, no. 5, pp. 579-595, Oct. 2004.
106. L. Ma, Z. Wang, C, Cai, and F. E. Alsaadi, Dynamic event-triggered state estimation for discrete-time singularly perturbed systems with distributed time-delays, *IEEE Transactions on Systems, Man, and Cybernetics: Systems*, vol. 50, no. 9, pp. 3258-3268, Sep. 2020.
107. F. Li, S. Xu, and B. Zhang, Resilient asynchronous $\mathscr{H}_\infty$ control for discrete-time Markov jump singularly perturbed systems based on hidden Markov model, *IEEE Transactions on Systems, Man, and Cybernetics: Systems*, vol. 50, no. 8, pp. 2860-2869, Aug. 2020.
108. W. Assawinchaichote, S. K. Nguang, and P. Shi, $\mathscr{H}_\infty$ output feedback control design for uncertain fuzzy singularly perturbed systems: An LMI approach, *Automatica*, vol. 40, no. 12, pp. 2147-2152, 2004.
109. S. K. Nguang, W. Assawinchaichote, and P. Shi, Robust $\mathscr{H}_\infty$ control design for fuzzy singularly perturbed systems with Markovian jumps: An LMI approach, *IET Control Theorem and Applications*, vol. 1, no. 4, pp. 893-908, Jul. 2007.
110. S. Xu and G. Feng, New results on $\mathscr{H}_\infty$ control of discrete singularly perturbed systems, *Automatica*, vol. 45, no. 10, pp. 2339-2343, Oct. 2009.
111. H. Shen, F. Li, Z.-G. Wu, J. H. Park, and V. Sreeram, Fuzzy-model-based nonfragile control for nonlinear singularly perturbed systems with semi-Markov jump parameters, *IEEE Transactions on Fuzzy Systems*, vol. 26, no. 6, pp. 3428-3439, Dec. 2018.
112. Y. Wang, Y. Gao, H. R. Karimi, H. Shen, and Z. Fang, Sliding mode control of fuzzy singularly perturbed systems with application to electric circuit, *IEEE Transactions on Systems, Man, and Cybernetics: Systems*, vol. 48, no. 10, pp. 1667-1675, 2018.
113. G.-H. Yang and J. Dong, Control synthesis of singularly perturbed fuzzy systems, *IEEE Transactions on Fuzzy Systems*, vol. 16, no. 3, pp. 615-629, 2008.
114. K.-J. Lin, Adaptive sliding mode control design for a class of uncertain singularly perturbed nonlinear systems, *International Journal of Control*, vol. 87, no. 2, pp. 432-439, 2014.
115. R. P. Roesser, A discrete state-space model for linear image processing, *IEEE Transactions on Automatic Control*, vol. 20, no. 1, pp. 1-10, Feb. 1975.

116. W. Marszalek, Two-dimensional state-space discrete models for hyperbolic partial differential equations, *Applied Mathematical Modelling*, vol. 8, no. 1, pp. 11-14, Feb. 1984.
117. T. Kaczorek, *Two-Dimensional Linear Systems*. Berlin: Springer, 1985.
118. L. V. Hien and H. Trinh, Stability analysis and control of two-dimensional fuzzy systems with directional time-varying delays, *IEEE Transactions on Fuzzy Systems*, vol. 26, no. 3, pp. 1550-1564, Jun. 2018.
119. J.-W. Wang, S.-H. Tsai, H.-X. Li, and H. K. Lam, Spatially piecewise fuzzy control design for sampled-data exponential stabilization of semilinear parabolic PDE systems, *IEEE Transactions on Fuzzy Systems*, vol. 26, no. 5, pp. 2967-2980, Oct. 2018.
120. Y. Luo, Z. Wang, G. Wei, and F. E. Alsaadi, Robust $\mathscr{H}_\infty$ filtering for a class of two-dimensional uncertain fuzzy systems with randomly occurring mixed delays, *IEEE Transactions on Fuzzy Systems*, vol. 25, no. 1, pp. 70-83, Feb. 2017.
121. M. A. Ghezzar, D. Bouagada, and M. Chadli, Influence of discretization step on positivity of a certain class of two-dimensional continuous-discrete fractional linear systems, *IMA Journal of Mathematical Control and Information*, vol. 35, no. 3, pp. 845-860, Sep. 2018.
122. L. Wu, H. Gao, and C. Wang, Quasi sliding mode control of differential linear repetitive processes with unknown input disturbance, *IEEE Transactions on Industrial Electronics*, vol. 58, no. 7, pp. 3059-3068, Jul. 2011.
123. W. S. Lu and A. Antoniou, *Two-dimensional Digital Filters*. New York: Marcel Dekker, 1992.
124. J. Lam, S. Xu, Y. Zou, Z. Lin, and K. Galkowski, Robust output feedback stabilization for two-dimensional continuous systems in Roesser form, *Applied Mathematics Letters*, vol. 17, no. 12, pp. 1331-1341, Dec. 2004.
125. C.-K. Ahn, L. Wu, and P. Shi, Stochastic stability analysis for 2-D Roesser systems with multiplicative noise, *Automatica*, vol. 69, no. 9, pp. 356-363, Jul. 2016.
126. C. Du, L. Xie, and C. Zhang, $\mathscr{H}_\infty$ control and robust stabilization of two-dimensional systems in Roesser models, *Automatica*, vol. 37, no. 2, pp. 205-211, Feb. 2001.
127. M. Wang, G. Feng, and J. Qiu, Finite frequency fuzzy output feedback controller design for Roesser-type two-dimensional nonlinear systems, *IEEE Transactions on Fuzzy Systems*, vol. 29, no. 4, pp. 861-873, Apr. 2021.
128. L. Li, W. Wang, and X. Li, New approach to $\mathscr{H}_\infty$ filtering of two-dimensional T-S fuzzy systems, *International Journal of Robust and Nonlinear Control*, vol. 23, no. 17, pp. 990-2012, Nov. 2013.
129. L. Li and W. Wang, Fuzzy modeling and $\mathscr{H}_\infty$ control for general 2D nonlinear systems, *Fuzzy Sets and Systems*, vol. 207, pp. 1-26, Nov. 2012.
130. X. Chen, J. Lam, H. Gao, and S. Zhou, Stability analysis and control design for 2-D fuzzy systems via basis-dependent Lyapunov functions, *Multidimensional Systems and Signal Processing*, vol. 24, no. 3, pp. 395-415, Sep. 2013.
131. Y. Luo, Z. Wang, G. Wei, and F. E. Alsaadi, Robust $\mathscr{H}_\infty$ filtering for a class of two-dimensional uncertain fuzzy systems with randomly occurring mixed delays, *IEEE Transactions on Fuzzy Systems*, vol. 25, no. 1, pp. 70-83, Feb. 2017.
132. L. Wu and H. Gao, Sliding mode control of two-dimensional systems in Roesser model, *IET Control Theory and Applications*, vol. 2, no. 4, pp. 352-364, Apr. 2008.
133. H. Adloo, P. Karimaghaee, and A. S. Sarvestani, An extension of sliding mode control design for the 2-D systems in Roesser model, *in Proceeding of the 48th IEEE Conference on Decision and Control*, Shanghai, China, Dec. 2009, pp. 7753-7758.

134. A. Argha, L. Li, and S. W. Su, A new approach to applying discrete sliding mode control to 2D systems, *in Proceeding of the 52nd IEEE Conference on Decision and Control*, Florence, Italy, Dec. 2013, pp. 3584-3589.
135. R. Yang and W. Zheng, Two-dimensional sliding mode control of discrete-time Fornasini-Marchesini systems, *IEEE Transactions on Automatic Control*, vol. 64, no. 9, pp. 352-364, Sep. 2019.
136. J. Qiu, W. Ji, and M. Chadli, A novel fuzzy output feedback dynamic sliding mode controller design for two-dimensional nonlinear systems, *IEEE Transactions on Fuzzy Systems*, vol. 29, no. 10, pp. 2869-2877, Oct. 2021.
137. W. Ji, J. Qiu, and H.-K. Lam, Fuzzy-affine-model-based sliding-mode control for discrete-time nonlinear 2-D systems via output feedback, *IEEE Transactions on Cybernetics*, vol. 53, no. 2, pp. 978-987, Feb. 2023.
138. U. E. Kocamaz, A. Göksu, H. Taşkın, and Y. Uyaroğlu, Synchronization of chaos in bonlinear finance system by means of sliding mode and passive control methods: A comparative study, *Information Technology and Control*, vol. 44, no. 2, pp. 172-181, Jun. 2015.
139. Q. Gao and J. Ma, Chaos and Hopf bifurcation of a finance system, *Nonlinear Dynamics*, vol. 58, pp. 209-216, Feb. 2009.

Index